Modern Spatiotemporal Geostatistics

George Christakos
Department of Geography
San Diego State University
San Diego, California

Dover Publications, Inc.
Mineola, New York

Bibliographical Note

This Dover edition, first published in 2012, is an unabridged republication of the work originally published under the auspices of the International Association for Mathematical Geology in 2000 by Oxford University Press, Inc., as Volume Six in the series, *Studies in Mathematical Geology*.

Library of Congress Cataloging-in-Publication Data

Christakos, George.
Modern spatiotemporal geostatistics / George Christakos.
p. cm. — (Dover earth science)
Originally published: Oxford ; New York : Oxford University Press, 2000.
Includes bibliographical references and index.
ISBN-13: 978-0-486-48818-9 (pbk.)
ISBN-10: 0-486-48818-7 (pbk.)
1. Earth sciences—Statistical methods. 2. Maximum entropy method. 3. Bayesian statistical decision theory. I. Title.

QE33.2.S82.C47 2012
550.72'7—dc23

2011053378

Manufactured in the United States by Courier Corporation
48818701
www.doverpublications.com

To my family and friends in space-time

PREFACE

"If an old rule of thumb from the publishing industry is to be believed, every equation included in a book halves the number sold." *Economist*, Jan. 2–8, 1999

Stochastic characterization of spatial and temporal attributes began as a collection of mathematical concepts and methods developed originally (mostly in the 1930's through 1950's) by A.N. Kolmogorov, H. Wold, N. Wiener, A.M. Yaglom, K. Itô, I.M. Gel'fand, L.S. Gandin, B. Matern, P. Whittle, and others. G. Matheron coined the term "geostatistics" to refer to these developments, brought them together, modified them in some cases, and then applied them systematically in the mining exploration context. Rapid commercial success allowed Matheron to establish the Fontainebleau Research Center in the late 1960's outside Paris. Later, geostatistical techniques were used in other disciplines as well, including hydrogeology, petroleum engineering, and environmental sciences. Geostatistics was introduced in the 1970's in Canada and the United States by geostatisticians including M. David, A.G. Journel, and R.A. Olea. References to selected publications of the above researchers may be found in the bibliography at the end of this book.

It is widely recognized that the techniques of classical geostatistics, which have been used for several decades, have reached their limit and the time has come for some alternative approaches to be given a chance. In fact, many researchers and practitioners feel that they may soon be faced with some kind of law of diminishing returns for geostatistics, inasmuch as the problems of the rapidly developing new scientific fields are becoming more complex, and seemingly fewer new geostatistical concepts and methods are available for their solution.

With these concerns in mind, this book is an introduction to the fundamentals of modern spatiotemporal geostatistics. Modern geostatistics is viewed in the book as a group of spatiotemporal concepts and methods which are the products of the advancement of the epistemic status of stochastic data analysis. The latter is considered from a novel perspective promoting the view that a deeper understanding of a theory of knowledge is an important prerequisite for the development of improved mathematical models of scientific mapping. A spatiotemporal map, *e.g.*, should depend on what we know about the natural variable it represents, as well as how we know it (*i.e.*, what sources of knowledge we selected and what kinds of methods we used to process knowledge). As is discussed in the book, modern geostatistical approaches can be developed that are consistent with the above epistemic framework. The main focus of the book is the Bayesian maximum entropy (BME) approach for studying spatiotemporal distributions of natural variables. As part of the modern spatiotemporal geostatistics paradigm, the BME approach provides a fundamental insight into the mapping problem in which the knowledge of a natural variable, not the variable itself, is the direct object of study. This insight plays a central role in numerous scientific disciplines. BME's rich theoretical basis provides guidelines for the adequate interpretation and processing of the knowledge bases available (different sorts of knowledge enter the modern geostatistics paradigm in different ways). It also forces one to determine explicitly the available physical knowledge bases and to develop logically plausible rules and standards for knowledge integration and processing. BME is formulated in a rigorous way that preserves earlier geostatistical results, which are its limiting cases, and also provides novel and more general results that could not be obtained by classical geostatistics. Indeed, a number of situations are discussed in the book in which BME's quest for greater rigor serves to expose new, hitherto ignored possibilities. In addition, the presentation of the quantitative results, with their full technical beauty, is combined with an effort to communicate across the various fields of natural science. Finally, an attempt has been made to ensure that the case studies considered in this book involve data that are publicly accessible, so that all hypotheses made and conclusions drawn can be critically examined and improved by others (in fact, a scientific model gains authority by withstanding the criticism of other scientists). Naturally, ideas and practical suggestions on how to efficiently apply BME theory will evolve as more case studies are done.

Metaphorically speaking, the aim of modern spatiotemporal geostatistics is to integrate effectively the powerful theoretical perspective of the "Reason of Plato" (who proposed a conceptual framework that dominated mathematical reasoning and philosophical thinking for thousands of years) with the practical thinking of the "Reason of Odysseus" (who was always capable of coming up with smart solutions to all kinds of practical problems he faced during his long journey). It has been said that "Plato shared his perspective with the Gods and Odysseus with the foxes." The modern geostatistician shares it with both! At this point, I must admit to using these great men as a provocative and authoritative means of setting things up and getting readers into the right mood.

In light of the above considerations, the Ariadne's thread running throughout the book is that the modern geostatistical approach to real-world problems is that of natural scientists who are more interested in a stochastic analysis concerned with both the *ontological* level (building models for physical systems) and the *epistemic level* (using what we know about the physical systems and integrating and modeling knowledge from a variety of scientific disciplines), rather than in the pure or naive *inductive* account of science based merely on a linear relationship between data and hypotheses and theory-free techniques that may be useful in other areas. In this sense, modern spatiotemporal geostatistics facilitates yet another kind of integration: the horizontal integration among disparate scientific fields. By processing a variety of physical knowledge bases, the BME approach brings together several sciences which are all relevant to the aspect of reality that is to be examined. For example, BME can become an integral component of the interdisciplinary attack on fundamental environmental health systems which involve physical variables, exposure mechanisms, biological processes, human anatomy and physiology parameters, and epidemiological indicators. The subject of horizontal integration is a source of great excitement among scientists; new ideas are generated incessantly. It is expected that as the domain of modern spatiotemporal geostatistics continues to expand in search of new conquests, a variety of mapping methods aiming at horizontal integration will be added to its arsenal.

The crux of this book was projected in Spring 1986 while I was a research scientist in what is now the Mathematical Geology Section of the Kansas Geological Survey. Some results were published in a paper in *Mathematical Geology* in 1990. Several other research obligations prevented me from working systematically on the subject for the following six or seven years. My involvement in BME analysis has been renewed in recent years, due to increased interest in developing a new conceptual and methodological framework for geostatistics, and aided by generous funding from the Army Research Office (Grants DAAG55-98-1-0289 and DAAH04-96-1-0100). To this financial benefactor I remain grateful.

In carrying out the project, I have benefited from comments made by my colleagues Patrick Bogaert, Dionissios Hristopulos, Ricardo Olea, John Davis, Jürgen Pilz, Tom Jones, and Hyemi Choi. Also my students, Marc Serre, Kyung-mee Choi, Alexander Kolovos, and Jordan Kovitz read my class notes and recommended improvements. In some cases, it was indeed my confrontation with youth that prompted a fresh look at the basis of geostatistics. These students continue to work in the field of modern geostatistics and are expected to contribute significantly to its further advancement. Finally, I am deeply indebted to Jo Anne DeGraffenreid, IAMG editor of the Oxford monograph series, *Studies in Mathematical Geology*; her continuous encouragement and editorial acumen have proven invaluable.

It has been said that in science, the quality of a scientist's work is closely related to the quality of those thinkers with whom he/she disagrees. I have personally benefited greatly from discussions, criticisms, and exchanges of ideas with theoretical opponents for whom I have the greatest respect. If this book is somehow critical of some of their ideas, it is because healthy disagreement—and not imitation—is the deepest sign of an abiding appreciation.

The book is dedicated to the unknown Pythagorean. He was a student at the famous Pythagorean school of ancient Greece (ca. 6th century B.C.), a young man who for the first time in history proved an irrefutable truth by the power of reason alone.* For his astonishing discovery that challenged the views of the Establishment of the time he was cursed and declared blasphemous, and then—legend has it—he was thrown into the sea. Thus perished in the dark waters of the Mediterranean, remaining forever unknown, the young man who brought us the light of reason...

George Christakos

Chapel Hill, 1999

---

* He proved that there is no quotient of integers whose square equals two (see, *e.g.*, the presentation of the Pythagorean School in Omnés, 1999).

TABLE OF CONTENTS

Modern
Spatiotemporal
Geostatistics

1

SPATIOTEMPORAL MAPPING IN NATURAL SCIENCES

"Science is built of facts, as a house is built of stones; but an accumulation of facts is no more a science than a heap of stones is a house." H. Poincaré

Mapping Fundamentals

The urge to *map* a natural pattern, an evolutionary process, a biological landscape, a set of objects, a series of events, *etc.*, is basic in every scientific domain. Indeed, the mapping concept is deeply rooted in the human desire for spatiotemporal understanding: What are the specific distributions of proteins in cells? What are the locations of atoms in biological molecules? What is the distribution of potentially harmful contaminant concentrations in the subsurface? What are the genetic distances of human populations throughout a continent? What are the prevailing weather patterns over a region? How large is the ozone hole? How many light-years do galaxies cover? Answers to all these questions—extending from the atomic to the cosmic—are ultimately provided by means of good, science-based spatiotemporal maps.

Furthermore, studies in the cognitive sciences have shown that maps are particularly suitable for the human faculty of *perception*, both psychological and neurological (Anderson, 1985; Gregory, 1990). These faculties can most efficiently recognize characteristic elements of information when it is contained in a map that helps us build visual pictures of the world. Every scientific discipline depends fundamentally on the faculty of perception in order to interpret a process, derive new insights, conceptualize and integrate the unknown.

What exactly *is* a spatiotemporal map? The answer to this question depends upon one's point of view, which is, in turn, based on one's scientific background and practical needs. From a geographer's point of view, a map is the visual representation of information regarding the distribution of a topographic variable in the spatiotemporal domain (*e.g.*, ozone distribution, radon concentration, sulfate deposition, disease rate). From an image analyst's perspective, a map is the reconstruction of some field configuration within a confined region of space/time. From a physical modeler's standpoint, a map is the output of a mathematical model which represents a natural phenomenon and uses observations, boundary/initial conditions, and other kinds of knowledge as input. While the viewpoints of the geographer and the image analyst are more descriptive, that of the physical modeler is more explanatory. Therefore, a variety of scenarios is possible regarding the way a physical map is produced and the meaning that can be assigned to it:

(*i.*) The map could be the outcome of statistical data analysis based on a set of observations in space/time.

(*ii.*) It could represent the solution of a mathematical equation modeling a physical law, such as a partial differential equation (pde) given some boundary/initial conditions.

(*iii.*) It could be the result of a technique converting physical measurements into images.

(*iv.*) It could be a combination of the above possibilities.

(*v.*) Or, the map could be any other kind of visual representation documenting a state of knowledge or a sense of aesthetics.

The following example illustrates some of the possible scenarios described above.

EXAMPLE 1.1: (*i.*) Studies of ozone distribution over the eastern United States that used data-analysis techniques include Lefohn *et al.* (1987), Casado *et al.* (1994), and Christakos and Vyas (1998). These studies produced detailed spatiotemporal maps, such as those shown in Figure 1.1. Interpreted with judgment (*i.e.*, keeping in mind the underlying physical mechanisms, assumptions, and correlation models), these maps identify spatial variations and temporal trends in ozone concentrations and can play an important role in the planning and implementation of policies that aim to regulate the exceedances of health and environmental standards. The use of data-analysis techniques is made necessary by the complex environment characterizing certain space/time processes at various scale levels (highly variable climatic and atmospheric parameters, multiple emission sources, large areas, *etc.*).

(*ii.*) While in these multilevel situations most conventional ozone distribution models cannot be formulated and solved accurately and efficiently, in some other, smaller scale applications, air-quality surfaces have been computed using pde modeling techniques. In particular, the inputs to the relevant air-quality models are data about emission levels or sources, and the outputs (ozone maps) represent numerical solutions of these models (*e.g.*, Yamartino *et al.*, 1992; Harley *et al.*, 1992; Eerens *et al.*, 1993).

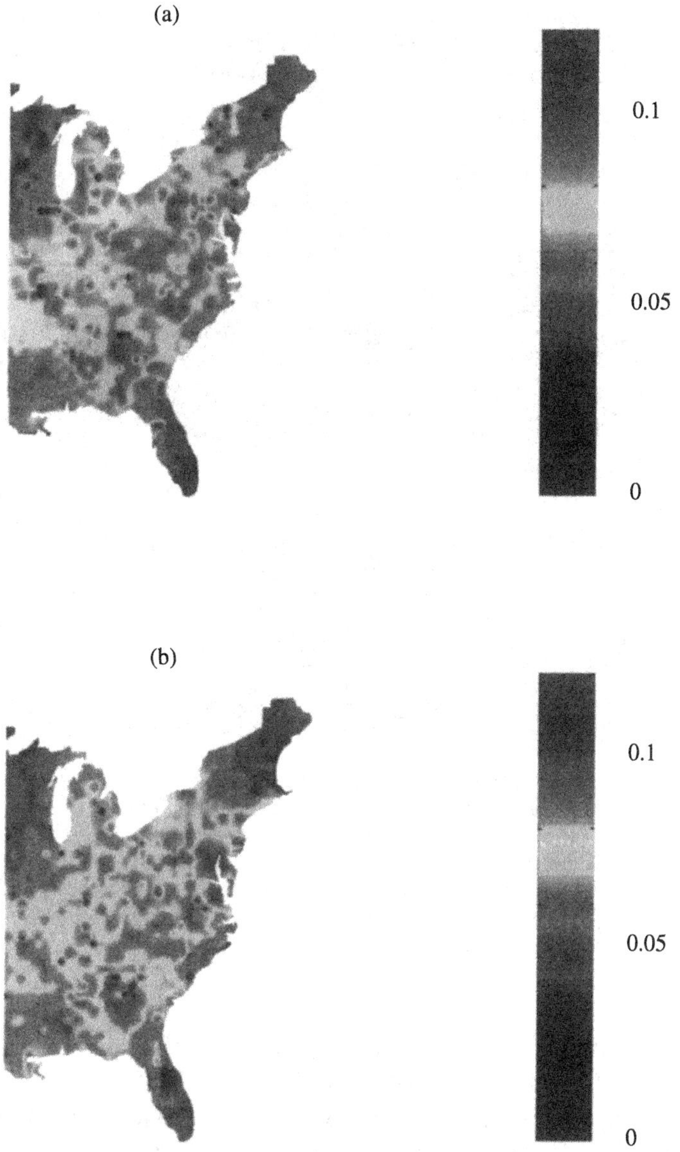

Figure 1.1. Maps of estimated maximum hourly ozone concentration (ppm) over the eastern U.S. on (a) July 15, 1995, and (b) July 16, 1995. From Christakos and Vyas (1998).

(*iii.*) Measuring the travel times of earthquake waves and using a seismic tomography technique, the Earth's core and mantle are mapped in Figure 1.2 (Hall, 1992). As is shown in this figure, the cutaway map that covers the area

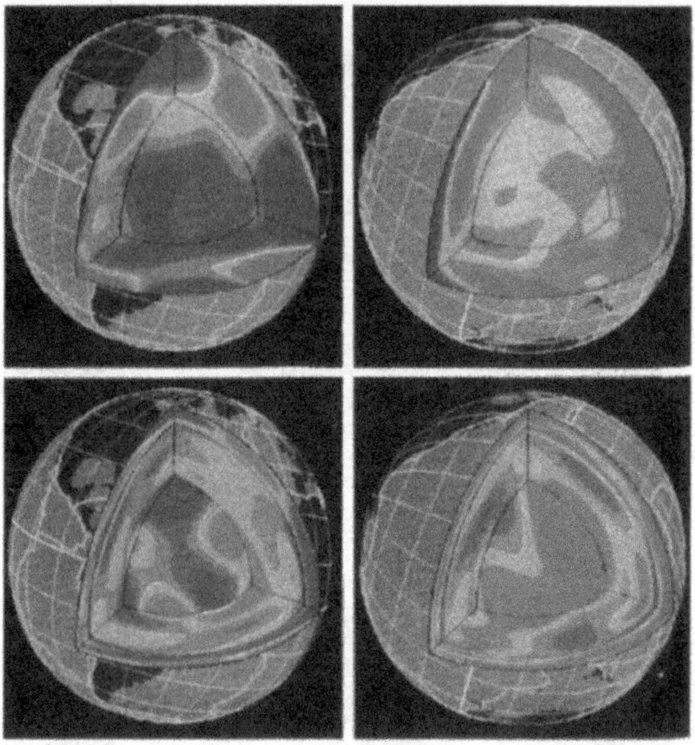

Figure 1.2. Maps of the Earth's core and mantle. The top row shows cutaway maps below the Atlantic (left) and the Pacific (right) oceans to a depth of 550 km. The bottom row shows cutaway maps of the Atlantic and Pacific oceans to a depth of 2,890 km. While the Atlantic maps reveal cold, dense, sinking material, the Pacific maps represent hot, buoyant, rising material. [From Dziewonski and Woodhouse, 1987; ©1987 by AAAS, reproduced with permission.]

of the mid-Atlantic ridge is completely different than the cutaway map that covers the area around the East Pacific Rise.

(*iv.*) Finally, the two-dimensional porous medium map plotted in Figure 1.3 consists of oil-phase isopressure contours for an anisotropic intrinsic permeability field. This map represents the solution of a set of partial differential equations and constitutive relations modeling two-phase (water–oil) flow in the porous medium (Christakos *et al.*, 2000b). □

The salient point of our discussion so far is properly expressed by the following postulate. (Postulates presented throughout the book should not be considered as self-evident truths, but rather as possible truths, worth exploring for their profusion of logical consequences. Indeed, a proposed postulate will be adopted only if its consequences are rich in new results and solutions to open questions.)

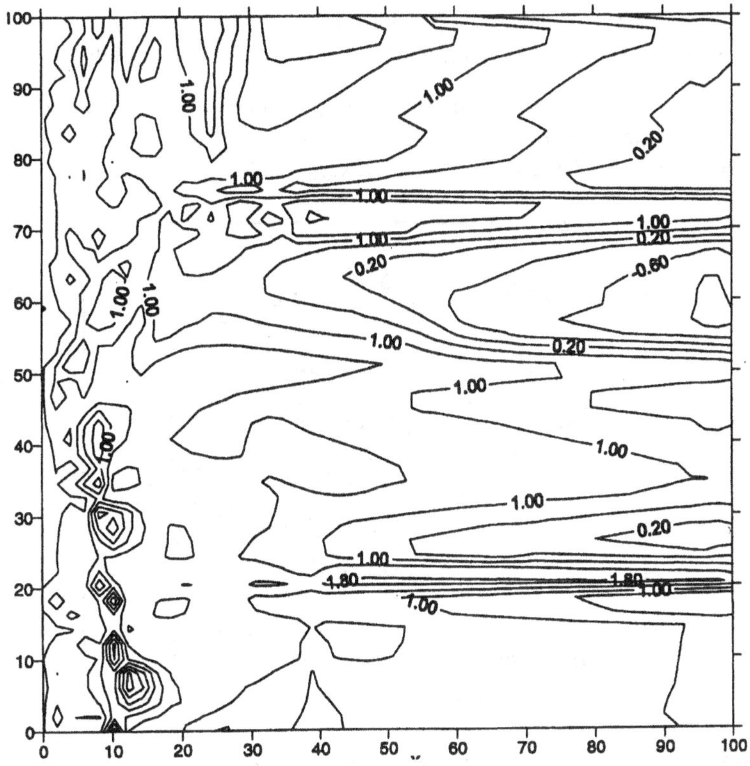

Figure 1.3. Map of oil-phase isopressure contours for an anisotropic intrinsic permeability field (pressure given in units of entry pressure). From Christakos *et al.* (2000b).

POSTULATE 1.1: In the natural sciences, a map is not merely a data-loaded artifact, but rather a visual representation of a scientific theory regarding the spatiotemporal distribution of a natural variable.

According to Postulate 1.1, a map is a representation of what we know (a theory) about reality, rather than a representation of reality itself. In view of this representation, scientific explanation and prediction are to some extent parallel processes: a cogent explanation of a specific map should involve demonstrating that it was predictable on the basis of the knowledge and evidential support available. Maps represent one of the most powerful tools by which we make sense of the world around us. In fact, once our minds are tuned to the concept of maps, our eyes find them everywhere.

Why is mapping indispensable to the natural sciences? If a convincing answer to this question is not offered by the discussion so far, the following examples can provide further assistance in answering the question by describing a wide range of important applications in which spatiotemporal mapping techniques play a vital role. The reality is that significant advances in various branches of science have made it possible to measure, model, and thus map a breathtaking range of spatiotemporal domains. Examples 1.2–1.5 below refer to the various uses of maps in agricultural, forestry, and environmental studies.

EXAMPLE 1.2: Thermometric maps (see Fig. 1.4) provide valuable information for a variety of atmospheric studies, agricultural activities, pollution control investigations, *etc.* (Bogaert and Christakos, 1997). □

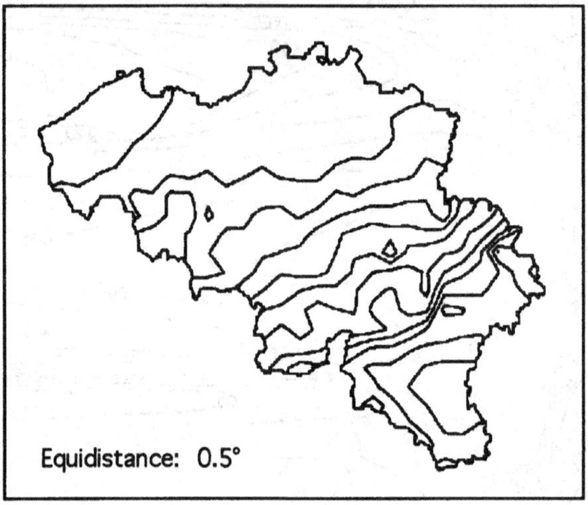

Figure 1.4. Map of the predicted maximum daily temperatures (°Celsius) over Belgium for one day of the year 1990. The equidistance between contours is 0.5°; the lowest level contour is 4.5° (SE part). From Bogaert and Christakos (1997).

EXAMPLE 1.3: In forestry, ground inventory provides important information on biodiversity that cannot be obtained by remote sensing (Riemann Hershey, 1997). A ground inventory, however, is expensive and labor intensive, especially when it covers large areas. Mapping techniques provide the means for estimating unsurveyed areas using a limited number of sample points in space/time. □

EXAMPLE 1.4: Assessment of environmental risk due to some pollutant often requires information regarding the pollutant's distribution on grids covering large spatial domains and multiple time instances (*e.g.*, Bilonick, 1985). This information can be provided most adequately by means of mapping techniques, which on the basis of a limited number of existing measurements and mathematical modeling lead to estimates of the pollutant at other locations and time periods. Also, in studies relating health status to pollutant distribution, an air-quality sampling network usually consists of fewer points in space than are available for health data sets (Briggs and Elliott, 1995). Mapping techniques must then be employed in order to derive pollutant estimates in wider area units. □

EXAMPLE 1.5: Using data from satellites orbiting the Earth, spatiotemporal maps of radioactivity in the atmosphere (Fig. 1.5) revealed unusually high energy emissions which made the detection of the nuclear incident at Chernobyl possible, prior to its official Soviet acknowledgment (Sadowski and Covington, 1987; Arlinghaus, 1996). □

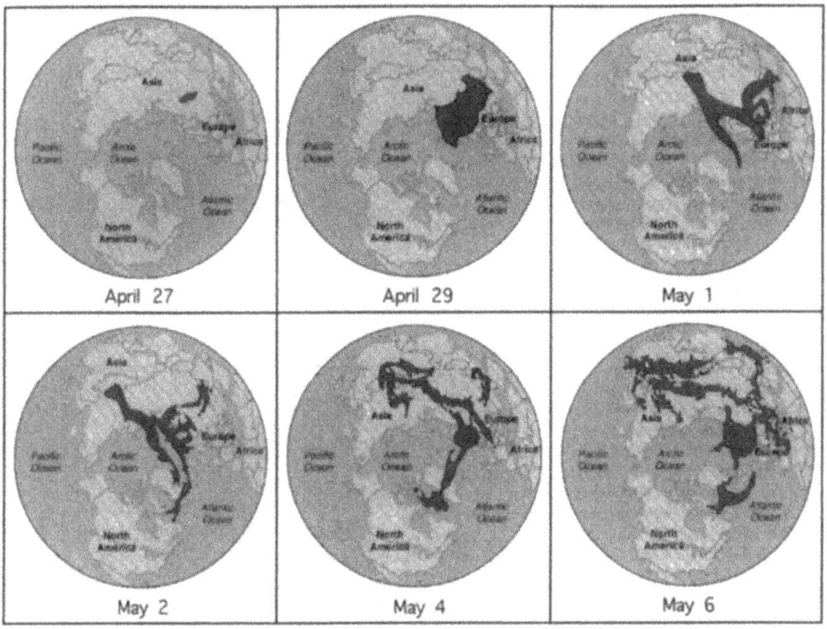

Figure 1.5. Maps of radioactivity present in the atmosphere following the Chernobyl accident (U.S. Air Force weather data and computer simulation by Lawrence Livermore National Laboratory; see Enger and Smith, 1995).

A map can offer more information than merely the distribution of the spatiotemporal variable it represents. The distribution of an air pollutant, *e.g.*, may be used in combination with an exposure-response model to predict the pollutant's impacts on human health and the ecosystem.

EXAMPLE 1.6: Figure 1.6 shows a health damage indicator map (expected number of representative receptors affected/km^2) expressing damage due to ozone exposure in the New York City–Philadelphia area on July 20, 1995; a sublinear exposure-response model was assumed (Christakos and Kolovos, 1999). Interpreted with judgment (*i.e.*, keeping in mind the assumptions made concerning exposure, biological and health response parameters, cohort characteristics of the representative receptor, *etc.*), such maps may offer valuable insight about the possible distributions of population health damage due to pollutant exposure. □

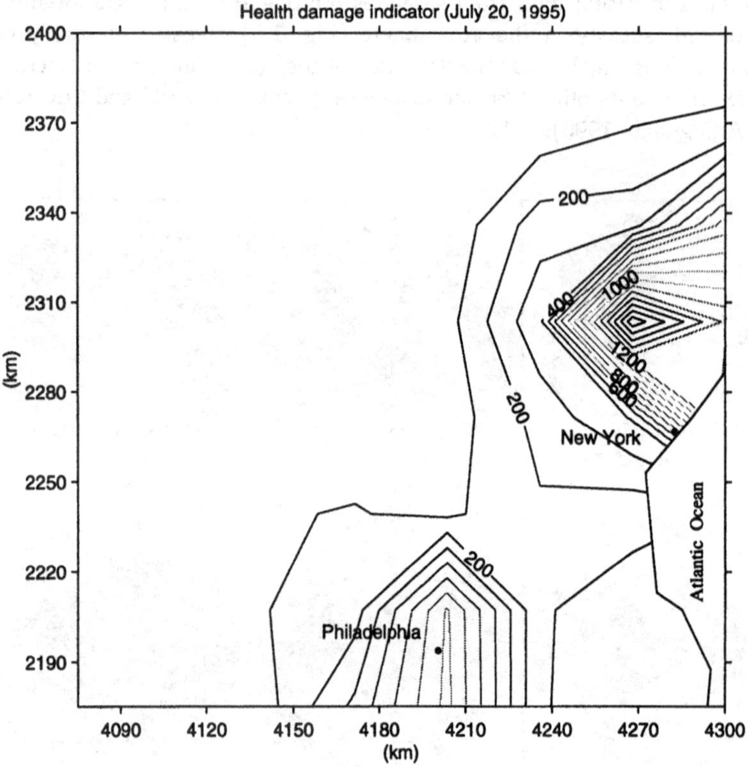

Figure 1.6. A health damage indicator map (number of representative receptors affected/km^2) showing damage due to ozone exposure in the New York City–Philadelphia area on July 20, 1995 (from Christakos and Kolovos, 1999).

Mapping applications also are abundant in the chemical, nuclear, and petroleum engineering fields.

EXAMPLE 1.7: Maps representing a type of material (*e.g.*, chemical element) and the amount (*e.g.*, concentration) of the material on a surface as a function of time are becoming increasingly important for determining inhomogeneities on and in solids (Schwedt, 1997). Nuclear waste facilities are interested in maps showing the migrations and activities of materials encapsulated in concrete barrels (Louvar and Louvar, 1998). The oil industry produces series of geological maps based on reflection seismic data for exploration and development purposes, *etc.* (Doveton, 1986). □

Many applications in which mapping plays a vital role can be found in medical sciences and in genetic engineering.

EXAMPLE 1.8: Simulated spatiotemporal cell fields representing human organs damaged by exposure to chemical agents and other pollutants are increasingly important in environmental health studies (Christakos and Hristopulos, 1998).

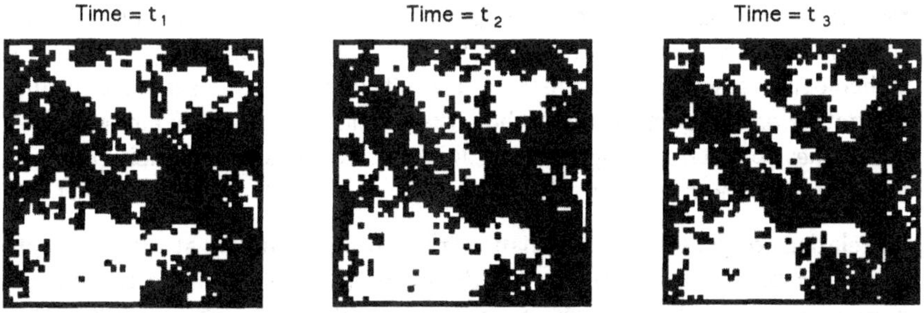

Figure 1.7. A simulated spatiotemporal cell field for a target organ. The affected cells are white and the normal cells are black; some repair is taking place, as well (from Christakos and Hristopulos, 1998).

Spatial maps simulating cell distribution at different times are shown in Figure 1.7. Note the change in number of normal cells (black) *vs.* affected cells (white) in space/time (some of the affected cells are repaired in time). These maps take into consideration the spatial and temporal correlations between cells. □

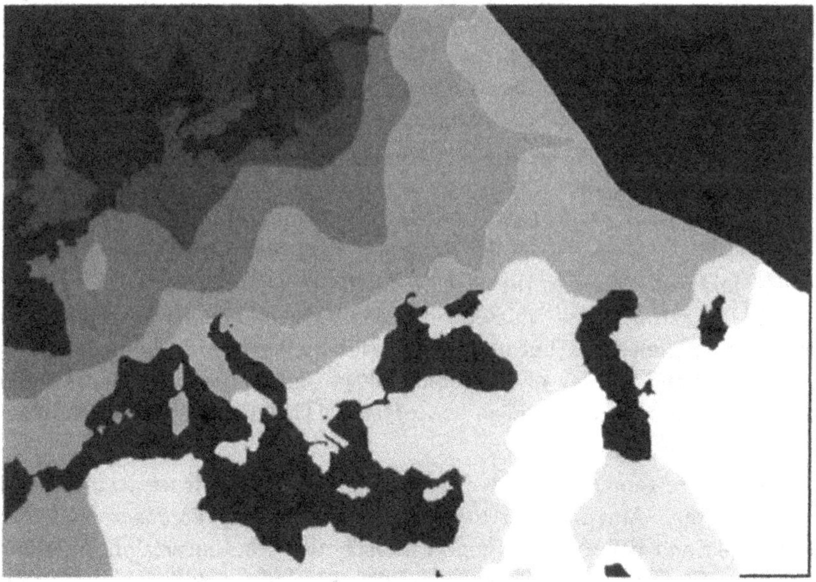

Figure 1.8. Genetic map of populations from the Near East to the European continent. [From Menozzi *et al.*, 1978; ©1978 by AAAS, reproduced with permission.]

EXAMPLE 1.9: Genetic distances between populations can be mapped based on gene frequencies in human beings. The map in Figure 1.8 shows a systematic pattern that slopes from the Near East and southeastern Europe to the north-eastern portions of the European continent. The more dissimilar the shade, the more dissimilar the genetic composition of the populations involved. The contours of the map closely match those of a map developed independently by archeologists studying the spread of agriculture from the Near East to hunting/gathering tribes of Europe. Thus, the conclusion that early farmers spread their genes (by intermarriage with local inhabitants), as well as their grains and agricultural know-how, is justified (Menozzi *et al.*, 1978; Wallace, 1992).

Correlations of gene-frequency maps with health parameters at the geographic level have been instrumental in the discovery of specific genetic adaptations. A map of the sickle-cell anemia gene, *e.g.*, showed a correlation with that of malaria, leading to the hypothesis that this gene may confer resistance to malaria. This hypothesis was later confirmed by more direct tests (*e.g.*, Cavalli-Sforza *et al.*, 1994). □

The maps discussed above, even when they do not offer the ultimate answer to fundamental questions related to the natural phenomena they represent, certainly suggest where answers should be sought. In this section we have selected a representative sample of maps that cover some of science's most fascinating frontiers. This selection is, though, by no means complete. Many other important mapping applications have been omitted (for more detailed accounts of various mapping projects—past, present and future—the interested reader is referred to, *e.g.*, Bagrow, 1985, and Hall, 1992).

Having demonstrated the importance of spatiotemporal maps in every branch of science, we can now address the next question: What constitutes a spatiotemporal mapping approach? Formally, a mapping approach consists of three main components:

1. The *physical knowledge* $\mathcal{K}$ available, including data sets, physical models, scientific theories, empirical functions, uncertain observations, justified beliefs, and expertise with the specified natural phenomenon.

2. The *estimator* $\hat{X}$ which denotes the mathematical formulation used to approximate the actual (but unknown) natural variable X.

3. The *estimates* $\hat{\chi}$ of the actual values χ generated from the estimator $\hat{X}$, usually on a regular grid in space/time. These grid values constitute a spatiotemporal *map*.

There are various methods that can be used to construct accurate maps in space/time. Among other things, a useful mapping approach should explain when and how one can cope rationally with the uncertainty of natural variables. For many years, because of their versatility, *classical geostatistics* methods emerged as the methods of choice for many spatial estimation applications (Matheron, 1965; Journel, 1989; Cressie, 1991; Kitanidis, 1997; Olea, 1999). However, when it comes to scientific mapping (*i.e.*, mapping that proceeds on the basis of scientific principles and laws), these methods suffer from certain well-documented limitations (restrictive assumptions and approximations are often used to compensate for the absence of a sound theoretical basis, a rigorous approach is lacking for incorporating important knowledge sources, physically inadequate space/time geometries are sometimes assumed, constraints are often imposed on the form of the estimator, extrapolation is not reliable beyond the range of the data, computational problems emerge in the practical implementation of some methods, *etc.*; see, also, discussions in the following sections). In this book an attempt is made to develop a group of what might be called *modern spatiotemporal geostatistics* concepts and methods. The goal of modern spatiotemporal geostatistics is to remove some limitations of the older methods, thus providing a significant improvement in the field of scientific mapping. As is described in the following postulate, the book's approach to spatiotemporal modeling will be that of the physical scientist.

POSTULATE 1.2: Modern spatiotemporal geostatistics is concerned with stochastic analysis that functions at both the ontological level (*i.e.*, building models for natural systems) and the epistemic level (*i.e.*, using what is known about the systems and how knowledge is integrated and processed from a variety of scientific disciplines), rather than with pure inductive procedures based on linear relationships between data and hypotheses using physical theory-free techniques.

This fundamental postulate—which will be discussed in more detail in the following section—is made necessary for the additional reason that, despite calls for closer interaction between scientific fields, some disciplines were never more narrow-minded and agoraphobic than today.

Some researchers may sound rather pessimistic when they argue that "we are slipping into a new form of darkness: one where it is popular, profitable, and politically expedient to suppress science" (Milloy and Gough, 1998, p. *vi*), but the phenomenon of people withdrawing in fear from novelty is not uncommon—especially when the new tools originate in disciplines other than their own. The current rigid system often discourages the vitally important cross-fertilization of concepts, models, and techniques between disciplines.

COMMENT 1.1: *The term "epistemic" in Postulate 1.2 above signifies the scientific study of knowledge, as opposed to the philosophical theory of knowledge, which is known as epistemology (Bullock* et al., *1977). A more extended definition of epistemic would include "constructing formal models of the processes by which knowledge and understanding are achieved, communicated, and integrated within the framework of scientific reasoning."*

The Epistemic Status of Modern Spatiotemporal Geostatistics: It Pays to Theorize!

What is the main distinction between the modern geostatistics ideas advocated in this book and classical geostatistics? To answer this question, one must look into the different scientific reasoning *frameworks* underlying these two fields.

Classical geostatistics was designed to fit into a *pure inductive* framework. This framework basically involves the following stages: (*i.*) piling up experimental data, (*ii.*) fitting mathematical functions to these data, and (*iii.*) piling up more experimental data and tests.

It is widely recognized, however, that there are some serious problems with such a framework. Indeed, modern developments in scientific reasoning have stressed deep-seated difficulties associated with pure inductivism (see, *e.g.*, Popper, 1962; Carnap, 1966; Harre, 1989; Chalmers, 1994; Dunbar, 1996; Newton, 1997). Scientific progress is not based merely on pure induction. It involves a significant amount of *theorizing*, as well. Data accumulation surely plays an important role in the growth of science. The data, however, are theory-dependent. Experiments involve planned, theory-guided interference with nature. Change in a theoretical viewpoint regarding a phenomenon results in a change of data. By not referring to theory to adjudicate, the use of pure induction to infer a law from the data leads to *indeterminate* results (discussed on p. 16 in the section entitled "Indetermination thesis"). It also fails to include *explanation* in the scientific process, and it lacks global prediction features (*i.e.*, extrapolation is not possible beyond the range of the data). Therefore, there is no scientific knowledge independent of theory.

This important shift in views concerning the appropriate scientific reasoning framework which occurred during the 20th century was reflected in the foregoing discussions of Examples 1.1–1.9. Indeed, a central theme of these examples was that, in the context of modern spatiotemporal geostatistics, a map is viewed as a representation of a scientific theory regarding the spatiotemporal distribution of the natural variable it represents (Postulate 1.1). According to this postulate, it is absolutely important to perform a deeper theoretical analysis of the mapping problem. A map without theoretical interpretation does not constitute a mature body of scientific information. And it should be expected that the more knowledge and diversified kinds of data we need to process, the less straightforward this theoretical representation will be.

Certainly, all scientific disciplines involve some form of induction in their early descriptive stages. However, as Dunbar (1996) emphasized in his treatise on the scientific method, any discipline that remains locked in this stage can do nothing except describe correlations in natural processes: it can never aspire to full scientific status by providing *explanation* and *understanding*. The latter are very important stages that involve theories and mathematical models developed from sets of hypotheses and assumptions.

EXAMPLE 1.10: Rigorous and clearly formulated theories are a prerequisite for precise observation statements and scientific predictions. That there exist elaborate theories (molecular physics, thermodynamics, *etc.*) presupposed by the observation statement, "the molecular structure of the fluid was affected by heating," should not need much arguing. The meaning of scientific terms used in experimental investigations depends on the role they play in a specific theory (*e.g.*, the term "entropy" has a different meaning in thermodynamics than in information theory; or, the term "covariance" has a different meaning in geostatistics than in relativity theory). □

In modern geostatistics, the epistemic process that leads to a spatiotemporal map is a combination of theoretical concepts, physical knowledge, assumptions, models, *etc.* that goes far beyond the pure inductive framework of accumulating and massaging observational data. Observational data are always expressed in the language of a scientific theory and will be as precise as is the theoretical and conceptual framework they use. Then, on the basis of a meaningful map, theoretical interpretation can lead to a useful picture of reality. This mapping paradigm is schematically expressed in Equation 1.1.

$$\left\{\begin{array}{l}\textit{Physical}\\ \textit{knowledge}\end{array}\right\} \xrightarrow{\textit{Theory}} \left\{\begin{array}{l}\textit{Spatiotemporal}\\ \textit{map}\end{array}\right\} \xrightarrow{\textit{Interpretation}} \left\{\begin{array}{l}\textit{Picture of}\\ \textit{reality}\end{array}\right\} \qquad (1.1)$$

As was anticipated by Postulate 1.1, theorizing plays a vital role in any stage of the scheme (Eq. 1.1). The sound theory and unifying principles of the paradigm make it possible to construct an informative map from the physical knowledge available, as well as to obtain a meaningful interpretation of the map. Given the important connections between scientific explanation (interpretation) and mapping (prediction), an ideal situation should consist of theory-driven improvements in mapping performance that can be explained within the context of our epistemic understanding. Ignoring the theoretical rationale underlying the mapping process can only damage our scientific interpretation of what the map represents. The lack of sound theoretical underpinnings and unifying principles is, perhaps, the key shortcoming of many cookbook approaches to data analysis. One should think of a geostatistical algorithm as the end result of an analysis that goes deeper into the fundamentals of a problem, rather than a collection of techniques and recipes without any clear underlying rationale.

In light of our discussion so far, the following definition of modern geostatistics seems reasonable (it is, however, a rather broad definition, the specific elements of which will later become more clear).

DEFINITION 1.1: Modern spatiotemporal geostatistics is a scientific discipline that arises from the advancement of the ontological and epistemic status of stochastic analysis, as described in Postulate 1.2 above.

In light of Definition 1.1, the problem domain is expanded to include the *observer* as well as the *observed*, so that the final space/time map is the result of the interaction between the two. The observer here is the *geostatistician* with his/her epistemic tools and knowledge bases (scientific theories, logical reasoning skills, engineering laws, *etc.*). The observed is the *natural world* with its ontological structure (physical phenomena, natural processes, biological mechanisms, *etc.*). Surely, Definition 1.1 is a broad one that leaves room for several ways out of the restrictive pure inductivist geostatistical framework that has been proven so ineffective in providing useful modeling tools for the rapidly developing new scientific fields. In this book we have chosen to focus on a specific group of modern geostatistics methods that have the following basic elements in common.

• *Spatiotemporal random field modeling*: The natural variable of interest is represented in terms of a spatiotemporal random field, which offers a general framework for analyzing data distributed in space/time. This framework is especially effective mathematically; it allows us to grasp difficult problems, improve our insight into the physical mechanisms and, thus, enhance our predictive capabilities.

• *Physical knowledge classification*: The two primary physical knowledge bases considered in spatiotemporal analysis and mapping are *general* knowledge (obtained from theories, physical laws, summary statistics, *etc.*) and *specificatory* knowledge (obtained through experience with the specific situation).

• *Epistemic paradigm*: Modern spatiotemporal geostatistics is underpinned by a cogent epistemic foundation which combines the world of empirical data with the world of theory and scientific reasoning. This is a powerful combination that leads to a distinctive methodology for the acquisition, interpretation, integration, and processing of physical knowledge.

The course of each one of these three topical elements is substantially influenced by each of the others, to the extent that they form a *net* or *web* of theoretical and empirical support for modern geostatistics, rather than simply converging upon it.

Why Modern Geostatistics?

The discussion of the previous section was partially motivated by the following question: Why should the data analysis community bother with modern spatiotemporal geostatistics when there exist already other alternatives which are fully developed, such as regression methods, spline functions, basis functions, and trend surface techniques? The answer to this question seems to be threefold:

1. A general answer is a matter of scientific progress: many of the above techniques—which have been used for several decades—have reached their limits, and it is time that novel methods be tried in spatiotemporal mapping applications. In fact, this development in the field of geostatistics is the *natural course* of all human constructions: the time comes when their limits are recognized and new methods need to be devised. The latter is a necessary step for the continuing vitality of a scientific field, and geostatistics should not be an exception.

2. Another answer to the above question is that the case for the existing methods would be *logically much stronger* if one could show that all the alternatives are less good or even inadequate. This is an important reason for examining other possibilities.

3. Finally, a more specific answer is that many of the existing methods suffer from a number of well-documented *limitations* which modern geostatistics makes a serious effort to eliminate. We have already mentioned some of these limitations. Due to its importance, the matter deserves further discussion,

particularly with respect to four issues: scientific content, indetermination thesis, spatiotemporal geometry, and sources of physical knowledge.

Scientific content

In spatial statistics the mapping process is viewed mostly as an exercise of mathematical optimization involving data-fitting techniques (regression, polynomial interpolation, spline functions, *etc.*). By ignoring the *scientific content* of the mapping process, purely instrumental data-processing techniques can seriously damage important scientific interpretations (*e.g.*, they can lead to unrealistic models of space/time correlation). If this is the case, one may soon be faced with some kind of *law of diminishing returns* for geostatistics, inasmuch as the problems of the rapidly developing new scientific disciplines are becoming more complex and seemingly fewer new geostatistical methods with a sound scientific rationale are available for their solution.

EXAMPLE 1.11: Mapping techniques based on spline functions seem attractive to some, for they show a relative lack of conceptual bias (Thiebaux and Pedder, 1987). These techniques have a conventional and purely instrumental character (they merely include conditions on continuity, smoothness, and closeness to data). Unfortunately, this lack of conceptual bias is usually accompanied by a notable lack of scientific content. Indeed, no knowledge of the structural and functional mechanisms of the natural process underlying the data is assumed. Similarly, shortcomings of the kriging mapping techniques include:

(*i.*) the inability to account for important knowledge bases (see "Sources of physical knowledge," p. 20), thus leading to maps which in many cases do not reflect the opinion of the experts (see Bardossy *et al.*, 1997);

(*ii.*) the lack of epistemic content (kriging's concern is merely how to deal with data, rather than how to interpret and integrate them into the understanding process);

(*iii.*) the restrictive assumptions and approximations used, as well as the computational problems (instability, high costs, *etc.*). See, *e.g.*, Dietrich and Newsam (1989) and Dowd (1992). □

Is has been argued (*e.g.*, Newton, 1997) that it is a characteristic of immature scientific fields to rely primarily on *taxonomy* (collecting, describing, and tabulating observational facts). This is particularly true for these fields at their early stages of development, at which time classical geostatistics is, indeed, a suitable tool. It usually takes a fierce struggle on the part of scientific modelers to end the hegemony of taxonomists, and to allow such fields to follow the theory-driven steps towards becoming a mature science. In the context of such an effort, the methods of modern spatiotemporal geostatistics are definitely more appropriate.

EXAMPLE 1.12: Biology is an example of a scientific field that was dominated for decades by the culture of taxonomy. This culture was, perhaps, necessary at the early stages of biology, but biology became a mature science only when it was able to go beyond taxonomy and make theoretical understanding its primary goal (Harre, 1989; Omnés, 1999). In modern scientific disciplines, such as bioinformatics, the emphasis progressively switches from the accumulation of data to its scientific interpretation (Baldi and Brunak, 1998). Also, in genetics most of the controversial issues regarding the genetic origin of complex human behavioral patterns are ultimately generated not by inadequate data, but rather by more difficult explanatory and interpretive issues such as: Can we attribute some features of the genes if they interact with the environmental (nongenetic) factors? or, Does linkage analysis establish anything more than correlation? (see Sarkar, 1998). □

In some cases, the reliability of claims presented as scientific can be an issue of life and death (*e.g.*, DNA maps in capital murder trials). The important issue here is the scientific content of these claims. Justice Blackmun of the U.S. Supreme Court wrote that in order for expert testimony to be of real assistance to the courts, "a valid scientific connection to the pertinent inquiry as a precondition to admissibility" is required, and the question is "whether reasoning and methodology properly can be applied to the facts in issue" (U.S. Supreme Court, 1993). In such cases the important factor is the extent to which a set of data can be related by a credible theory to the situation at hand. Viewed from the wrong perspective lacking the support of a sound scientific theory, even the most accurate data may imply false conclusions. As Foster and Huber (1999, p. 24) point out in their treatise, *Scientific Knowledge and the Federal Courts*: "the history of science records many instances of precise and accurate measurements being piled up around false conclusions," and "facts acquire meaning from the theory (express or implied) in which they are presented, and in turn they determine what conclusions might be drawn from a theory." This being the situation within the legal system, it could be a tough case for the proponents of a theory-free geostatistics to demonstrate the "practicality" of their recipes if the scientific content of the claims is not considered reliable by the Federal courts.

Indetermination thesis

The main idea of theory-free analysis (also known as physical model-free analysis or naive empiricism) is to "let the data speak for themselves"; in other words, "let the data tell us what the mathematical model is." Theory-free analysis constitutes a fallacy that is a consequence of the pure inductive framework mentioned in the previous section. Indeed, what this kind of analysis ignores is the fundamental fact of scientific reality—called the *indetermination thesis*—that, while the mapping "model → data" is one-to-one (*i.e.*, given a specific set of physical conditions, a model produces one data set), the mapping "data → model" is one-to-many (a data set may be represented by numerous models). As a matter of fact, this is the way the incompleteness of any system of models or hypotheses manifests itself, in accordance with Gödel's theorem on the incompleteness of any system of axioms (Nagel and Newman, 1958).

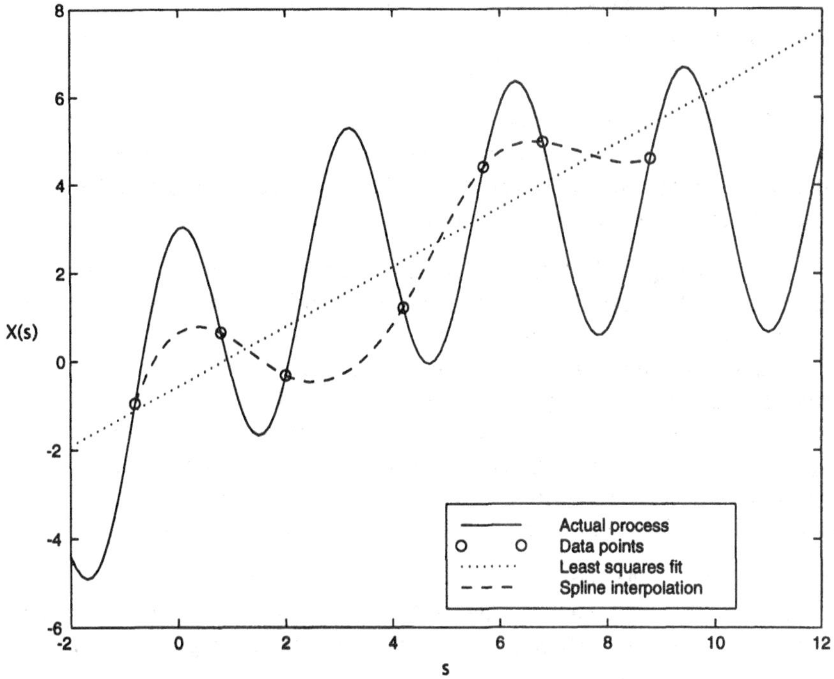

Figure 1.9. Actual process *vs.* least squares and spline models.

EXAMPLE 1.13: There are many curves that honor the same set of data. So, when the data speak, all they do is give us many choices. In the simple case of Figure 1.9, two different models (least squares and the spline functions) honor the data points but, nevertheless, they provide a quite misleading picture of the actual physical process, $X(s) = 3\cos(2s) + s\ \exp[-0.1\,s]$, which is also shown in Figure 1.9. □

COMMENT 1.2: *Poincaré (1952, p. 146) made the following remark regarding curve-fitting: "We draw a continuous line as regularly as possible between the points given by observation. Why do we avoid angular points and inflexions that are too sharp? Why do we not make our curve describe the most capricious zigzags? It is because we know beforehand, or think we know, that the law we have to express cannot be so complicated as all that."*

In sciences with little or no mathematical basis (*e.g.*, psychology), the indetermination thesis can pose severe difficulties, for one must choose among a very large number of possibilities. Seeking to reduce problems to forms to which their traditional curve-fitting tools are most easily applicable, some descriptive epidemiology studies are often led to produce data fits that merely satisfy statistical criteria, at the expense of the attainable mechanistic or biologic realism that is required to use the results in risk assessment. Physical sciences, on the other hand, contain rich enough mathematical guidelines to let us focus on a few choices. As a matter of fact, the case of the indetermination thesis implies that one should use a more scientifically sound method in one's effort to choose an appropriate model, as is suggested by the following postulate.

POSTULATE 1.3: In modern spatiotemporal geostatistics, the rational approach for choosing the appropriate model from the data is by means of a theory that represents the physical knowledge available and satisfies plausible logical principles of scientific reasoning.

The meaning of Postulate 1.3 is that, in addition to conforming to empirical evidence, there exist physical knowledge (scientific theories, laws, *etc.*) and epistemic criteria (explanatory power, informativeness, *etc.*) that are capable of singling out one model from a class of empirically equivalent models (or, at least, capable of narrowing the choices considerably). In this sense, Postulate 1.3 is closely related to the issue of scientific content addressed in the previous section. As the following example demonstrates, instead of fitting various mathematical models to the available data, one could fit the model that is justified on the basis of a physical theory that produces the specific set of data.

EXAMPLE 1.14: Assume that a set of temporal burden data $X(t_i)$, $i = 1, \ldots, m$ on a target human organ subjected to pollutant exposure is available. If the pure induction (theory-free) approach is adopted, several valid covariance functions

$$c_k(t_1, t_2), \quad k = 1, 2, \ldots \tag{1.2}$$

could be fitted to the burden data by means of a conventional data-fitting technique. The question then arises: Which model best describes the temporal burden correlations? The physics of the situation offers a realistic answer to this question. The burden $X(t)$ on a target organ due to exposure obeys the stochastic first-order kinematics

$$\frac{dX(t)}{dt} + \lambda X(t) = U(t) \tag{1.3}$$

where λ is the transfer rate out of the organ determined from the data, $U(t)$ is the uptake rate (zero mean, white-noise process with a unit variance), and $X(0) = 0$ is the initial condition. On the basis of the law (Eq. 1.3), the theoretical burden covariance is given by

$$c_x(t_1, t_2) = (2\lambda)^{-1} \left[\exp(-\lambda\, |t_1 - t_2|) - \exp(-\lambda\, t_1 - \lambda\, t_2)\right] \tag{1.4}$$

The process of fitting the covariance model (Eq. 1.4) to the burden data has physical content, which was not the case with the arbitrary functions $c_k(t_1, t_2)$, $k = 1, 2, \ldots$ of Equation 1.2 above. □

Other applications of the important Postulate 1.3 will be discussed in the following chapters (see *e.g.*, Chapter 7, Example 7.5, p. 143).

COMMENT 1.3: *Most studies of the scientific method (*e.g., *Lastrucci, 1967; Kosso, 1998) show that the most ardent "believer in data" eventually realizes that these data are valid and meaningful because they rest upon some theory of significance and credibility, even though he/she may not have reasoned through the surface of the data to their origin in the theoretical notion which supports them. In the words of Wang (1993, p. 168): "... a safe bet is that, without a good theory, even if the data are thrown in the face of the 'data analyst,' the chance that he will find anything significant is next to zero."*

Spatiotemporal geometry

Spatiotemporal structures differ, as do the *geometries* which describe them. Euclidean formulas, *e.g.*, are appropriate for geometric figures on flat surfaces, but generally are inappropriate for describing intrinsically geometric relations on curved surfaces. Several other examples will be discussed in Chapter 2.

Classical geostatistics routinely employs the Cartesian coordinate system equipped with a Euclidean metric (distance), and considers time merely as an additional dimension. Within such a framework, space and time have an existence all on their own, independent of the existence of physical processes and objects. In many applications, however, Cartesian geometry is not a realistic model of the physical situation, which requires a higher level of understanding of the spatiotemporal domain. In many cases, this understanding involves non-Cartesian coordinate systems and non-Euclidean metrics. Also, the concepts of absolute space and time—independent of the underlying physical processes—usually do not exist in the real world. Applications discussed in Schäfer-Neth and Stattegger (1998) as well as in Christakos *et al.* (2000a) have shown that the mapping techniques of classical geostatistics—which rely on Cartesian coordinate systems or use a physically inappropriate Euclidean metric—can lead to inaccurate maps which imply false conclusions (see Chapter 2, Examples 2.31 and 2.32, p. 67). The inadequacy of classical geostatistics to handle these situations requires the development of new methods, as is suggested by the following postulate.

POSTULATE 1.4: Modern spatiotemporal geostatistics recognizes that spatiotemporal geometry is not a purely mathematical affair and relies on physical knowledge in order to decide which mathematical geometry best describes reality.

Postulate 1.4 essentially expresses the view that the spatiotemporal geometry of modern geostatistical applications is intimately connected with the laws of the physical domain and cannot exist independent of them. As a result, in practical applications it is necessary to investigate whether the traditional geometrical structures need to be replaced by a physically more meaningful spatiotemporal geometry. The appropriate coordinate system should allow, *e.g.*, representations of the spatiotemporal geometry on the basis of the underlying natural symmetry. In some cases the use of a specific system depends on mathematical convenience. The local properties of space and time (*e.g.*, curvature), as well as rules imposed by the specific natural process (*e.g.*, diffusion) play a significant role in the choice of the appropriate metric.

Sources of physical knowledge

Yet another argument in favor of the development of modern spatiotemporal geostatistics may be offered. Most people consider geostatistics to be an *applied science*. As such, its intent should be to produce marketable products, capitalizing on the stores of *basic knowledge* that have accumulated thus far in a richly productive century. But no one really believes these stores of knowledge consist merely of observational facts. Moreover, it is inconceivable that cookbook recipes and techniques are capable of taking important basic knowledge into account. The real meaning of "basic knowledge" in the definition of applied science surely goes far beyond observational facts to include several other forms of knowledge, such as physical laws, scientific theories, empirical rules, phenomenological relationships, and factual statements (Lewis, 1983).

In most practical applications, classical geostatistics does not possess the means which will allow it to account for important sources of scientific knowledge, put together information in new ways, and find creative solutions to changing conditions (*e.g.*, most of the classical methods are regression-type data-processing techniques). The quality of decisions generated by various decision-support systems (DSS; *e.g.*, a water resources DSS or a health management DSS) depends heavily on the proper use of high-quality physical and medical knowledge (Foster *et al.*, 1993; Birkin *et al.*, 1996; van Bemmel and Musen, 1997). As a consequence, the development of modern geostatistics requires methods which are capable of accounting for various sources of scientific knowledge and weighing all the evidence available in a rigorous and systematic manner. Ideally, knowledge bases (including scientific literature and expert opinions from the specified domain) should be systematically organized so that they are accessible electronically and interpretable by computer. New methods should make possible the interdisciplinary cross-fertilization of concepts, models, and research tools. Modern geostatistics should also recognize the growing sophistication of research tools, and particularly their ability to reveal the modes of action of natural phenomena, environmental pollutants, toxicokinetic and biological processes, epidemiologic patterns, *etc*. Furthermore, novel methods should include explanation as part of the scientific process and incorporate global prediction features.

The non-Procrustean spirit

The above arguments should be viewed in the context of constructive criticism aimed at the improvement and continuing vitality of geostatistics. There is, however, a possibility that suggestions will be misinterpreted or even rejected by the Establishment. In fact, the tendency of the Establishment to reject criticism of traditional methods has been called "obscurantism" by Whitehead (1969, p. 43), who writes in a memorable passage, "Obscurantism is the refusal to speculate freely on the limitations of traditional methods. It is more than that: it is the negation of the importance of such speculation, the insistence of incidental dangers."

It is not appropriate to consider geostatistics from the Procrustean viewpoint of obscurantism (Procrustes was the ancient mythical giant who chopped off the legs of travelers when they did not fit his bed). For geostatistics to flourish, it must be allowed to use new concepts and tools freely, unconstrained by preconceived notions of what geostatistics ought to be. Therefore, the developments proposed by modern spatiotemporal geostatistics should be viewed in a *non-Procrustean* spirit. Certainly, the epistemic approach to modern geostatistics discussed in this book is not the only possibility. The book is more like a "call for research," encouraging a multidisciplinary conception of modern geostatistics directed toward novel ideas and models which consider the advances of numerous scientific disciplines where geostatistical ideas can be applied.

Bayesian Maximum Entropy Space/Time Analysis and Mapping

The study of the scientific status of spatiotemporal modeling and mapping forms a group of methods that can be placed in a unified framework demonstrating its significance to the development of modern geostatistics (Definition 1.1). From the epistemic viewpoint, spatiotemporal mapping is a combination of both functions of scientific reasoning, namely, the examination of hypotheses regarding a natural variable and the determination of estimates for the values of that variable. A particularly important member of this group of methods is *Bayesian maximum entropy* (BME) spatiotemporal analysis and mapping (Christakos, 1990, 1991a, 1992, 1998a, b).

COMMENT 1.4: *Before proceeding any further, let us make a brief note about the term "BME." The epistemic framework of the new approach involves the concepts of Bayesian conditionalization (see Chapter 4, Eq. 4.9, p. 96) and entropy (Chapter 5, Eq. 5.2, p. 105), thus the acronym "BME." For readers who would like to review Bayesian analysis and entropic concepts, good references are the books by Howson and Urbach (1993) and Robert (1994) and the collection of papers by Jaynes (1983). It should be kept in mind, however, that the introduction of Bayesian thinking in modern spatiotemporal geostatistics is epistemically and physically motivated, rather than merely promoting a statistical methodology. Also, as we shall see below, the modern geostatistics framework is very general and should not be restricted by such acronyms. Indeed, the Bayesian concept can be generalized significantly by means of the knowledge processing rules of mathematical logic, and information measures other than entropy may be considered.*

Due to its strong epistemic component, BME focuses on levels of spatiotemporal analysis as they relate to understanding. This is a powerful approach of scientific reasoning, with particularly appealing features in the mapping of natural variables (see BME applications in Christakos, 1992, 1998c; Christakos and Li, 1998; Choi *et al.*, 1998; Serre *et al.*, 1998; Bogaert *et al.*, 1999; Serre and Christakos, 1999a; D'Or, 1999). BME features are discussed in detail in the following sections. To whet the reader's appetite, a brief summary is provided here.

BME features

From the perspective of modern geostatistics, some of the most appealing features of BME are as follows (see also discussion on p. 249 *ff*):

• It satisfies sound epistemic ideals and incorporates physical knowledge bases in a rigorous and systematic manner.

• BME does not require any assumption regarding the shape of the underlying probability law; hence, non-Gaussian laws are automatically incorporated.

• It can be applied as effectively in the spatial as in the spatiotemporal domains and can model nonhomogeneous/nonstationary data.

• BME leads to nonlinear estimators, in general, and can obtain well-known kriging estimators as its limiting cases.

• It is easily extended to functional (block, temporal averages, *etc.*) and vector natural variables, and it allows multipoint mapping (*i.e.*, interdependent estimation at several space/time points simultaneously), which most traditional mapping techniques do not offer.

• By incorporating physical laws into spatiotemporal mapping, BME has global prediction features (*i.e.*, extrapolation is possible beyond the range of observations).

• It is computationally efficient due to the availability of closed form analytical expressions for certain mapping distributions, *etc.*

To the above features, one should add BME's beauty which, regretfully, cannot be captured by words, but can be known only by those who come sufficiently close to it (which is, of course, the case with any kind of beauty). In BME analysis, spatiotemporal mapping is viewed as a *generalization* process that expresses the important relationship of theory to data. Depending on the amount of data available, the generalization process may take two forms:

(*i.*) In many applications only a few data are available. Mapping is then a process that seeks to create a spatiotemporal pattern by enlarging upon the limited amount of data available. For the purpose of such an enlargement, physical knowledge that comes from sources other than direct measurements becomes extremely important.

(*ii.*) In several other applications, a huge collection of data must be confronted, little of which is really relevant to the problem. In this case, mapping seeks a meaningful pattern that will make sense out of shapeless heaps of data. The best way to deal with the situation is to develop a deeper theoretical understanding and adapt the data-processing techniques to the new theories.

In both cases considered above, the theory aims at deriving *operational* concepts, *i.e.*, concepts that concern potentially observable quantities and processes which can be expressed in terms of efficient computer algorithms.

A geostatistician's first duty is to be *creative*. The second duty is to be *skeptical*. The traits of creativity and skepticism are necessary complements in science. In principle, the *limits* of BME analysis and mapping are of an epistemic as well as an ontological nature. This is, of course, the case of all predictive theories where epistemic limits are associated with our inability to collect enough data, with poor understanding of the underlying physical mechanisms, or with limited computational capabilities. Ontological limits, on the other hand, could be due to the inherent complexity of the phenomenon being mapped, the lack of causal relations and well-defined patterns, *etc.* While modern geostatisticians should take conventional criticisms into consideration, they should not be constrained unnecessarily by them.

The Integration Capability of Modern Spatiotemporal Geostatistics

The practice of geostatistics is changing nowadays. Changes result from rapid technological developments and globalization, as well as from the fact that science is becoming more interdisciplinary. *Horizontal* integration represents a striking phenomenon of convergence in science which leads to new, highly interdisciplinary fields, many of which lie at the frontier of current research. This compels researchers and practitioners alike to be aware of developments and challenges in fields other than their own. The methods of modern spatiotemporal geostatistics can play a vital role in the horizontal integration among disparate scientific disciplines. By integrating a variety of knowledge bases and case-specific goals and objectives, the BME model of modern geostatistics can generate highly informative spatiotemporal maps that improve our understanding and decision-making processes. This function of the BME model may be symbolically represented as follows

$$\underbrace{\int_{\substack{\textit{General}\\ \textit{knowledge}}} \int_{\substack{\textit{Specificatory}\\ \textit{knowledge}}} \int_{\substack{\textit{Goals \&}\\ \textit{Objectives}}}}_{\textit{Interdisciplinary}} \textit{BME} \longrightarrow \begin{matrix}\textit{Spatiotemporal}\\ \textit{maps}\end{matrix} \longrightarrow \begin{cases}\textit{Understanding}\\ \textit{Decision making}\end{cases} \tag{1.5}$$

where the symbol "$\int$" denotes integration. The horizontal integration facilitated by Equation 1.5 brings together several sciences which are all relevant to the problem under consideration and, thus, generates an improved picture of reality. Equation 1.5 also acknowledges the importance of studying the internal relations between the various components of the problem. Furthermore, the cross-disciplinary interactions inspired by this approach present exciting opportunities for fruitful communication between experimentalists and modelers. The implication of Equation 1.5 is that BME can become a vital component of an *interdisciplinary* attack on complex problems. The integration capability of the BME model is emphasized throughout the book. Some examples related to environmental health systems are discussed in Chapter 9 (see "BME in the Context of Systems Analysis," p. 181).

The "Knowledge-Map" Approach

The following chapters elaborate further on each of the BME concepts and the features briefly discussed in this introductory chapter. We will also address important practical issues arising in the application of modern spatiotemporal geostatistics. Particularly, the presentations in this book may be viewed as a set of "knowledge-maps" that gradually become more detailed.

1. With the first knowledge-map (which includes Chapters 2–4, in which the ontological and epistemic foundations of modern geostatistics are discussed), the reader should obtain a general picture of how the BME group of methods works, without a full grasp of all finer points.

2. The second knowledge-map is more detailed, consisting of the rigorous mathematical formulation of the BME model (Chapters 5–8). The mathematics of BME are rather straightforward, which is a useful thing in computational applications.

3. The third knowledge-map deals with extensions of the BME model to functional, vector, and multipoint mapping situations (Chapters 9–11). In Chapter 12, the BME model is compared to some popular mapping methods both on theoretical grounds and by means of numerical applications.

4. The final "map" that provides a complete understanding of BME is, of course, for the reader to make: **One should try one's own applications!**

As a theory of physical knowledge-based spatiotemporal analysis and mapping, BME comprises two distinct parts, a formal part focused on mathematical structure and logical process (formulation and solution of mapping equations, organization of logical connections, *etc.*) and an interpretive part concerned with application of the formal part in real-world situations (choice of natural language, physical meaning of mathematical terms, methodology, *etc.*). As is usual in scientific inquiry, some geostatisticians choose to concentrate their efforts on the formal part while others prefer to address interpretative aspects. The two parts influence each other considerably and are equally important to the development and growth of modern spatiotemporal geostatistics; but, we must maintain a distinction between their different goals. Much misunderstanding has been caused by not making the distinction sufficiently clear.

2

SPATIOTEMPORAL GEOMETRY

"The basis of metrical determination must be sought outside the manifold in the binding forces which act upon it." G.F.B. Riemann

A More Realistic Concept

As geostatistics expands its domain in search of new concepts and applications, the return to its foundations will continue, each process nourishing the other. The complete understanding of any line of scientific reasoning calls for examining the underlying hypotheses and exposing fallacies, whether of the factual or conceptual sort. Chapter 1 discusses the goals of the epistemic analysis that lies at the heart of modern spatiotemporal geostatistics. Among the goals are the exposition and correction of *three fallacies* of classical geostatistics:

1. Spatiotemporal structure is a purely geometrical affair.
2. The data always speak for themselves.
3. Estimation is an exercise of mathematical optimization.

These fallacies exist as a set of hidden assumptions implicitly held for a number of years by many classical geostatisticians. Epistemic analysis can advance the state of geostatistics by removing misconceptions that have blocked its path. This chapter demonstrates how modern geostatistics succeeds in correcting the fallacy that spatiotemporal structure is purely geometrical. This is accomplished by promoting a more realistic concept involving knowledge of the underlying physical processes. The geometry placed upon space/time should establish a structure within which physical data, laws, theories, *etc.* make sense and can be represented with consistency, free of contradiction, and can be tested and used to make predictions in space/time.

The response of modern spatiotemporal geostatistics to the other misconceptions listed above is the subject of later chapters. As we shall see in Chapter 3, the second fallacy is corrected by demonstrating the importance of scientific theories, in addition to empirical evidence: data do not always speak for themselves. Mathematics is an instrument of analysis, not a constituent of things. It does not describe the behavior of natural phenomena, but only our knowledge of that behavior. The real challenge in today's multidisciplinary scientific arena is not merely to develop mathematical techniques to deal with numerical data, but also to develop the theoretical means for interpreting and integrating these data—as well as other important sources of physical knowledge—into the process of understanding. The analysis in Chapter 4 suggests a powerful response to this challenge, thus making possible the correction of the third fallacy that estimation is an exercise of mathematical optimization.

The Spatiotemporal Continuum Idea

Physical theories parameterized by space and time variables are considered more basic than those that are not (this is why, *e.g.*, mechanics is considered more basic than thermodynamics). Table 2.1 depicts the considerable ranges of these variables (*i.e.*, spatial distances and time intervals) encountered in natural studies (Ridley, 1994). Spatiotemporal continuity implies an integration of space with time and is a fundamental property of the mathematical formalism of natural phenomena. In most applied natural sciences, the following postulate is of vital importance for the purposes of spatiotemporal analysis and mapping.

POSTULATE 2.1: Space/time $\mathcal{E}$ is viewed as a set of points that are associated with a continuous spatial arrangement of events combined with their temporal order.

Events represent attributes related to a natural phenomenon or process (*e.g.*, contaminant concentration, hydraulic head, temperature, fluid velocity). Within the space/time continuum $\mathcal{E}$, space represents the order of coexistence of events and time represents the order of their successive existence. This is, in essence, the view of space/time held by the majority of scientists today.

While the concept of *identifiable points* in a continuum is most important, it is often taken for granted because it seems so obvious. Each identifiable point in a space/time continuum $\mathcal{E}$ is associated with an event, but $\mathcal{E}$ does not merely serve as a repository of events. In addition, many other considerably more complicated objects and processes can be described within $\mathcal{E}$. A few examples follow, in an attempt to throw some light on these basic spatiotemporal continuum concepts.

EXAMPLE 2.1: The space/time path of an individual entity P (*e.g.*, a particle) in $\mathcal{E}$ is represented by a set of points forming a *space/time trajectory*—also called a *world-line* (see Fig. 2.1). In other words, a world-line defines the spatial position of P at every instant of time with respect to a reference system (a system of coordinates that will be defined mathematically in a following section).

Just as points and curves constitute the basic elements of Euclidean geometry, events and world-lines are the basic elements of spatiotemporal geometry.

Table 2.1. Ranges of spatial distances and time intervals.

Meters		Seconds	
10^{25}	Size of Universe		
10^{20}	Distance to galactic center	10^{20}	
			Age of Universe
			Age of Earth
10^{15}	Distance to nearest star	10^{15}	Age of present-day continents
			Age of man
10^{10}	Distance to Sun	10^{10}	Lifespan of man
	Radius of Earth		
			Orbital period of Earth (1 yr)
10^{5}		10^{5}	Rotation period of Earth (1 day)
			Lifetime of free neutron
1	Man	1	Reaction time of man
10^{-5}	Living cell	10^{-5}	Lifetime of muon
	DNA molecule		
10^{-10}	Size of atom	10^{-10}	Response time of electronic device
10^{-15}	Size of proton	10^{-15}	Period of visible light wave
			Period of X-ray
		10^{-20}	
			Period of γ-ray
			Lifetime of unstable particle
		10^{-25}	

However, as we shall see later in this chapter, while the rules of Euclidean geometry are given axiomatically, the rules of spatiotemporal geometry are statements about the physical world. On the basis of events and world-lines, other useful space/time entities may be defined, as follows. A *position-line* is a set of points in $\mathcal{E}$ representing events associated with P that have the same spatial position with respect to the reference system (*e.g.*, the world-line

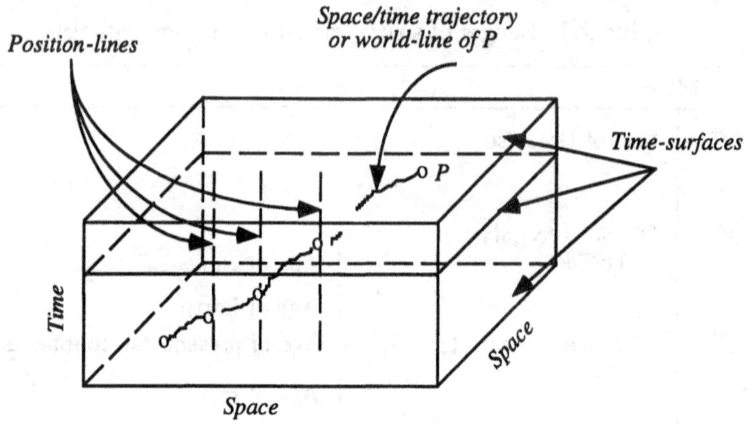

Figure 2.1. The dynamics of an entity in space/time $\mathcal{E}$.

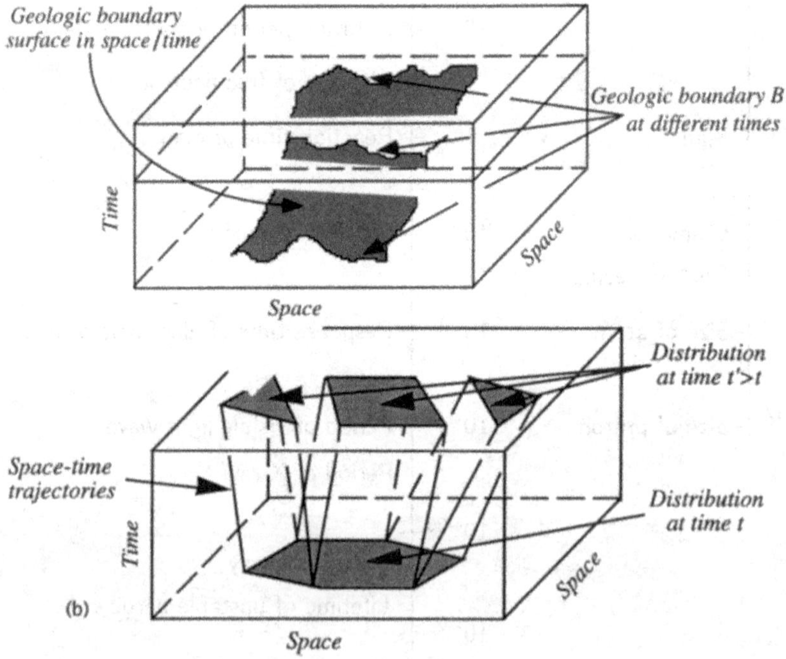

Figure 2.2. (a) The dynamics of the geologic boundary B in space/time $\mathcal{E}$. (b) Time evolution of contaminant distribution.

of a particle not moving in space is a position-line). A *time-surface* is a set of points in $\mathcal{E}$ representing a spatial arrangement of events having the same temporal coordinate with respect to the reference system. Therefore, a time-surface (sometimes technically called a *space-like hypersurface*) is not a single space that exists at all times, but rather a "space-at-an-instant" within the space/time continuum. □

EXAMPLE 2.2: In Figure 2.2a the evolution of a geologic boundary B at different times is represented by a set of lines. This set forms a *space/time surface* in $\mathcal{E}$. In Figure 2.2b, a two-dimensional contaminant distribution is broken into three parts, sweeping out a three-dimensional volume in space/time. □

The message conveyed by the above examples is that in most scientific studies we are seeking to establish a theory of relationships between events. This leads to our next postulate.

POSTULATE 2.2: Since events are associated with points of the space/time continuum $\mathcal{E}$, relationships between events are essentially relationships between the points of $\mathcal{E}$.

Beyond the continuum with its individual points, the *spatiotemporal structures* differ, as do the *geometries* which describe them. Generally four essential characteristics of spatiotemporal geometry may be identified:

(*i.*) Geometric objects (individual points, lines, planes, vectors, and tensors) that give geometry its *objective* features.

(*ii.*) Measurable properties of objects and spaces (angles, ratios, curvature, and distance or metric) that give geometry its *quantitative* features.

(*iii.*) Modes of comparison (equal, less than, greater than) that give geometry its *comparative* features.

(*iv.*) Spatiotemporal relationships (inside, outside, between, before, after) that give geometry its *relative* features.

COMMENT 2.1: *It is noteworthy that in their physical application these four characteristics of spatiotemporal geometry may be considered independently. For example, the spatiotemporal metric and the coordinate system used to describe that metric could be independent of each other. As we shall see below, this case has significant consequence in spatiotemporal analysis.*

EXAMPLE 2.3: Traditionally, geometries are divided into Euclidean and non-Euclidean. Table 2.2 summarizes some of the main features of these geometries in the case of two-dimensional space. Euclidean geometry was invented around 300 B.C. Euclid, in his epoch-making work, *Elements* (see Heath, 1956 [1908]), reduces the whole of geometrical science to an axiomatic form in which all propositions and theorems are deduced from a small number of axioms and postulates. Non-Euclidean geometries are derived by modification of one or more of the five basic Euclidean postulates. Well-known non-Euclidean geometries have been constructed by modifying Euclid's infamous fifth postulate (given a straight line and a point not on that line, there is one and only one line through that point parallel to the given line). In particular, the geometry of Bolyai and Lobachevski asserts the existence of more than one parallel, and the geometry of Riemann denies the existence of parallels altogether (Faber,

Table 2.2. Some Euclidean and non-Euclidean geometries in two-dimensional space.

Geometry	Euclidean	Riemannian	Bolyai–Lobachevskian
Surface	Plane	Sphere	Saddle
Parallels	1	0	Many
Curvature	0	>0	<0
Angular sum for triangles	180°	>180°	<180°
Ratio of circumference/diameter of circle	π	$<\pi$	$>\pi$

1983). The geometry of Riemann is exemplified by a spherical surface and that of Bolyai–Lobachevski by a saddle surface. □

Implicit in a spatiotemporal geometry are certain hypotheses concerning the way space and time operate. The goal of modern geostatistics is to find what kind of spatiotemporal structure we can choose on $\mathcal{E}$. Now, in other words, given that the set of events $\mathcal{E}$ forms the basic set in space/time, the problem is to equip modern spatiotemporal geostatistics with a mathematical structure that captures the significant physical relations. This important issue is addressed by the following postulate.

POSTULATE 2.3: Since a set of physical relationships between events is associated with a set of geometrical relationships between points in $\mathcal{E}$, a spatiotemporal structure is imposed on $\mathcal{E}$ by means of these physical relationships.

Starting from the assumption that relationships between events express physical knowledge $\mathcal{K}$ (laws of nature, scientific theories, empirical correlations, *etc.*; see Chapter 3) and that relationships between points are geometrical, the implication of Postulate 2.3 is that the geometrical structure we decide to use on space/time $\mathcal{E}$ has a strong influence on which kinds of knowledge $\mathcal{K}$ we can consider. Therefore, while all of the geometries are on an equal footing from a logical point of view, they are not on an equal footing *epistemically*.

Another important factor in the choice of a geometry on the space/time continuum $\mathcal{E}$ is the concept of $\mathcal{E}$ visualized *intrinsically* (or *internally*) *vs.* $\mathcal{E}$ visualized *extrinsically* (or *externally*). To visualize $\mathcal{E}$ internally is to imagine the kinds of experiences we would have if we were living in such a space/time.

To visualize $\mathcal{E}$ externally is to view it from a higher dimensional space that includes it.

EXAMPLE 2.4: Non-Euclidean geometries are more common than one might think. When our existence is confined to the surface of the Earth, we have a two-dimensional world view (*e.g.*, points on the surface are unambiguously located by two coordinates—longitude and latitude). The surface can be internally represented by a non-Euclidean geometry of the Riemannian type (which violates certain major assumptions of the Euclidean geometry; see Table 2.2). If we are able to leave the surface of the Earth and move to outer space, we can view Earth's surface from a three-dimensional standpoint that can be externally described in terms of Euclidean geometry (Table 2.2). However, we cannot step outside of the three-dimensional world into a four-dimensional space in order to visualize externally either a three-dimensional Euclidean space or a three-dimensional non-Euclidean space. Instead, our visualization of three-dimensional space can only be internal, *i.e.*, from the standpoint of beings confined within that space. □

The epistemic conclusions to be drawn from the above analysis are summarized by the final postulate of this section, as follows.

POSTULATE 2.4: The choice of an appropriate geometry to describe space/time continuum $\mathcal{E}$ depends on whether one adopts an intrinsic or an extrinsic visualization of $\mathcal{E}$.

The four fundamental postulates underlying the concept of space/time continuum $\mathcal{E}$ are illustrated in Figure 2.3. The continuum concept is paramount in representing the evolution of natural variables which assume values at any point in space/time, thus requiring continuously varying spatiotemporal coordinates. The *operational* importance of $\mathcal{E}$ is its bookkeeping efficiency that permits an ordered recording of physical measurements and the establishment of links between measurements by means of physical theories and mathematical expressions.

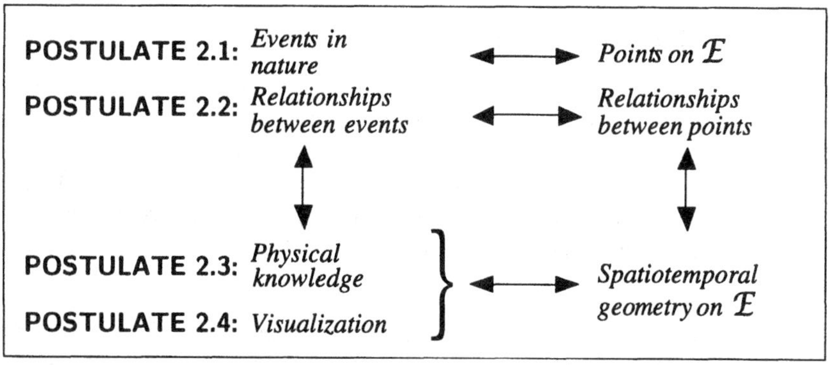

Figure 2.3. The four fundamental postulates of space/time continuum $\mathcal{E}$.

Viewed separately, both space and time are continua. Thus, in general, they share all the properties possessed by the abstract notion of a continuum. But there are important differences, as well. Time has certain *extra-continua* physical properties not shared with any other continuum, by virtue of which time is specifically time and not just a continuum (*e.g.*, recursivity is not a property that continua have in general; recursivity is, indeed, an extra-continua property of time, but not of space). The same is true with certain properties of space. These extra-continua physical properties—some of them known, some of them not—may contribute considerably to the behavior of the space/time system as a whole. Indeed, the difference between the degree of importance that space and time play in natural processes may depend on their extra-continua properties.

When we bring space and time together, the extra-continua physical properties of space integrate with those of time, producing a *holistic* space/time in which the whole is greater than the sum of its parts. In such a holistic environment, the spatiotemporal connections and cross-effects could control natural variations. We are not, *e.g.*, concerned merely about the distance between two geographical locations l_1 and l_2, but rather about the distance between the location l_1 at a specified time t and the location l_2 at another time t'. In the case of many natural variables, some aspects of time have repetitive or cyclical features which result because of holistic relationships between the spatial and temporal domains (not because time actually repeats). The cyclic behavior of precipitation profiles, *e.g.*, is due to the occurrence of certain spatiotemporal climatic processes. Also, the intimate connection between space and time is embodied in the astronomical unit of distance: the light year (*i.e.*, the distance traveled by light in one year).

Postulates 2.1–2.4 express rather broad, qualitative features of the space/time continuum $\mathcal{E}$. In order for $\mathcal{E}$ to be useful in real-world applications it must be equipped with *numerical information* and *operational concepts*. With this in mind we will examine these postulates in more detail in the following sections.

The Coordinate System

Postulate 2.3 involves a combination of three components: (*i.*) an *axiomatic* component (*i.e.*, a set of geometric objects, axioms, relations, and their logical consequences); (*ii.*) an *analytical* component (points represented by coordinate systems, relations expressed in terms of algebraic equations, *etc.*); and (*iii.*) an *empirical* component (a means of investigating which combination of axiomatic and analytical geometrical constructions best describes the observed facts).

The introduction of a *coordinate system* is essential in making a decision about how to assign "addresses" to different points in the space/time continuum $\mathcal{E}$. Generally, a point $\boldsymbol{p}$ in $\mathcal{E}$ can be identified by means of two separate numerical entities: the spatial coordinates $s = (s_1, \ldots, s_n) \in S \subset R^n$ and the temporal coordinate t along the temporal axis $T \subset R^1$, so that

$$p = (s, t) \in S \times T \tag{2.1}$$

In Equation 2.1 the union of space and time is defined in terms of their Cartesian product $S \times T$. Every point in space R^n is like every other point—a property called *homogeneity*. The existing observational evidence requires the introduction of up to three spatial dimensions (*i.e.*, $n = 1$, 2, or 3). In two- or three-dimensional space, the dimensions are equivalent to one another, a property sometimes called *isotropy*. Isotropy is a simple but precious property because it allows us to set up a suitable coordinate system. Time is also homogeneous—origin indifferent. In many applications it is sufficient to investigate the temporal evolution after an initial time instant in the not-too-distant past. Then, the initial time is set equal to zero and $T \subset [0, \infty)$. Exceptions are processes with long-range correlations (*e.g.*, the fractional Brownian motion).

COMMENT 2.2: *It is possible that unsolved physical problems in the future will oblige us to introduce a fourth spatial dimension. The fourth spatial dimension would be of a different character than the others, one that presumably will be in contrast to our intuitive ideas about space* (e.g., *it will question the idea of isotropy). Perhaps, while the existence of a fourth dimension external to our world cannot be directly experienced by us, we may nevertheless need to infer it from the geometry of our world. This suggestion should not be surprising, since the human mind can arrive at valid conclusions pertaining to things and events not directly within our perceptual world or evolutionary history.*

It can be seen from the above analysis that the "address" of a point in space/time is characterized by $n + 1$ numbers: n ($= 1$, 2, or 3) for the *spatial* coordinates plus 1 for the *temporal* coordinate. There are several methods for specifying the $n + 1$ numbers, each of them associated with a different coordinate system. It is mathematically important that any appropriate method ensure continuity and unambiguity in the assignment of "addresses." In physical applications it is typical to choose a coordinate system that works as simply as possible.

Space, though an obvious component of natural variation, requires careful definition. Equation 2.1 suggests more than one way to define the spatial location of a point, depending upon the choice of spatial coordinates $s = (s_1, \ldots, s_n)$. Essentially, the only constraint on the coordinate system implied by Equation 2.1 is that it possess n independent quantities available for denoting spatial position. We start with the so-called *general curvilinear* system of coordinates for the spatial component of Equation 2.1. This system is defined as follows.

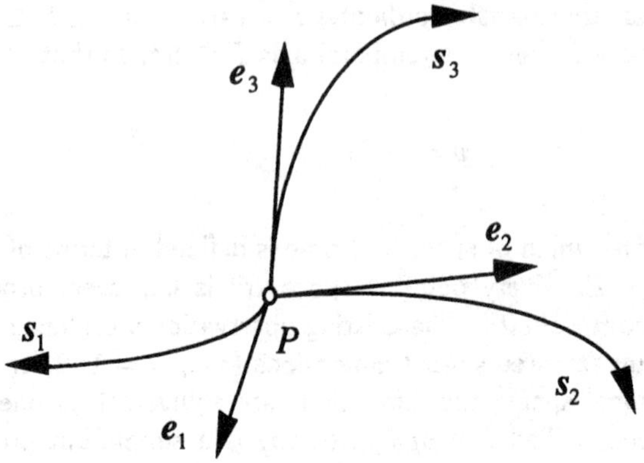

Figure 2.4. Curvilinear coordinates in three-dimensional space.

DEFINITION 2.1: The general curvilinear coordinate system $\{s_i\}$, $i = 1, \ldots, n$ associated with a point P is the set of oriented n coordinate curves that are the intersections of the n coordinate spaces (s_i is a constant) through the point P.

The orientation of the coordinate curves is established at each point P by drawing a set of basis vectors $\{e_i\}$, $i = 1, \ldots, n$ tangent to the coordinate curves. The basis vectors generally are nonorthogonal to one another and *local* (since they may change from point to point). When the basis vectors are orthogonal to one another we are dealing with an *orthogonal* curvilinear coordinate system. Such systems include the polar coordinate system (in R^2), as well as the cylindrical and spherical coordinate systems (both in R^3). The coordinate systems most widely used in geostatistics are orthogonal. When the coordinate spaces are planes in R^3 and, hence, the coordinate lines are straight lines, we are dealing with a *rectangular* coordinate system (in which the basis vectors do not vary from point to point).

EXAMPLE 2.5: A curvilinear system of coordinates $\{s_1, s_2, s_3\}$ is shown in Figure 2.4. The three coordinate surfaces (s_i is a constant, $i = 1, 2, 3$) through a point P determine the three coordinate curves (or lines, or directions). At each point P we draw basis vectors $\{e_1, e_2, e_3\}$ tangent to the coordinate curves. □

An interesting classification of the coordinate systems available in modern spatiotemporal geostatistics can be made in terms of the following two major groups:

1. The *Euclidean* group of coordinate systems, which assumes that the Euclidean geometry is valid (the Euclidean postulates and theorems apply). A widely used group of Euclidean coordinate systems are derived as special cases of the orthogonal curvilinear systems (see the following section on Euclidean coordinate systems).

2. The *non-Euclidean* group of coordinate systems, which assumes that the Euclidean geometry is not necessarily valid (certain of the Euclidean axioms, postulates, or theorems are violated). This group includes the *Gaussian* and *Riemannian* coordinate systems (see "Non-Euclidean coordinate systems" on p. 38).

The general curvilinear system introduced above can be used to coordinate events and processes in association with many kinds of geometries. Before proceeding with the detailed discussion of the coordinate systems used in modern spatiotemporal geostatistics, an issue worth emphasizing is the possibility of using global as well as local coordinate systems. In many geostatistical applications a single *global* coordinate system covers the whole space/time continuum of interest. But it is also possible that a global coordinate system is inappropriate in several other applications and *local* coordinate systems should be used instead. In other words, we may need to work with coordinate systems which cover only a portion of space/time and offer an internal visualization of the domain (these systems are sometimes called *coordinate patches*). The above notions are important in the development of certain non-Euclidean coordinate systems described in the following sections.

Euclidean coordinate systems

The most common Euclidean coordinate systems are essentially special cases of the orthogonal curvilinear systems mentioned above. In $n = 2$ or 3 spatial dimensions, the Euclidean group includes: (*i.*) The *rectangular* (*Cartesian*) coordinate system ($n = 2$ or 3; $s_1, \ldots, s_n$). (*ii.*) *Non-rectangular* coordinate systems (Fig. 2.5), such as the *polar* ($n = 2$; r, θ), the *cylindrical* ($n = 3$; r, θ, s_3), and the *spherical* system ($n = 3$; ρ, φ, θ).

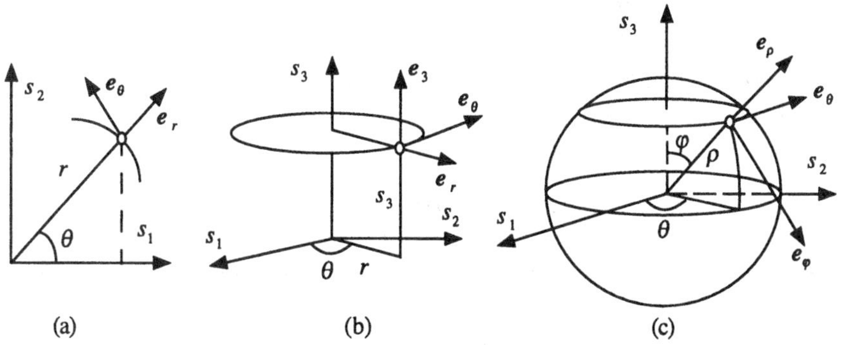

Figure 2.5. Polar (a), cylindrical (b), and spherical (c) coordinate systems.

The rectangular coordinate system is the most commonly used system in geostatistical applications. Its extension in the spatiotemporal domain is described in the following definition.

DEFINITION 2.2: In the rectangular Euclidean coordinate system, $\boldsymbol{s} = (s_1, \ldots, s_n)$ and t are the orthogonal projections of a geometrical point P_i on the spatial axes and the temporal axis, respectively; *i.e.*, the following mapping is defined

$$P_i \rightarrow (\boldsymbol{s}_i, t_i) = (\boldsymbol{s}, t)_i = \boldsymbol{p}_i \tag{2.2}$$

Alternatively, starting from the spatial coordinates $\boldsymbol{s}_i \in S$ and the time instant $t_j \in T$, a geometrical point P_{ij} can be defined in Euclidean space/time as

$$(\boldsymbol{s}_i, t_j) = \boldsymbol{p}_{ij} \rightarrow P_{ij} \tag{2.3}$$

EXAMPLE 2.6: In Figure 2.6 an illustration is given of the two approaches to defining a point in the rectangular $R^2 \times T$ domain. While in Equation 2.3 a point is denoted by a pair (i, j) of space and time labels, in Equation 2.2 a unified space/time label (i) is used. Equation 2.3 is more convenient when explicit reference is made to spatial locations and time instants; Equation 2.2 is more efficient in other cases because it provides a simpler notation. □

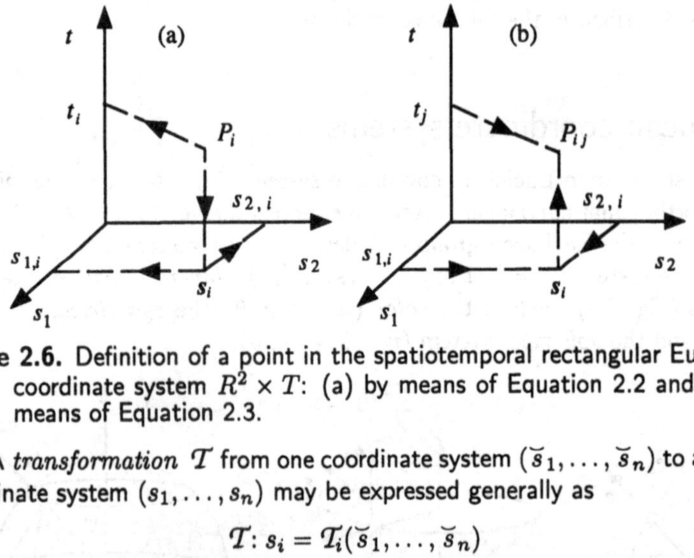

Figure 2.6. Definition of a point in the spatiotemporal rectangular Euclidean coordinate system $R^2 \times T$: (a) by means of Equation 2.2 and (b) by means of Equation 2.3.

A *transformation* $\mathcal{T}$ from one coordinate system $(\breve{s}_1, \ldots, \breve{s}_n)$ to another coordinate system $(s_1, \ldots, s_n)$ may be expressed generally as

$$\mathcal{T}\colon s_i = \mathcal{T}_i(\breve{s}_1, \ldots, \breve{s}_n) \tag{2.4}$$

$i = 1, \ldots, n$. Modern geostatistics regards transformations (Eq. 2.4) in two ways: as active and passive transformations. *Active* transformations refer to the same sort of coordinate system, but are associated with a change of origin, change of axis, and change of unit.

EXAMPLE 2.7: In Figure 2.7 the initial Cartesian system $(\breve{s}_1, \breve{s}_2)$ in R^2 is spatially translated and rotated, thus leading to another Cartesian system (s_1, s_2). This transformation is expressed as

$$\mathcal{T}\colon s_1 = \breve{s}_1 \cos\theta + \breve{s}_2 \sin\theta - a, \quad s_2 = -\breve{s}_1 \sin\theta + \breve{s}_2 \cos\theta - b \tag{2.5}$$

where the parameters a, b, and θ are shown in Figure 2.7. □

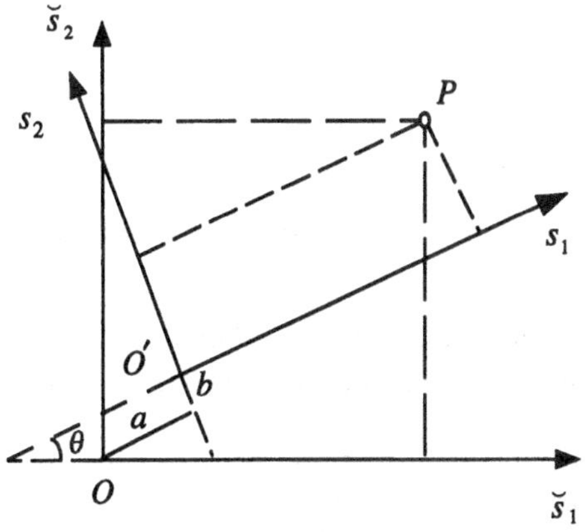

Figure 2.7. Transformation in R^2 (combination of translation with a rotation) that leads from one Cartesian coordinate system to another.

Passive transformations, on the other hand, relate different sorts of coordinate systems (*e.g.*, a Cartesian system and a polar system). An arbitrary orthogonal system $\{s_i\}$ may be expressed by means of a passive transformation of the general form (Eq. 2.4) where, for future notational convenience, the $(\breve{s}_1, \ldots, \breve{s}_n)$ denote the rectangular coordinates. In the special case that $\mathcal{T}$ is a linear transformation, the s_i are called *affine* coordinates. A few example transformations follow.

EXAMPLE 2.8: In the Euclidean polar coordinate system, $n = 2$ and $\boldsymbol{s} = (s_1, s_2) = (r, \theta)$, with $r > 0$, in which case the following transformations are established

$$\left.\begin{aligned} &\mathcal{T}\colon (s_1 = \sqrt{\breve{s}_1^2 + \breve{s}_2^2} = r,\ s_2 = \tan^{-1}(\breve{s}_2/\breve{s}_1) = \theta) \\ &\mathcal{T}^{-1}: (\breve{s}_1 = s_1 \cos s_2,\ \breve{s}_2 = s_1 \sin s_2) \end{aligned}\right\} \tag{2.6}$$

The inverse for $s_2 = \theta$ above is valid in the first and fourth quadrants of the $\breve{s}_1\,\breve{s}_2$ plane, while other solutions can be obtained over the remaining two quadrants (likewise for the θ coordinate in the cylindrical and spherical coordinate systems below). In cylindrical coordinates, $n = 3$ and $\boldsymbol{s} = (s_1, s_2, s_3) = (r, \theta, s_3)$, in which case the following transformations hold true

$$\left.\begin{aligned} &\mathcal{T}\colon (s_1 = \sqrt{\breve{s}_1^2 + \breve{s}_2^2} = r,\ s_2 = \tan^{-1}(\breve{s}_2/\breve{s}_1) = \theta,\ s_3 = \breve{s}_3) \\ &\mathcal{T}^{-1}: (\breve{s}_1 = s_1 \cos s_2,\ \breve{s}_2 = s_1 \sin s_2,\ \breve{s}_3 = s_3) \end{aligned}\right\} \tag{2.7}$$

In spherical coordinates, $n = 3$ and $\boldsymbol{s} = (s_1, s_2, s_3) = (\rho, \varphi, \theta)$, which are related to the rectangular coordinates by

$$\left.\begin{array}{l} \mathcal{T}\colon (s_1 = \sqrt{\breve{s}_1^2 + \breve{s}_2^2 + \breve{s}_3^2} = \rho,\ s_2 = \cos^{-1}(\breve{s}_3/\sqrt{\breve{s}_1^2 + \breve{s}_2^2 + \breve{s}_3^2}) = \varphi, \\ \qquad s_3 = \tan^{-1}(\breve{s}_2/\breve{s}_1) = \theta) \\ \mathcal{T}^{-1}\colon (\breve{s}_1 = s_1 \sin s_2 \cos s_3,\ \breve{s}_2 = s_1 \sin s_2 \sin s_3,\ \breve{s}_3 = s_1 \cos s_2) \end{array}\right\} \tag{2.8}$$

A summary of the main features of the orthogonal curvilinear coordinate systems is provided later in this chapter in Table 2.3 (p. 48). □

COMMENT 2.3: *The spherical coordinates (Eq. 2.8) are similar to the coordinates used in geographical studies* (e.g., *Langran, 1992; Bonham-Carter, 1994). In geographical coordinates,* $|\theta|$ *denotes the longitude (meridian angle) and is called east or west longitude according to whether* θ *is positive or negative.* $|\pi/2 - \varphi|$ *denotes the latitude (equatorial angle) and is called north or south latitude according to whether* $\pi/2 - \varphi$ *is positive or negative.*

The coordinate system used in a physical application can have important consequences. As shown later in Example 2.32 (p. 67), geostatistical analysis in terms of spherical *vs.* Cartesian coordinates can lead to very different variographies and spatial maps.

Non-Euclidean coordinate systems

Euclidean geometry and the associated coordinate systems are not appropriate for several types of physical space. As emphasized in Postulate 2.4 and Example 2.4, when we seek an internal visualization of arbitrary curved surfaces, the geometry is generally non-Euclidean and local descriptions of the domain may be considered (this is the case, *e.g.*, with spherical and saddle surfaces). The two-fold requirement (*i.e.*, local and internal characterization of the space/time domain) has led to the development of non-Euclidean coordinate systems which generally are not tied to rectangular coordinates. Among the most well-known non-Euclidean coordinate systems are the Gaussian system and its generalization, the Riemannian coordinate system. We will study these two systems next.

The *Gaussian* coordinate system was developed for the internal visualization of two-dimensional surfaces (*i.e.*, so that we can derive the coordinates of any point on the surface by means of measurements carried out without leaving the surface and moving into a third dimension into which the surface is embedded). In the Gaussian coordinate system, the network of parallel lines of the Euclidean plane is replaced by an arbitrary dense network of ordered curves (Fig. 2.8).

Fixing the value of one of the variables, u_1 or u_2, produces a curve on the surface in terms of the other variable, which remains free. In this way, a parametric network of two one-parameter families of curves on the surface is created, so that just one curve of each family passes through each point

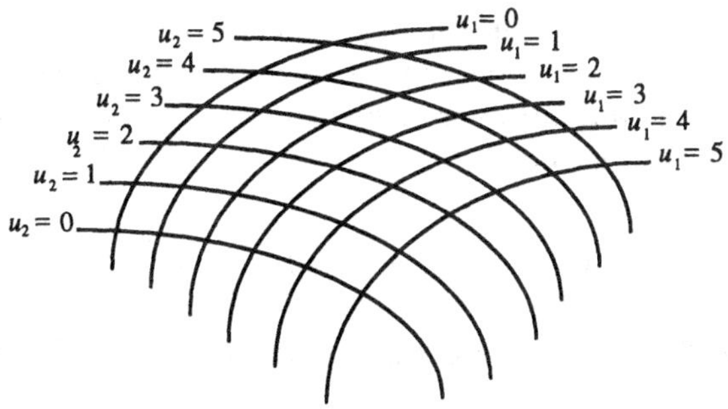

Figure 2.8. Gaussian coordinate system.

defined by the pair (u_1, u_2). Neither the u_1 curves nor the u_2 curves are equally spaced; the important idea of Gauss is that the spacing between them is not a consideration at all. The u_1 curves never cross other u_1 curves, and the u_2 curves never cross other u_2 curves. The u_1 curves intersect the u_2 curves, but not necessarily at right angles. This intersecting grid allows us to locate points, but not to measure distances between them directly. Indeed, unlike Euclidean coordinates, the idea of distance is not essential to Gaussian coordinates. The values of u_1 and u_2 are just numbers that assign an order to the network curves. So, when we say that the Gaussian coordinates of a point are $(3, 2)$, we provide no information about distance from the origin or any other point.

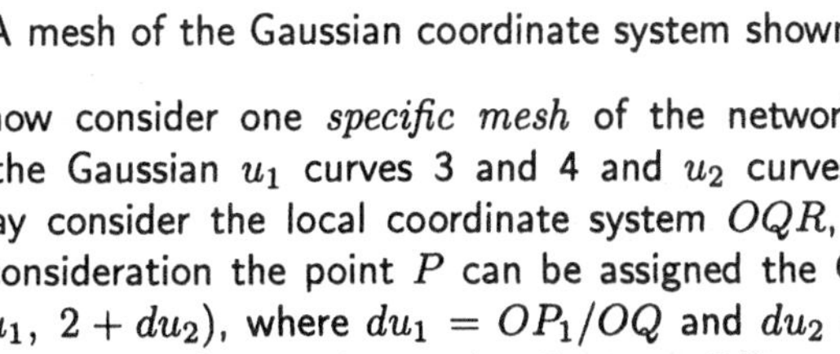

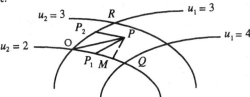

Figure 2.9. A mesh of the Gaussian coordinate system shown in Figure 2.8.

Let us now consider one *specific mesh* of the network in Figure 2.8 bounded by the Gaussian u_1 curves 3 and 4 and u_2 curves 2 and 3 (Fig. 2.9). We may consider the local coordinate system OQR, and within the mesh under consideration the point P can be assigned the Gaussian coordinates $(3 + du_1,\ 2 + du_2)$, where $du_1 = OP_1/OQ$ and $du_2 = OP_2/OR$ are ratios (which, however, do not give us the distances OP_1 and OP_2). If from actual measurements the lengths of OQ and OR are found to be ζ_{11} and ζ_{22}, respectively, the corresponding distances can be given by $(OP_1,\ OP_2) = (\zeta_{11}\, du_1,\ \zeta_{22}\, du_2)$. This discussion leads to the following definition.

DEFINITION 2.3: In the Gaussian coordinate system, the coordinates of a point P on a curved surface are defined by

$$(s_1,\ s_2) = (u_1 + du_1,\ u_2 + du_2) \tag{2.9}$$

where u_1, u_2 is a pair of real-valued parameters that serve as registration numbers or identification marks for the curves of the local mesh under consideration, and du_1, du_2 vary between 0 and 1. Within this mesh the corresponding distances are

$$(ds_1,\ ds_2) = (\zeta_{11}\, du_1,\ \zeta_{22}\, du_2) \tag{2.10}$$

where ζ_{11} and ζ_{22} are parameters obtained from physical measurements.

The parameters ζ_{11} and ζ_{22} do not change as long as we stay within the specific mesh, although they may change when we move from one mesh to another (see also the following section, "Metrical Structure"). The coordinates of any point on the surface covered by the Gaussian system (Fig. 2.8) are then known precisely if we know the values of ζ_{11} and ζ_{22} for every mesh. These values are obtained from physical measurements carried out by always remaining on the surface and never going outside of it.

EXAMPLE 2.9: As we saw above, Gaussian coordinates are useful when it is desirable to establish a coordinate system without leaving the surface under study, *i.e.*, internally. Consider a sphere, say, the Earth. If we think of the space inside and outside the Earth, we have a three-dimensional continuum with straight-line *geodesics* (the shortest distances between points) and we can use Euclidean geometry: a Cartesian coordinate system with origin at the Earth's center and axes along three mutually perpendicular diameters; and we have to refer to external points, lines, and planes. This would be very inconvenient, since it would necessitate taking measurements below the Earth's surface near its hot center, flying out into the atmosphere, *etc*. Things could be simplified considerably by having our coordinate system right on the surface of the Earth. In this case, straight lines will be replaced by arcs, for these are the geodesics. A triangle will consist of three intersecting arcs, and the sum of its angles will be greater than 180°. Longitude and latitude and all geographical facts related to them are defined in terms of the Earth's surface, *i.e.*, they are all internal aspects of the surface. Therefore, on the surface of the Earth we have a two-dimensional continuum with a non-Euclidean geometry of the Gaussian type. □

The preceding analysis reveals an important aspect regarding the choice of a coordinate system to describe natural phenomena. This aspect concerns the distinction between extrinsic *vs*. intrinsic coordinate systems (which is related to a similar distinction between visualization viewpoints—see Postulate 2.4) and is summarized in the following postulate.

POSTULATE 2.5: In certain situations understanding may be enhanced if the conception of points extrinsic (external) to physical space can be avoided and, instead, the analysis is performed in terms of intrinsic (internal) properties of the physical space.

The generalization of the Gaussian coordinate system to higher dimensional spaces requires the use of the concept (proposed by Riemann) of a *manifold*, which is a natural extension of Euclidean space (Riemann originally introduced the concept of a continuous manifold as a continuum of elements, such that a single element is defined by n continuous variable magnitudes; this definition includes the analytical conception of space in which each point is defined by n coordinates). In simple terms, an n-dimensional manifold is a set of points such that each point can serve as the origin of a local coordinate system (also called a *coordinate patch*) that covers only a portion of the space and offers an internal visualization of the surface. A set of coordinate patches that covers the whole space of interest is called an *atlas*. Since it requires two Gaussian coordinates to locate a point on an ordinary surface, the surface is called a *two-dimensional manifold*. If Cartesian coordinates are used, a relation among three of them is needed to describe such a manifold. Riemann extended Gauss' intrinsic geometry of two-dimensional surfaces ($n = 2$) to n-dimensional manifolds with $n > 2$, leading to the *Riemannian* coordinate system, as follows.

DEFINITION 2.4: The Riemannian coordinate system in an n-dimensional manifold consists of a network of u_i-curves ($i = 1, \ldots, n$) so that a one-to-one correspondence is established between each point P of the manifold and the n-tuples $(u_1, \ldots, u_n)$.

The Riemannian coordinates consistently assign to each point on a manifold a unique n-tuple; *i.e.*, they individuate, but neither relate nor measure. If explicit relations or measures are required by the physical situation of concern, extra constructions are introduced, like the metrical structure of a small n-dimensional region around each point (see the following section). In physical applications, one may choose to change or transform the parametric net $u_1, \ldots, u_n$ to suit some specified objective. Then, one may have to work with restricted regions (or patches) of a manifold to meet the requirements for such a transformation of parameters. In a Riemannian system, the jobs of Cartesian lines are divided between the coordinate patches fixing positions and the curves connecting points on a manifold. A rigorous mathematical presentation of the Riemannian theory of space, which is utilized in *differential geometry*, may be found in Chavel (1995).

Other useful non-Euclidean coordinate systems include the so-called *geodesics coordinate system*. According to this system, a point P on a surface is defined by choosing an origin O and then measuring the angle ϕ_P and the distance $|OP|$ to point P by means of the corresponding geodesic (Fig. 2.10). Depending on the shape of the surface, some constraints may apply. In the case of the Earth, *e.g.*, a maximum limit should be set for the distance such that $|OP| \leq c/2$, where c is the circumference of the Earth (considered as a sphere). In addition, the *Glebsch coordinates*, the *Boozer–Grad coordinates*, the *Hamada coordinates*, and the *toroidal coordinates* (*e.g.*, D'haeseleer *et al.*, 1991) are systems of coordinates with certain particular physical properties

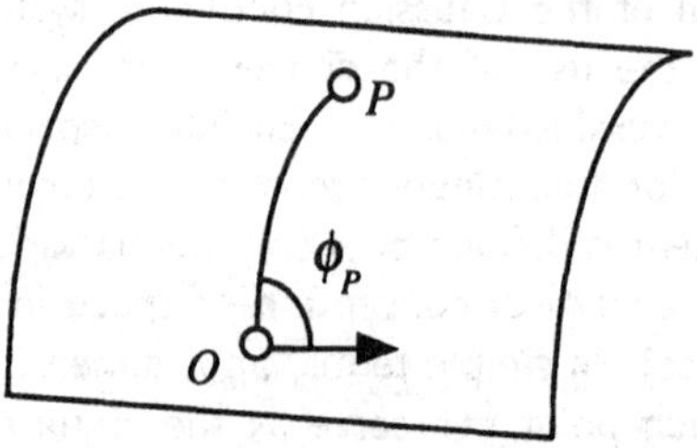

Figure 2.10. An illustration of the geodesics coordinate system.

(*e.g.*, toroidal coordinates are such that the equation of a magnetic-field line is that of a straight line in these coordinates). This section concludes with a postulate that summarizes an important conclusion from our examination of spatiotemporal analysis thus far and foreshadows results to be obtained in the following sections.

POSTULATE 2.6: The geostatistician should choose a coordinate system (Euclidean or non-Euclidean) that provides a physically meaningful representation of the situation and works as simply as possible.

In other words, in many geostatistical applications, the physical interpretation of the subject matter (*i.e.*, "In what physical situations do we use which system of coordinates?") may be equally as important as the computational convenience resulting from a particular choice of the coordinate system.

Metrical Structure

Central among the quantitative features of a spatiotemporal geometry is its *metrical structure*, that is, a set of mathematical expressions that define distances. Metrical properties (or distance relations) must be "added" once we have a set of points in space/time; the same set of points may be compatible with many metrical properties.

DEFINITION 2.5: A spatiotemporal metric is a function defined for a coordinate system such that the spatiotemporal distance between any two points in that system is determined from their coordinates.

Distance expressions cannot always be defined unambiguously. The expression for the metric of any continuum is dependent on two entirely different factors: (*i.*) a "relative" factor—the particular *coordinate system*; and (*ii.*) an "absolute" factor—the *nature of the continuum* itself (whether it forms a plane, a sphere, or an ellipsoid; the physical laws governing the natural variables occurring within the continuum, *etc.*).

In modern spatiotemporal geostatistics, it is usually convenient to consider two prime metrical structures: one is the *separate* metrical structure and the other is the *composite* metrical structure. Both metrical structures are discussed below in considerable detail.

Separate metrical structures

These metrics are convenient for many geostatistical applications because they treat the concept of distance in space and time separately. The main idea is introduced by the following definition.

DEFINITION 2.6: The separate metrical structure includes a spatial distance $|ds| \in R^1_{+,0}$ and an independent time interval $dt = t_2 - t_1 \in T$, so that

$$|dp| : (|ds|\,,\ dt) \in R^1_{+,0} \times T \tag{2.11}$$

In Equation 2.11, the structures of space and time are posited independently. Thus "distance" has meaning only at a fixed point in space (when it means "time elapsed") or at a fixed time (when it means "distance between spatial locations"). The formulation (Eq. 2.11) includes two celebrated space/time structures: Newtonian and Galilean. *Newtonian* space/time is defined by the following properties: between any two points in space/time (events) $\boldsymbol{p}$ and $\boldsymbol{p}'$ there exist a Euclidean spatial distance $|d\boldsymbol{s}|$ and a temporal interval dt. An alternative conception to Newtonian space/time is *Galilean* space/time, which is defined so that between any two points $\boldsymbol{p}$ and $\boldsymbol{p}'$ there always exists a temporal interval dt, but only if the two points are simultaneous is there any spatial distance $|d\boldsymbol{s}|$ between them.

The distance $|d\boldsymbol{s}|$ can take on different meanings depending upon the particular topographic space under consideration. The following examples explore several Euclidean and non-Euclidean spatial distances in two and three dimensions.

EXAMPLE 2.10: In the Euclidean plane and in ordinary three-dimensional Euclidean space, $|d\boldsymbol{s}|$ is defined as the length of the line segment between the spatial locations $\boldsymbol{s}_1$ and $\boldsymbol{s}_2 = \boldsymbol{s}_1 + d\boldsymbol{s}$, *i.e.*, the *Euclidean* distance in a rectangular coordinate system is defined by the *Pythagorean* formula

$$|d\boldsymbol{s}| = \sqrt{\sum\nolimits_{i=1}^{n} ds_i^2} \tag{2.12}$$

Frequently, the distance concept most useful in a particular application is non-Euclidean. In some physical situations it makes sense to define the *absolute* distance between the spatial locations P_1 and P_2 with coordinates $\boldsymbol{s}_1$ and $\boldsymbol{s}_2 = \boldsymbol{s}_1 + d\boldsymbol{s}$, respectively, as follows (see, also, Fig. 2.11)

$$|d\boldsymbol{s}| = \sum\nolimits_{i=1}^{n} |ds_i| \tag{2.13}$$

Distance as defined in Equation 2.13 could represent the length of the shortest path traveled by a particle that moves from P_1 to P_2, if the particle is constrained by the physics of the situation to move along the sides of the grid; or, it could represent the shortest path that a car must travel to get from point

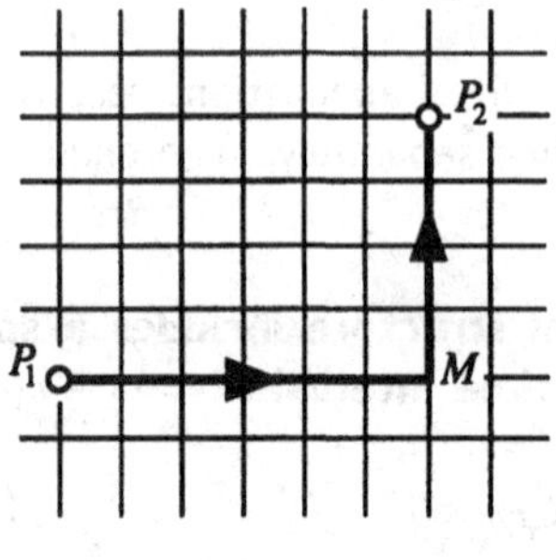

Figure 2.11. Distance in the sense of Equation 2.13.

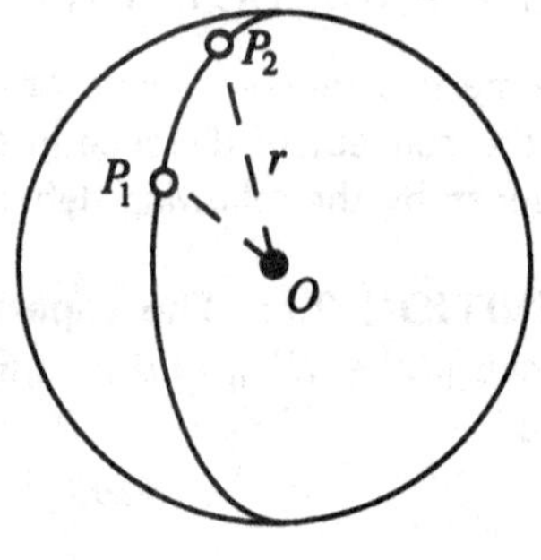

Figure 2.12. Distance in the sense of Equation 2.15.

P_1 to P_2 in a city. Therefore, this distance (Eq. 2.13) may be more appropriate for processes that actually occur on a discrete grid or network of some sort (however, this may not be true for the simulation of continuous processes by means of numerical grids, since in this case the grid is only a convenient modeling device and does not change the spatiotemporal metric).

Yet another distance $|ds|$ is defined by

$$|ds| = \max\,(|ds_i|\,;\ i = 1, \ldots, n) \tag{2.14}$$

In the case of Figure 2.11, Equation 2.14 gives $|ds| = |P_1 M|$. When dealing with distances between two geographical points P_1 and P_2 on the surface of the Earth (consider a sphere with center O and radius r, as shown in Fig. 2.12), the *arc* distance $|ds|$ is defined as the length of the smaller arc of the great circle joining these two points, *i.e.*,

$$|ds| = r\ d\varphi \tag{2.15}$$

where $d\varphi$ is the radian measure of the angle $P_1 O P_2$.

Finally, Euclidean metrics are not usually the most suitable measures of distance in fractal spaces (*e.g.*, processes taking place within porous media are more adequately represented by fractal rather than Euclidean geometry). There is not a general expression of the metric in fractal spaces, which rather depends on the physics of the situation (Christakos *et al.*, 2000a). In fact, formulating explicit metric expressions (such as Eq. 2.12) is not always possible in fractal spaces, since the physical laws may not be available in the form of differential equations. Geometric patterns in fractal spaces are self-similar (or statistically self-similar in the case of random fractals) over a range of scales (Mandelbrot, 1982; Feder, 1988); self-similarity implies that fractional (fractal) exponents characterize the scale dependence of geometric properties. A common example is the percolation fractal generated by random occupation of sites or bonds on a discrete lattice. Length and distance measures on a percolation cluster, denoted by $\ell(r)$, scale as power laws with the Euclidean (linear) size of the

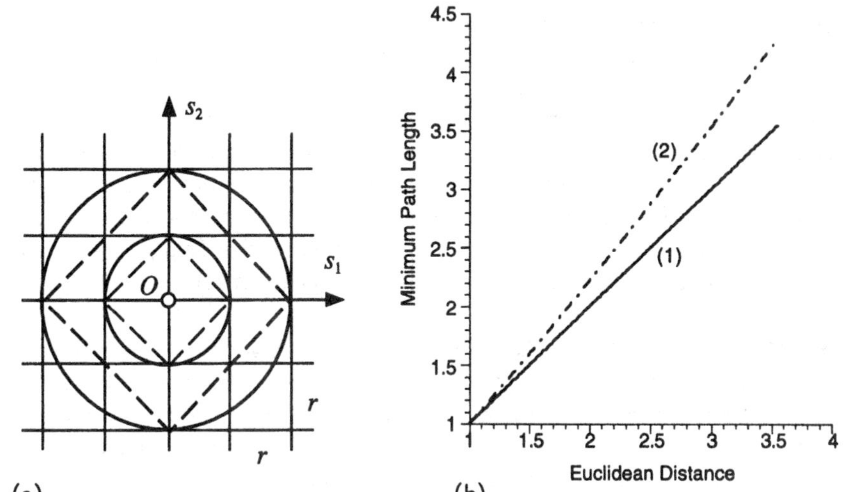

(a) (b)

Figure 2.13. (a) Iso-covariance contours assuming a Euclidean metric (Eq. 2.12)—continuous lines—and an absolute metric (Eq. 2.13)—dashed lines. (b) Minimum path length between two points at a Euclidean distance r in Euclidean space (curve 1) and in a space with fractal length dimension $d_o = 1.15$ (curve 2).

cluster. Fractals are homogeneous functions such that $\ell(br) = b^{d_o}\ell(r)$, where r is the appropriate Euclidean distance, d_o is the fractal exponent for the specific property, and b is a scaling factor. In two dimensions, *e.g.*, the perimeter of a cluster's hull scales as $\ell_h \propto \alpha^{d_h}$ for a cluster of linear dimension α (d_h is the corresponding fractal exponent); in contrast, the perimeter of Euclidean objects varies linearly with the length α. □

The following example gives an illustration of how some of the metrics considered above can lead to quite different geometric properties of space.

EXAMPLE 2.11: In the geostatistical analysis of spatial *isotropy* in R^2 (Olea, 1999), one needs to define iso-covariance contours. In Figure 2.13a it is shown that, while in the case of the metric in Equation 2.12 these contours are circles of radius r, $2r$, *etc.*, in the case of the metric in Equation 2.13 the iso-covariances are squares with sides $\sqrt{2}r$, $2\sqrt{2}\,r$, *etc.* In Figure 2.13b (Christakos *et al.*, 2000a) we show the minimum path length between two points separated by a Euclidean distance r in Euclidean space and in a fractal space with $d_o = 1.15$. The path length in the Euclidean space is a linear function of the distance between the two points, for all types of paths. The fractal path length increases nonlinearly, because the fractal space is nonuniform and obstacles to motion occur at all scales. □

We should distinguish carefully a mere change of coordinates from an actual change of geometry. One can use, *e.g.*, non-rectangular coordinates on a flat Euclidean plane as well. Generally, what distinguishes a flat Euclidean geometry is the *existence* of coordinates satisfying the Pythagorean formula. The curvilinear coordinate systems presented above (polar, cylindrical, and spherical) are associated with the Euclidean metric in Equation 2.12—since they are connected via Equation 2.4 with rectangular coordinates and Euclidean space. However, these same systems can be formally adopted in a non-Euclidean space if there are physical and/or mathematical reasons for doing so. This implies that the spatiotemporal metric and the coordinate system used to describe that metric are *independent* of each other (*e.g.*, we may have a Euclidean coordinate system with a non-Euclidean metric).

EXAMPLE 2.12: On the Euclidean plane one can always transform any coordinate system into coordinates that satisfy the Pythagorean metric (Eq. 2.12), *e.g.*, by transforming polar coordinates into Cartesian coordinates. By contrast, on a curved non-Euclidean surface it is not possible to perform such a transformation, since Cartesian coordinates do not exist. □

Several of the spatial distances—Euclidean and non-Euclidean—can be summarized in terms of the following definition.

DEFINITION 2.7: The Riemannian distance is defined as

$$|ds| = \sqrt{\sum\nolimits_{i,j=1}^{n} g_{ij}\, ds_i\, ds_j} \tag{2.16}$$

where g_{ij} are coefficients that generally are dependent on the spatial location.

Equation 2.16 may be viewed as Riemann's generalization of Pythagoras' metric (Eq. 2.12)—although one may wonder if Pythagoras would have recognized this. A Riemannian coordinate system (Definition 2.4) together with a metric (Definition 2.7) determine a *Riemannian space*. The following two examples present some Euclidean and non-Euclidean metrics as special cases of Equation 2.16.

EXAMPLE 2.13: The Euclidean metric in a rectangular coordinate system—Equation 2.12—is a special case of Equation 2.16 for $g_{ii} = 1$ and $g_{ij} = 0$ $(i \neq j)$. In a polar coordinate system, the Euclidean metric is obtained from Equation 2.16 for $n = 2$ and $g_{11} = 1$, $g_{22} = s_1^2$, $g_{ij} = 0$ $(i \neq j)$. For $n = 3$ and $g_{11} = g_{33} = 1$, $g_{22} = s_1^2$, $g_{ij} = 0$ $(i \neq j)$, Equation 2.16 provides the metric in a cylindrical system. In a spherical coordinate system, the Euclidean metric is obtained from Equation 2.16 for $n = 3$ and $g_{11} = 1$, $g_{22} = s_1^2$, $g_{33} = [s_1 \sin(s_2)]^2$, $g_{ij} = 0$ $(i \neq j)$. □

As is demonstrated in the following example, as far as the Gaussian or Riemannian geometries are concerned, the physical surface leaves its stamp on the metric.

EXAMPLE 2.14: In the case of a two-dimensional curved physical surface (*e.g.*, the surface of a hill), the Riemannian distance (Eq. 2.16) reduces to the Gaussian distance $|OP|$ of Figure 2.9; *i.e.*, for $n = 2$, Equation 2.16 gives

$$|ds| = \sqrt{g_{11}\,ds_1^2 + g_{22}\,ds_2^2 + 2g_{12}\,ds_1 ds_2}$$

where, for notational generality, the u_1 and u_2 are replaced by s_1 and s_2, respectively; the values of g_{11}, g_{22}, and $g_{12} = g_{21}$ are found from actual measurements (P_1M is the projection of P_1P on OQ). The *internal* geometry of the surface is determined completely if the values of g_{11}, g_{22}, and g_{12} are known for every local mesh (recall that the g's are generally functions of the spatial location coordinates s_i, $i = 1, 2$). Suppose that $s_1 = \theta$ and $s_2 = \pi/2 - \varphi$ represent the longitude and latitude of a point on a spherical surface (Fig. 2.12) whose radius r is one unit of length; then, $g_{11} = \cos^2 s_2$, $g_{12} = g_{21} = 0$, and $g_{22} = 1$.

If the surface is a hyperbolic paraboloid, one can choose Gaussian coordinates such that $g_{11} = 1 + 4s_1^2$, $g_{12} = g_{21} = -4s_1\,s_2$, and $g_{22} = 1 + 4s_2^2$.

A surface of constant curvature a is determined by the following metric coefficients $g_{11} = g_{22} = [1+0.25\,a(s_1^2+s_2^2)]^{-2}$ and $g_{12} = g_{21} = 0$. The metric coefficients $g_{11} = k^2(k^2+s_2^2)(k^2+s_1^2+s_2^2)^{-2}$, $g_{22} = k^2(k^2+s_1^2)(k^2+s_1^2+s_2^2)^{-2}$, and $g_{12} = g_{21} = -s_1\,s_2(k^2+s_1^2+s_2^2)^{-2}$ have been used to characterize a certain surface of curvature k^{-2}. Note that as k tends to infinity, this metric converges to the Pythagorean metric. □

The tensor $\boldsymbol{g} = (g_{ij})$ is called the metric tensor. Although from the viewpoint of differential geometry the metric tensor gives infinitesimal length elements, the mathematical form of Equation 2.16 may be used to define finite distances as well (see, *e.g.*, the section entitled "Correlation analysis and spatiotemporal geometry" on p. 61). The choice of $\boldsymbol{g}$ should satisfy certain physical and mathematical conditions. A possible set of mathematical conditions is described in the comment below.

COMMENT 2.4: *The metric tensor* $\boldsymbol{g}$ *should be: (i.) of differentiability class* C^2 *(all second-order derivatives of* g_{ij} *exist and are continuous); (ii.) symmetric* ($g_{ij} = g_{ji}$)*; (iii.) nonsingular* ($|g_{ij}| \neq 0$)*; and (iv.) such that the Riemannian distance* $|ds|$ *is invariant with respect to a change of coordinates.*

EXAMPLE 2.15: Consider the transformation from a special $\{s_i\}$ coordinate system to the rectangular $\{\breve{s}_i\}$ coordinate system defined by $\breve{s}_1 = s_1$ and $\breve{s}_2 = s_1 + \ln s_2$. The Jacobian is

$$J = \begin{bmatrix} 1 & 0 \\ 1 & (s_2)^{-1} \end{bmatrix}$$

Table 2.3. Common orthogonal curvilinear coordinates.

Curvilinear	Rectangular	Polar	Cylindrical	Spherical
s_1	s_1	r	r	ρ
s_2	s_2	θ	θ	φ
s_3	s_3	–	s_3	θ
g_{11}	1	1	1	1
g_{22}	1	r^2	ρ^2	r^2
g_{33}	1	–	1	$(r \sin\varphi)^2$
$\mathbf{e}_1$	$\mathbf{e}_1$	$\mathbf{e}_r$	$\mathbf{e}_r$	$\mathbf{e}_\rho$
$\mathbf{e}_2$	$\mathbf{e}_2$	$\mathbf{e}_\theta$	$\mathbf{e}_\theta$	$\mathbf{e}_\varphi$
$\mathbf{e}_3$	$\mathbf{e}_3$	–	$\mathbf{e}_3$	$\mathbf{e}_\theta$

Therefore, the Euclidean metric tensor is

$$g = J^T J = \begin{bmatrix} 2 & (s_2)^{-1} \\ (s_2)^{-1} & (s_2)^{-2} \end{bmatrix}$$

The distance for the $\{s_i\}$ system is given by

$$|ds| = \sqrt{2\, ds_1^2 + (s_2)^{-2} ds_2^2 + 2(s_2)^{-1}\, ds_1\, ds_2}$$

and the conditions of Comment 2.4 are satisfied. □

A summary of the commonly used orthogonal curvilinear (Euclidean) coordinate systems is provided in Table 2.3. As mentioned, Euclidean metrics are special cases of Equation 2.16. For a metric of the form of Equation 2.16 to be considered Euclidean, a transformation of coordinates must exist such that Equation 2.16 can be put in Cartesian form (Eq. 2.12). In Euclidean space, the general arc-length formula—which is valid in various coordinate systems—is given by

$$s(v) = \int_a^v \sqrt{\left| \sum\nolimits_{i,j=1}^{n} g_{ij} \frac{ds_i}{du} \frac{ds_j}{du} \right|}\, du \tag{2.17}$$

where $s(v)$ gives the arc-length of the curve $s_i = s_i(v)$, $1 \leq i \leq n$, and $a \leq v \leq b$. Equation 2.17 yields

$$(ds/dv)^2 = \sum\nolimits_{i,j=1}^{n} g_{ij} (ds_i/dv)(ds_j/dv)$$

and by introducing the differential $ds_i \equiv [ds_i(v)/dv]\, dv$ and $ds^2 = |ds|^2$, we find Equation 2.16. Thus, in Euclidean space, Equation 2.16 is equivalent to Equation 2.17. Equation 2.17 is independent of the particular parameterization of the curve.

EXAMPLE 2.16: The Euclidean metric in a rectangular coordinate system—see Equation 2.12—is obtained from Equation 2.17 by setting $g_{ij} = \delta_{ij}$, *etc.* □

COMMENT 2.5: *Assume that a transformation can be established from a given coordinate system* $\{s_i\}$ *to a rectangular coordinate system* $\{\breve{s}_i\}$, *and let* $J = (\partial \breve{s}_i / \partial s_i)$ *be the Jacobian matrix of this transformation. Then, the matrix* $\boldsymbol{g} = (g_{ij})$ *of the Euclidean metric in the* $\{s_i\}$ *coordinate system can be expressed as* $\boldsymbol{g} = J^T J$.

EXAMPLE 2.17: Consider the transformation of systems from cylindrical $\{s_i\}$ coordinates to rectangular $\{\breve{s}_i\}$ coordinates in Equation 2.7. The Jacobian is

$$J = \begin{bmatrix} \cos s_2 & -s_1 \sin s_2 & 0 \\ \sin s_2 & -s_1 \cos s_2 & 0 \\ 0 & 0 & 1 \end{bmatrix} \tag{2.18}$$

and, hence, the Euclidean metric for cylindrical coordinates is given by

$$g = J^T J = \begin{bmatrix} 1 & 0 & 0 \\ 0 & (s_1)^2 & 0 \\ 0 & 0 & 1 \end{bmatrix} \tag{2.19}$$

i.e., $g_{11} = g_{33} = 1$, $g_{22} = s_1^2$, and $g_{ij} = 0$ $(i \neq j)$, which is the same result as that obtained in Example 2.13. Of course, other (non-Euclidean) metrics could be considered for the cylindrical coordinate system, as well. □

COMMENT 2.6: *Suppose that a transformation between a system of generalized curvilinear coordinates* $\{s_i\}$ *and an underlying system of rectangular coordinates* $\{\breve{s}_i\}$ *exists. Then, the metric coefficients can be expressed as*

$$g_{ij} = \sum\nolimits_{k=1}^{n} \frac{\partial \breve{s}_k}{\partial s_i} \frac{\partial \breve{s}_k}{\partial s_j} \tag{2.20}$$

In Equation 2.16 above we used the metric coefficients g_{ij} to represent the coefficients in the distance $|ds|$ for a general coordinate system. The metric coefficients g_{ij} play an important role in several other *operations* of spatiotemporal geometry related to the modeling of natural phenomena, such as the determination of a suitable metric (the issue is discussed also in the section entitled "Restrictions on spatiotemporal geometry imposed by physical laws" on p. 56). The gradient, Laplacian, and divergence operators for a scalar field $X(\boldsymbol{p})$ and a vector field $\boldsymbol{X}(\boldsymbol{p})$ can be expressed in terms of general curvilinear coordinates as follows

$$\nabla X = \sum\nolimits_i (\sqrt{g_{ii}})^{-1} \frac{\partial X}{\partial s_i} e_i \tag{2.21}$$

$$\Delta X = (\sqrt{g})^{-1} \sum_i \sum_j \frac{\partial}{\partial s_i} \left[\sqrt{g}\, g^{ij} \frac{\partial X}{\partial s_j} \right] \tag{2.22}$$

and

$$\nabla \cdot \boldsymbol{X} = (\sqrt{g})^{-1} \sum_i \frac{\partial}{\partial s_i} \left[\sqrt{g\, g_{ii}^{-1}}\, X_i \right] \tag{2.23}$$

where g^{ij} are elements of the inverse of the matrix $\boldsymbol{g} = (g_{ij})$, and $g = |\boldsymbol{g}|$.

EXAMPLE 2.18: Consider the special case of orthogonal curvilinear coordinates. Then, $g_{ij} = 0\ (i \neq j), = 1\ (i = j)$; $g^{ij} = 0\ (i \neq j), = g_{ii}^{-1}\ (i = j)$; $\boldsymbol{g} = diag\,(g_{ii})$ and $g = \prod_i g_{ii}$. The form of Equation 2.21 remains the same, but Equations 2.22 and 2.23 are now written as follows

$$\Delta X = (\sqrt{g})^{-1} \sum_i \frac{\partial}{\partial s_i} \left[\sqrt{\prod_{j \neq i} g_{jj}}\, \frac{\partial X}{\partial s_i} \right] \tag{2.24}$$

and

$$\nabla \cdot \boldsymbol{X} = (\sqrt{g})^{-1} \sum_i \frac{\partial}{\partial s_i} \left[\sqrt{\prod_{j \neq i} g_{jj}}\, X_i \right] \tag{2.25}$$

where $g = \prod_i g_{ii}$. □

Composite metrical structures

In the case of composite metrical structure, a higher level of physical understanding of space/time is assumed which may involve theoretical and empirical facts of the natural sciences. According to this approach, the basis of metrical determination should be sought outside the abstract geometric objects in the physical processes that act on them. The composite metrical structure approach is described by the following definition.

DEFINITION 2.8: In the composite metrical structure, space and time parameters are connected by means of an analytical expression, *i.e.*,

$$|d\boldsymbol{p}| = g(ds_1, \ldots, ds_n, dt) \tag{2.26}$$

where g is a function determined from the knowledge available (topography, physical laws, *etc.*).

EXAMPLE 2.19: Consider a point P in the space/time continuum $R^2 \times T$ with coordinates $\boldsymbol{p} = (s_1, s_2, t)$, as in Figure 2.14. A natural variable varying within this continuum is written as $X(\boldsymbol{p}) = X(s_1, s_2, t)$.

If the separate metrical structure were used, the distance $|\overrightarrow{OP}|$ would be defined indirectly in terms of two independent entities—space and time—forming the pair $(|\boldsymbol{s}|, t)$, where the distance $|\boldsymbol{s}|$ may have one of the spatial forms discussed in Examples 2.10 or 2.13 above. If, however, the composite metrical

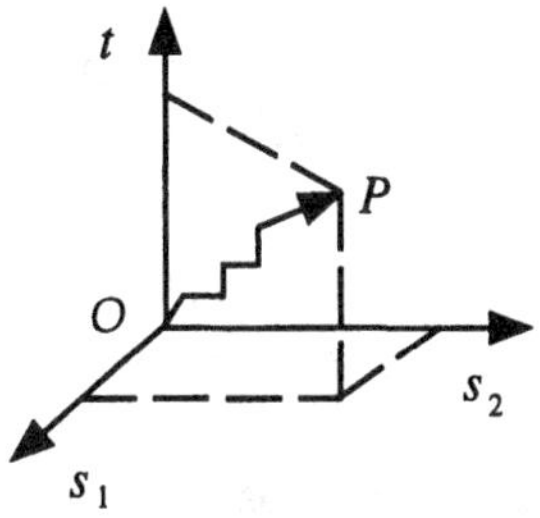

Figure 2.14. A point in the space-time domain $R^2 \times T$.

structure is used, the function g should be determined by means of the physical knowledge available about the natural variable $X(s_1, s_2, t)$.

The wavy line in Figure 2.14 is used to denote that the spatiotemporal vector $\vec{OP} = \mathbf{p}$ does not, in general, have a Euclidean metrical structure, *i.e.*, the function g in Equation 2.26 is not necessarily of the form $\sqrt{s_1^2 + s_2^2 + t^2}$. In fact, the latter form may have no physical meaning. We will continue the discussion of Figure 2.14 in Example 2.26. □

COMMENT 2.7: *The situation with the physical meaning of vector* $\vec{OP} = \mathbf{p}$ *in Figure 2.14 raises an important—although sometimes ignored—logical connection between a vector and its components, in general. In many cases, it is the vector itself that carries the fundamental meaning. In some other cases, however, while its components are physically meaningful, the vector itself has no physical meaning independent of its components. An example of the former situation is the velocity vector, which has a meaningful geometric representation in terms of its coordinates (directional velocities) and it also carries a fundamental meaning by itself (*i.e., *even if its coordinates are erased from the geometric representation). An example of the reverse situation is a vector that characterizes an individual so that its three components represent "age," "weight," and "height." If these components are erased from the geometric representation, nothing is signified by the remaining vector, which becomes totally meaningless.*

A special case of Equation 2.26 is the space/time generalization of the spatial distance in Equation 2.16 that leads to the spatiotemporal *Riemannian* metric defined below.

DEFINITION 2.9: The spatiotemporal Riemannian metric is defined as

$$|d\mathbf{p}| = \sqrt{\sum\nolimits_{i,j=1}^{n} g_{ij}\, ds_i\, ds_j + g_{00} dt^2 + 2dt \sum\nolimits_{i=1}^{n} g_{0i}\, ds_i} \qquad (2.27)$$

where the metric coefficients g_{ij} $(i,\ j = 1, \ldots, n)$ **are functions of the spatial location and time.**

A well-known spatiotemporal metric of the Riemannian type is discussed in the following example.

EXAMPLE 2.20: In the relativistic representation of space/time a composite (Riemannian) metric (Eq. 2.27) is defined as follows (Chavel, 1995)

$$|dp| = \sqrt{\sum_{i=1}^{3} ds_i^2 - c^2 dt^2} \tag{2.28}$$

The distance (Eq. 2.28) is obtained directly from Equation 2.27 by letting $g_{00} = -c^2$, $g_{ii} = 1$ ($i = 1, 2, 3$; c is the speed of light) and $g_{ij} = 0$ ($i \neq j$). This space/time distance is known as the *Minkowski* metric (for a physical interpretation, see Example 2.23 below). □

Riemann has shown that Equation 2.27 provides a sufficient but not necessary specification that satisfies the fundamental requirements of a metric (Weber, 1953). The metric (Eq. 2.27) can be used for an extrinsic (external) as well as an intrinsic (internal) characterization of space/time, where the g_{ij} coefficients are obtained from the available physical knowledge (measurements, laws, *etc.*) and are, in general, functions of space and time. In the case of the external geometry model, a point is defined in space/time using Euclidean coordinates. In the case of the intrinsic geometry model, the Riemannian coordinates of a point can be defined without knowledge of distances. Then, spatial distances and time lags are calculated within small local meshes using an expression of the form of Equation 2.27, where the g_{ij} coefficients may vary from one mesh to the other. These coefficients essentially convert the increments of Riemannian coordinates in each small local mesh into spatial distances and time lags.

Some comments on physical spatiotemporal geometry

By way of a summary, in modern geostatistics the spatiotemporal geometry is generally represented by the ordered pair

$$(\mathcal{E}, |dp|) \tag{2.29}$$

where $\mathcal{E}$ is a space/time continuum, each point of which is associated with a set of coordinates (Euclidean or non-Euclidean, Cartesian or non-rectangular, *etc.*), and $|dp|$ denotes a suitable space/time metric (separate or composite, Euclidean or Riemannian, *etc.*). As we saw above, Newtonian space/time requires two distinct structures to be specified: $|ds|$ and dt—see Equation 2.11. Minkowskian space/time, on the other hand, involves one structure: the metric in Equation 2.28.

As far as the representation of physical knowledge is concerned, the existing spatiotemporal geometries display some important differences. Traditional geometries roll together metrical and other spatiotemporal concepts, which may hamper the development of physics. In classical geostatistics, *e.g.*, the spatiotemporal continuum is routinely covered by a single coordinate system which implies a specified metrical structure (*i.e.*, rectangular coordinates with a Euclidean metric defined by the inner product) and no attempt is made to examine whether the geometry assumed is consistent with the physics of the domain. As a result, the geostatistical analysis could be based on unjustified assumptions about space/time which may lead to unrealistic maps of the phenomena under consideration (see section on the spatiotemporal random field concept beginning on p. 59). Physically more meaningful geometries are those that clearly distinguish the various space/time concepts so that scientific theories can employ only the necessary ones. In modern spatiotemporal geostatistics, *e.g.*, depending on the physical features of the domain, the metric and the coordinate system used to describe that metric may be independent of each other and not necessarily of the Euclidean form. This allows considerable flexibility in the choice of the geometry that best represents physical reality.

The separate metrical structure of Equation 2.11, in particular, would be suitable to represent our commonsense view of space as a container (within which all events take place) and time as an absolute entity (that registers the successive or simultaneous occurrences of these events); space and time exist independently of natural processes and laws, as a kind of theater in which the natural processes and laws enact their drama. On the other hand, the basic idea underlying Equation 2.26 is that the theater (the space/time continuum) is intimately linked to its actors (natural processes and laws) and cannot exist independently of them.

In several natural applications the separate metrical structure (Eq. 2.11) will suffice. In other applications, however, the more involved composite structure (Eq. 2.26) may be necessary. In the latter case, considering the several existing spatiotemporal geometries that are mathematically distinct but *a priori* and generically equivalent, we must ask ourselves: Which mathematical spatiotemporal metrical structure (*i.e.*, what sort of function g) describes reality, inasmuch as there is no need to single out the Euclidean metrical structure? The answer to this question should take into account the fact that mathematics describes the possible geometric spaces, and physical knowledge determines which one of them corresponds to real space. According to this approach the question is subject to empirical investigation. It is the space of physical experience and intuition that gives rise to the concepts of metric and measurement and presents the notion and possibility of spatiotemporal relationships, such as Equation 2.26. In other words, a system of pure axiomatic geometry may not suffice, if geometry is to be applied to physical space/time. What is required in many cases is to establish a *relationship* between the geometric concepts of the abstract system with the natural processes of the physical system.

How can such a relationship between the abstract geometry with the physical system be established? Again, any answer to this question must take into consideration the fact that mathematical geometry is purely logical, whereas physical geometry is empirical. In other words, the following postulate makes sense.

POSTULATE 2.7: The nature of the mathematical geometry that best describes space/time should be disclosed in terms of the empirical

investigation of space/time as a whole. The term "empirical" includes all available physical knowledge bases.

According to Postulate 2.7, the basis of metrical determination should be sought outside the abstract geometric objects in the natural processes that act upon them. Postulate 2.7 implies that one's physical knowledge $\mathcal{K}$ (observational data, correlation functions, physical laws, empirical relationships, *etc.*; see Chapter 3) must make sense within one's view of the structure of space/time. In this context, Postulate 2.7 is closely related to Postulates 2.3, 2.4, and 2.6 considered earlier. Before proceeding with a more detailed analysis of the metrical structures, it is necessary to introduce the concept of "field," which is essential in the geostatistical modeling of physical data and processes.

The Field Idea

Another fundamental idea of geostatistics is the idea of a *field* that associates mathematical entities—such as the scalar, vector, or tensor—with spatiotemporal points. A field may be viewed as a model, *i.e.*, as a mathematical construction for representing the distribution of natural variables in space/time, which leads us to the following postulate.

POSTULATE 2.8: A field presupposes a continuum $\mathcal{E}$ of spatiotemporal points and then attributes values of a natural variable to these points. Specification of the values at all points in $\mathcal{E}$ specifies a realization of the field.

From the viewpoint of physical modeling, fields have a number of interesting features. There are many kinds of fields—some represent materialistic variables (*e.g.*, the spatiotemporal distribution of a material in soil or the concentration of a contaminant in water), others express non-materialistic regions of influence through which values of natural variables can affect each other (*e.g.*, the Earth's gravitational field or the electromagnetic field). It is also possible that field-based representations of reality may involve a hierarchy of fields. There is no unique field representing every aspect of reality. Instead, one field describes one characteristic of reality and another field some other characteristic (a detailed discussion of the field concept and its various applications can be found in Christakos and Hristopulos, 1998).

Restrictions on spatiotemporal geometry imposed by field measurements and natural media

The crux of Postulates 2.7 and 2.8 is that there is no space/time without natural fields. When we seek to develop a self-consistent geostatistical analysis we must take into consideration the physical characteristics of the domain and the diverse systems of measurement. The examples below show that the spatiotemporal distance may depend on the properties of the physical *medium* and the particular *experimental* procedure that we decide to use in order to obtain measurements of the natural field.

EXAMPLE 2.21: One cannot decide by means of purely geometrical notions how the distance d_{12} between the points $\boldsymbol{p}_1$ and $\boldsymbol{p}_2$ at a 3-km spatial lag and a 2-day time lag in a porous medium compares with the distance $d_{12'}$ between the points $\boldsymbol{p}_1$ and $\boldsymbol{p}'_2$ at a 2-km spatial lag and a 3-day time lag. Which spatiotemporal distance is larger can be decided only by means of a physical process as follows: An experiment is performed during which a tracer is released at $\boldsymbol{p}_1$. If the tracer is detected at $\boldsymbol{p}_2$ but not at $\boldsymbol{p}'_2$, then $d_{12} < d_{12'}$ *with respect to* the particular medium and experimental setup. Also, measuring distance by means of fluid-tracer dispersion can lead to very different results than measuring it by means of electromagnetic propagation. □

EXAMPLE 2.22: Consider two geostatisticians A and B who both use the same Euclidean system, say, rectangular coordinates. As a result of physical considerations, the two geostatisticians use different metrical systems. Geostatistician A employs the usual linear Euclidean metric system $s_{i,A} = \eta_{i,A}\, u$ (with a unit distance u), and geostatistician B uses the nonlinear metric system $s_{i,B} = \exp[\eta_{i,B}]\, v$ (with a unit distance v); $\eta_{i,A}$ and $\eta_{i,B}$ are the number of distance units for A and B, respectively. Since both systems are Euclidean, the spatial coordinates of a specified point P in R^3 provided by the two geostatisticians are related as follows

$$s_{i,B} = \varepsilon_i\, s_{i,A}, \quad i = 1,\ 2,\ 3 \tag{2.30}$$

where ε_i are real-valued coefficients. By adjusting the unit distances of the two geostatisticians along the s_1 axis so that $s_{1,B} = s_{1,A}$, the remaining coordinates are altered in the same ratio; then, Equation 2.30 yields

$$s_{1,B} = s_{1,A}, \quad s_{2,B} = \varepsilon_1^{-1}\varepsilon_2\, s_{2,A}, \quad s_{3,B} = \varepsilon_1^{-1}\varepsilon_3\, s_{3,A} \tag{2.31}$$

In other words, since the two geostatisticians use different *metric systems*, there is no way to adjust their measurements for all three coordinates. For this reason, the two geostatisticians obtain very different representations of objects and processes in space/time. The sphere $\sum_{i=1}^{n} s_{i,A}^2 = \rho^2$ of the first geostatistician, *e.g.*, is for the second geostatitician an ellipsoid $\sum_{i=1}^{n} \varepsilon_i^{-2}\, s_{i,B}^2 = \varepsilon_1^{-2}\, \rho^2$; the angles measured by the two geostatisticians are different; *etc.* □

The situation presented in the following example has interesting applications in the domain of *relativistic* geostatistics and has already been mentioned in Example 2.20.

EXAMPLE 2.23: In space/time, distances are physically measured most efficiently using *light pulses*. Let a flash of light be emitted from point P_1 at time t and reach point P_2 at time $t + dt$. The spatial distance $|ds|$ between the two points is given by

$$|ds|^2 = c^2\, dt^2 \tag{2.32}$$

where c is the speed of light. If ds_i ($i = 1,\ 2,\ 3$) are the orthogonal projections of ds, we can write the space/time metric as in Equation 2.28. Equation 2.28 attributes a physical meaning to the quantity $|dp|$, even if the points are so chosen that the corresponding $|dp|$ does not vanish (Einstein, 1994 [1954]). In particular, we say that the space/time distance is "space-like" when $|dp|$ is real, "time-like" when $|dp|$ is imaginary, and "light-like" or "null" when $|dp|$ is zero. Space-like distances occur between events with a spatial separation less than the distance the light pulse travels between the times of their occurrence; time-like distances occur between events with a spatial separation that exceeds the distance that the light pulse can travel in the time between them; null distances represent events in space/time that could just be connected by a light pulse. Mathematically, the distance (Eq. 2.28) is the *Minkowski* metric determined from Equation 2.27 by letting $g_{00} = -c^2$, $g_{ii} = 1$ $(i = 1, \ldots, n)$, and $g_{ij} = 0$ $(i \neq j)$. □

Restrictions on spatiotemporal geometry imposed by physical laws

We have seen that we can learn about the nature of space/time by studying the characteristics of the physical system. Nature does not, of course, allow the natural processes to vary in an arbitrary manner, but imposes constraints in the form of physical laws. Consider a physical law governing the distribution of a natural field $X(\boldsymbol{p})$ that is generally expressed as

$$X(\boldsymbol{p}) = \mathcal{L}[\boldsymbol{v},\ BC,\ IC;\ \boldsymbol{p}] \tag{2.33}$$

where $\boldsymbol{v} = (v_1, \ldots, v_k)$ are known physical coefficients, BC and IC are given boundary and initial conditions, and $\mathcal{L}[\cdot]$ is a known mathematical functional. Typically, Equation 2.33 is regarded as determining the values of the field $X(\boldsymbol{p})$ from the known coefficients $\boldsymbol{v}$, the BC and IC, and the space/time coordinates. When it comes to the spatiotemporal metrical structure, however, there are several ways in which the physical law (Eq. 2.33) can help one's effort to determine the appropriate metric form. Some of these ways are reviewed next.

In certain applications, Equation 2.33 leads to the following explicit expression for the metric

$$|d\boldsymbol{p}| = |\boldsymbol{p}' - \boldsymbol{p}| = g[\chi',\ \chi,\ \boldsymbol{v},\ BC,\ IC] \tag{2.34}$$

where χ and χ' are X-values at points $\boldsymbol{p}$ and $\boldsymbol{p}'$, respectively; and $g[\cdot]$ has a functional form associated with $\mathcal{L}[\cdot]$. As far as the metrical structure is concerned, the interpretation of Equation 2.34 can be different from that of Equation 2.33. Given any metric $|d\boldsymbol{p}|$, Equation 2.34 will not be satisfied automatically. In fact, Equation 2.34 will serve to cut down the number of possible metric models. Equation 2.34 is thus regarded as determining the metric $|d\boldsymbol{p}|$ of the space/time geometry from the natural field values at $\boldsymbol{p}$ and $\boldsymbol{p}'$, the coefficients $\boldsymbol{v}$, the BC, and the IC. Here, then, is a sense in which Equation 2.34 restricts the space/time geometry. On imposing Equation

2.34 into geostatistical analysis, one cannot specify the spatiotemporal metric and the natural field values independently. Rather these two entities must be connected via Equation 2.34.

In some other applications, the form of the metric is obtained indirectly from the field equations. This may happen in the case in which the solution of the physical law can be written as

$$X(\boldsymbol{p}) = \breve{X}[g(s_1, \ldots, s_n, t)] \tag{2.35}$$

The solution (Eq. 2.35) puts restrictions on the geometrical features of space/time and suggests a metric of the form $|\boldsymbol{p}| = g(s_1, \ldots, s_n, t)$, where $|\boldsymbol{p}|$ defines the space/time distance from the origin. In fact, in the section, "Separate metrical structures" (p. 43), we discussed the formulation of physical equations in terms of general curvilinear coordinates involving the metric coefficients g_{ij}. It is possible that the physical law could lead to a solution that offers information about the coefficients g_{ij} of the metric. These possibilities are demonstrated with the help of the following examples.

EXAMPLE 2.24: Consider the natural field $X(s_1,\ s_2)$ whose spatial distribution is governed by the physical law

$$s_1\, \partial X/\partial s_2 - s_2\, \partial X/\partial s_1 = 0 \tag{2.36}$$

The solution of Equation 2.36 is expressed as $X(s_1,\ s_2) = \breve{X}\ [g(s_1,\ s_2)] = \breve{X}\left(\sqrt{s_1^2 + s_2^2}\right)$. Hence, a spatial metric suggested by the physical law above is the Euclidean $|\boldsymbol{s}| = \sqrt{s_1^2 + s_2^2}$. On the other hand, the physical equation $s_1\, \partial X/\partial s_1 - s_2\, \partial X/\partial s_2 = 0$ has a solution of the form $X(s_1,\ s_2) = \breve{X}(s_1\, s_2)$, which means that the above physical equation may be associated with the metric $|\boldsymbol{s}| = s_1\, s_2$. □

EXAMPLE 2.25: The governing flow equations for phases α (= water and oil) in a porous domain are (Christakos, *et al.*, 2000b)

$$d\zeta_\alpha/d\ell_\alpha + \phi(\boldsymbol{e}_\alpha,\ K_\alpha)\, \zeta_\alpha = 0 \tag{2.37}$$

where $\boldsymbol{e}_\alpha$ is the direction vector of the α-flowpath trajectory, ζ_α is the magnitude of the pressure gradient along $\boldsymbol{e}_\alpha$, K_α are intrinsic permeabilities, and ϕ is a function of $\boldsymbol{e}_\alpha$ and K_α. The solution of the flow equation above is of the form $\zeta_\alpha = \zeta_\alpha(\,|\boldsymbol{s}|\,)$, where the metric $|\boldsymbol{s}| = \ell_\alpha$ is the distance along the α-flowpath.

EXAMPLE 2.26: Consider the $R^2 \times T$ space/time continuum considered in Example 2.19. The question is: How can we determine a suitable spatiotemporal metric? According to the preceding discussion, the physical knowledge available can be very helpful in this respect. Let us suppose that the natural field $X(s_1,\ s_2,\ t)$ is governed in $R^2 \times T$ by the flux-conservative law

$$\partial X/\partial t + \boldsymbol{v} \cdot \nabla X = 0 \tag{2.38}$$

where $\boldsymbol{v} = (v_1,\ v_2)$ is an empirical velocity to be determined from the data. The physical law (Eq. 2.38) puts certain restrictions on the geometrical features of the space/time domain. Indeed, by considering a coordinate transformation from the rectangular Euclidean system (s_i) to the system of coordinates defined by $\breve{s}_i = s_i - v_i t$, the solution of Equation 2.38 has the following form

$$X(s_1,\ s_2,\ t) = \breve{X}(s_1 - v_1 t,\ s_2 - v_2 t) \tag{2.39}$$

i.e., it depends on the space/time vector $\boldsymbol{p} = \boldsymbol{s} - \boldsymbol{v}t$. This dependence implies that in the rectangular coordinate system, a natural field with the physics described by Equation 2.39 may be assigned a metric of the form of Equation 2.27 above, where $n = 2$, $g_{00} = (v_1^2 + v_2^2)$, $g_{11} = g_{22} = 1$, $g_{10} = g_{01} = -2v_1$, $g_{20} = g_{02} = -2v_2$, and $g_{12} = g_{21} = g_{01} = g_{10} = g_{02} = g_{20} = 0$; *i.e.*, we have

$$|d\boldsymbol{p}| = \sqrt{ds_1^2 + ds_2^2 + (v_1^2 + v_2^2)\,dt^2 - 2(v_1 ds_1 + v_2 ds_2)dt} \tag{2.40}$$

Equation 2.40 demonstrates how the physical law determines the geometrical metric through the empirical vector $\boldsymbol{v} = [v_1,\ v_2]^T$. Also, a relation between the mathematical geometry and the physical parameters can be established by solving the law with respect to $\boldsymbol{v}$, which gives

$$\boldsymbol{v} = \begin{bmatrix} v_1 \\ v_2 \end{bmatrix} = - \begin{bmatrix} \partial^2 X/\partial s_1^2 & \partial^2 X/\partial s_1 \partial s_2 \\ \partial^2 X/\partial s_2 \partial s_1 & \partial^2 X/\partial s_2^2 \end{bmatrix}^{-1} \begin{bmatrix} \partial^2 X/\partial t \partial s_1 \\ \partial^2 X/\partial t \partial s_2 \end{bmatrix} \tag{2.41}$$

Equation 2.41 expresses the empirical vector $\boldsymbol{v}$ in terms of the natural field $X(s_1,\ s_2,\ t)$-values in space/time. Also, later in this chapter (see "Correlation analysis and spatiotemporal geometry," p. 61) we will discuss how the covariance functions can be instructive in determining the appropriate geometry in a spatiotemporal continuum. □

COMMENT 2.8: *The following notational remark is important for future reference, as well. In vector calculus, the vector* $\boldsymbol{x} = (x_1, \ldots, x_n)$ *is considered as a column vector*

$$\boldsymbol{x} = \begin{bmatrix} x_1 \\ \vdots \\ x_n \end{bmatrix}$$

when matrix or vector multiplications are involved (see, e.g., Marsden and Tromba, 1988).

The Complementarity Idea

In matters of scientific investigation, we need to consider alternatives where facts are unknown. In view of uncertainty and imperfect knowledge, actuality is, indeed, surrounded by an infinite realm of possibilities. The spatiotemporal distributions of most natural variables are not sharply defined but, instead, they have an uncertain or indeterminate structure. A theoretical as well as a practical need to account for this uncertainty gives rise to the complementarity idea in the following postulate.

POSTULATE 2.9: Complementarity considers a multiple parallel processing of field realizations which are diverse yet necessary for a complete understanding of the phenomenon of interest.

In light of complementarity, uncertainty manifests itself as an ensemble of *possible* field realizations that are all in agreement with what is known about the phenomenon of interest. These realizations are different but complementary facets of the description of the actual phenomenon. In other words, the different realizations are not contradictions, rather, they are complementary aspects of a seamless unity.

The concept of complementarity includes the actual (but usually unknown to us) realization of the natural field as well as several non-actual but possible field realizations. While the actual realization is located in the physical world, the possible realizations are located in a conceptual (or logical) world. A "possible field realization" is not the same as a "conceivable field realization." Just as there exist possible realizations that are not conceivable (on the basis of our current knowledge, technological abilities, *etc.*), there also exist conceivable realizations that are not possible because they violate rules of logic, natural laws, *etc.* In science, we consider only *physically* possible realizations, *i.e.*, realizations which are characterized by the same physical laws, data, *etc.* as the actual realization of the natural field.

COMMENT 2.9: *Complementarity may offer an interesting interpretation of the cause–effect concept: In mathematical logic, the use of* material conditionals *(*i.e.*, propositions of the form "if* χ_1*, then* χ_2*" or "*$\chi_1 \rightarrow \chi_2$*," where* χ_1 *is the antecedent and* χ_2 *is the consequent) requires us to admit various possible realizations besides the actual (but unknown) one. A conditional then may be true not in terms of how things are, but of how they would be in an appropriate field realization. A conditional* $\chi_1 \rightarrow \chi_2$ *is true in a realization if* χ_2 *is true in the same realization in which* χ_1 *is also true. In other words, an event* χ_1 *may be considered as causing an event* χ_2 *if both* χ_1 *and* χ_2 *occur in the observed realization, but in the vast majority of the other realizations in which* χ_1 *does not occur,* χ_2 *does not occur either. We revisit material conditionals in Chapter 4.*

Putting Things Together: The Spatiotemporal Random Field Concept

Spatiotemporal random field (S/TRF) modeling of natural phenomena has led to considerable successes over the last few decades. Conceptually, the S/TRF model is a combination of the three fundamental ideas (the spatiotemporal continuum, field, and complementarity) of the preceding sections. Here we will use large and small Roman characters to denote random fields and random variables, respectively; Greek characters will be used to denote realizations (data values, *etc.*). The following S/TRF definition is often used in modern spatiotemporal geostatistics.

DEFINITION 2.10: An S/TRF $X(p)$ is a collection of complementary field realizations χ associated with the values of a natural variable at points $p = (s,t)$ of a spatiotemporal continuum $S \times T$. Mathematically, the S/TRF $X(p)$ is the mapping

$$X(p) : S \times T \rightarrow L_q(\Omega, F, P) \tag{2.42}$$

where Ω is the sample space that includes all possible field realizations, F is a family of realizations, $P(\chi) \in [0,1]$ is a probability associated with each realization, and $L_q(\Omega, F, P)$, $q \geq 1$ denotes the norm on the probability space (Ω, F, P).

L_2-norms are usually considered in geostatistics. By "complementary field realizations" are meant all physically possible realizations of the natural variable. An S/TRF $X(p)$ is, therefore, a collection of realizations for the distribution of the natural variable in space/time. The S/TRF can be viewed also as a collection of correlated random variables, say, $\boldsymbol{x}_{map} = (x_1, \ldots, x_m, x_k)$ at the space/time points $\boldsymbol{p}_{map} = (p_1, \ldots, p_m, p_k)$. A realization of the S/TRF at these points is denoted by the vector $\boldsymbol{\chi}_{map} = (\chi_1, \ldots, \chi_m, \chi_k)$. We assume that the S/TRF takes values in the space of real numbers, since this assumption represents the majority of applications in the natural sciences. A detailed discussion of recent developments in the mathematical S/TRF theory may be found in Christakos and Hristopulos (1998).

COMMENT 2.10: *It must be clear to the reader that here we use the symbol $\boldsymbol{\chi}_{map}$ because our goal later in the book will be to obtain spatiotemporal maps displaying estimates at points $\boldsymbol{p}_k$ of the unknown values χ_k of the natural variable from its observed values $\chi_1, \ldots, \chi_m$. In the same context, the elements of the vector $\boldsymbol{p}_{data}$ are the data points $\boldsymbol{p}_i$ $(i = 1, \ldots, m)$ and the vector $\boldsymbol{p}_{map}$ has as elements the points $\boldsymbol{p}_i$ $(i = 1, \ldots, m, k)$.*

The complete characterization of an S/TRF is provided by the multivariate probability density function (pdf) $f_{\mathcal{K}}$ defined as

$$\text{Prob}\left[\chi_1 \leq x_1 \leq \chi_1 + d\chi_1, \ldots, \chi_m \leq x_m \leq \chi_m + d\chi_m, \chi_k \leq x_k \leq \chi_k + d\chi_k\right]$$

$$= f_{\mathcal{K}}(\boldsymbol{\chi}_{map};\ \boldsymbol{p}_{map})\, d\boldsymbol{\chi}_{map} \tag{2.43}$$

where the subscript $\mathcal{K}$ denotes the physical knowledge used to derive the pdf (physical knowledge bases are studied in Chapter 3). Notice that the notation $f_{\mathcal{K}}(\boldsymbol{\chi}_{map};\ \boldsymbol{p}_{map})$ in the rest of the book has been kept only when necessary (*e.g.*, when differentiation with respect to space and/or time takes place); in all other situations, $\boldsymbol{p}_{map}$ has been dropped and the simpler notation $f_{\mathcal{K}}(\boldsymbol{\chi}_{map})$ is used. A generally incomplete—yet in many practical applications satisfactory—characterization of the S/TRF is provided by a limited set of *statistical moments* (also called, simply, space/time *statistics*), which are defined as follows,

$$\overline{g(\boldsymbol{x}_{map})} = \overline{g}(\boldsymbol{p}_{map}) = \int d\boldsymbol{\chi}_{map}\, g(\boldsymbol{\chi}_{map})\, f_{\mathcal{K}}(\boldsymbol{\chi}_{map};\ \boldsymbol{p}_{map}) \tag{2.44}$$

where $g(\cdot)$ is some known function. [Notice the difference between $g(\boldsymbol{\chi}_{map})$, which is a function of the realization values, and its expectation $\overline{g}(\boldsymbol{p}_{map})$, which is a function of the space/time points.] There is, generally, no need to specify limits of integration in Equation 2.44, since if certain ranges of $\boldsymbol{\chi}_{map}$ are impossible, the pdf will be zero, removing contributions from these ranges.

EXAMPLE 2.27: If we let $g(\chi_1) = \chi_1$, Equation 2.44 provides the mean

$$\overline{g(x_1)} = \overline{g}(\boldsymbol{p}_1) = \overline{x_1}$$

of the S/TRF. If we let $g(\chi_1, \chi_2) = (\chi_1 - \overline{x_1})(\chi_2 - \overline{x_2})$, the (centered) covariance function

$$\overline{g(x_1, x_2)} = \overline{(x_1 - \overline{x_1})(x_2 - \overline{x_2})} = c_x(\boldsymbol{p}_1, \boldsymbol{p}_2)$$

between the points $\boldsymbol{p}_1$ and $\boldsymbol{p}_2$ is obtained; *etc.* □

COMMENT 2.11: *S/TRF characterization in terms of Equation 2.44 is considered "incomplete" (or general or vague), in the sense that several random fields exist that share the same space/time moments. Also, statistical moments can be defined for more than one random field simultaneously, in which case Equation 2.44 should involve data or map vectors for all these fields (see,* e.g.*, the multivariable or vector formulation of BME in Chapter 9).*

Correlation analysis and spatiotemporal geometry

In geostatistical applications, spatiotemporal correlation functions are usually part of the available physical knowledge bases. These functions could be derived from a physical law or fitted to the data. Commonly used correlation functions include the ordinary covariance, the variogram (sometimes also called semivariogram), and the generalized covariance. Ordinary covariance analysis, *e.g.*, can be helpful in determining the physically appropriate spatiotemporal geometry. Let us suppose that the form of a finite metric λ is sought such that

$$c_x(h_1, \ldots, h_n, \tau) = \breve{c}_x(\lambda) \tag{2.45}$$

Taking the derivatives of this equation, we find $\partial c_x/\partial h_i = (d\breve{c}_x/d\lambda)\,(\partial\lambda/\partial h_i)$ and $\partial c_x/\partial\tau = (d\breve{c}_x/d\lambda)\,(\partial\lambda/\partial\tau)$. The last two equations imply that the metric λ is related to the covariance of the natural field through the following set of equations

$$\left.\begin{aligned} \frac{\partial c_x/\partial h_i}{\partial c_x/\partial h_j} &= \frac{\partial\lambda/\partial h_i}{\partial\lambda/\partial h_j} \\ \frac{\partial c_x/\partial h_i}{\partial c_x/\partial\tau} &= \frac{\partial\lambda/\partial h_i}{\partial\lambda/\partial\tau} \end{aligned}\right\} \tag{2.46}$$

where $i,\ j = 1, \ldots, n$.

One may consider the metric as a transformation $\lambda = \mathcal{T}(h_1, \ldots, h_n, \tau)$ of the original coordinate system, where $\mathcal{T}$ has a Riemannian structure and the forms of the coefficients g_{ij} are sought on the basis of physical and mathematical facts. Let λ be expressed as

$$\lambda = \sqrt{\sum_{i,j=1}^{n} g_{ij}\, h_i\, h_j + 2\tau \sum_{i=1}^{n} g_{0i}\, h_i + g_{00}\tau^2} \tag{2.47}$$

While the finite space/time distance (Eq. 2.47) has the same form as the infinitesimal Riemannian distance (Eq. 2.27), the g_{ij} do not necessarily coincide with the metric coefficients of Equation 2.27. In Equation 2.47, the g_{ij} are assumed to denote functions of the spatial and lag distances rather than the local coordinates, as was the case of the definition in Equation 2.27; *i.e.*, $g_{ij} = g_{ij}(h_i, h_j)$, $g_{0i} = g_{0i}(\tau, h_i)$, $i, j = 1, \ldots, n$ and $g_{00} = g_{00}(\tau)$. Then, Equation 2.46 yields the expressions

$$\left.\begin{aligned}
&\frac{\partial c_x/\partial h_{i'}}{\partial c_x/\partial h_{j'}} = \\
&\frac{\sum_{j=1}^{n} g_{i'j}\, h_j + \sum_{j=1,\, j\neq i'}^{n} (\partial g_{i'j}/\partial h_{i'})\, h_{i'}\, h_j + 0.5\,(\partial g_{i'i'}/\partial h_{i'})h_{i'}^2 +}{\sum_{i=1}^{n} g_{ij'}\, h_i + \sum_{i=1,\, i\neq j'}^{n} (\partial g_{ij'}/\partial h_{j'})\, h_i\, h_{j'} + 0.5\,(\partial g_{j'j'}/\partial h_{j'})h_{j'}^2} \\
&\qquad \frac{g_{0i'}\,\tau + (\partial g_{0i'}/\partial h_{i'})\,\tau\, h_{i'}}{+ g_{0j'}\,\tau + (\partial g_{0j'}/\partial h_{j'})\,\tau\, h_{j'}} \\
&\frac{\partial c_x/\partial h_{i'}}{\partial c_x/\partial \tau} = \\
&\frac{\sum_{j=1}^{n} g_{i'j}\, h_j + \sum_{j=1,\, j\neq i'}^{n} (\partial g_{i'j}/\partial h_{i'})\, h_{i'}\, h_j + 0.5\,(\partial g_{i'i'}/\partial h_{i'})h_{i'}^2}{\sum_{i=1}^{n} g_{0i}\, h_i + \sum_{i=1}^{n} (\partial g_{0i}/\partial \tau)\,\tau\, h_i +} \\
&\qquad \frac{+ g_{0i'}\,\tau + (\partial g_{0i'}/\partial h_{i'})\,\tau\, h_{i'}}{g_{00}\,\tau + 0.5\,(\partial g_{00}/\partial \tau)\tau^2}
\end{aligned}\right\} \tag{2.48}$$

The determination of the coefficient g_{ij} is not always an easy task. In most cases it requires additional assumptions that are based on theoretical and experimental facts (*e.g.*, the local properties of space and time, and rules imposed by the specific natural processes). This is, clearly, an important issue that has been discussed already in previous sections.

If the g_{ij} are space and time independent (*e.g.*, the analysis is focused on a local mesh), Equation 2.48 reduces to the simpler expressions below ($i, j = 1, \ldots, n$)

$$\left.\begin{aligned} \frac{\partial c_x/\partial h_i}{\partial c_x/\partial h_j} &= \frac{\sum_{j=1}^{n} g_{ij}\, h_j + g_{0i}\, \tau}{\sum_{i=1}^{n} g_{ij}\, h_i + g_{0j}\, \tau} \\ \frac{\partial c_x/\partial h_i}{\partial c_x/\partial \tau} &= \frac{\sum_{j=1}^{n} g_{ij}\, h_j + g_{0i}\, \tau}{\sum_{i=1}^{n} g_{0i}\, h_i + g_{00}\, \tau} \end{aligned}\right\} \tag{2.49}$$

A healthy dose of intuition and a deeper understanding of the physical situation will also be of great value in determining g_{ij}. Indeed, as we saw in a previous section ("The Field Idea," p. 54), in light of the physical equations, the spatiotemporal geometry and the natural variables are not independent of one another; rather they are connected via these equations. As a result of this connection, the behavior of the natural phenomenon is tied to space/time itself. Consider the following example.

EXAMPLE 2.28: Consider a covariance in $R^1 \times T$ that satisfies the physical equation

$$\frac{\partial c_x/\partial h}{\partial c_x/\partial \tau} = \frac{h}{v^2\, \tau} \tag{2.50}$$

where $h = \Delta s_1 = s_1' - s_1$, $\tau = \Delta t = t' - t$ $(s_1' > s_1,\ t' > t)$, and v is an empirical parameter. Equation 2.50 was derived from the empirical evidence available, the scientific laws governing the corresponding natural field, *etc.* The question is, what is the form of the metric λ so that $c_x(h,\ \tau) = \breve{c}_x(\lambda)$? In light of Equations 2.49 and 2.50, when $(h,\ \tau) \xrightarrow{\mathcal{T}} \lambda$ the metric coefficients satisfy the relationship

$$\frac{g_{11} h + g_{01}\, \tau}{g_{01} h + g_{00}\, \tau} = \frac{h}{v^2\, \tau} \tag{2.51}$$

Hence, a geometric metric that satisfies the last relationship—and is thus consistent with the physical Equation 2.50—is of the form of Equation 2.47 with $n = 1$, $g_{00} = v^2$, $g_{01} = 0$, and $g_{11} = 1$; *i.e.*,

$$\lambda = \sqrt{h^2 + v^2 \tau^2} \tag{2.52}$$

Equation 2.52 demonstrates how the covariance coefficients determine the spatiotemporal metric. It also provides a sufficient, although not necessary, specification for a metric satisfying the physical conditions expressed by Equation 2.50. Note that for a function to be an appropriate covariance model—in addition to being a function of the metric (Eq. 2.52)—it must also satisfy certain permissibility conditions (Christakos, 1992; also discussed in the following section). A function which is a permissible covariance and has a metric of the form in Equation 2.52 is, *e.g.*, the model $c_x(h,\ \tau) = c_0 \exp(-h^2 - a^2\tau^2/b^2)$, where a and b are empirical correlation coefficients. □

COMMENT 2.12: *The empirical parameter v in Example 2.28 relates space and time intrinsically and prevents the dissolution of the spatiotemporal structure into independent spatial and temporal components. In this case, the spatial distance and the time interval are replaced by the so-called* proper *time interval between two space/time points defined as* $d\eta^2 = \tau^2 + h^2 v^{-2}$ *(a proper spatial interval may be defined in a similar fashion).*

In light of the preceding analysis, the choice of a spatiotemporal geometry in geostatistical applications must avoid discrepancies between the "natural" geometry—as revealed by the physical equations and data—and the appropriate mathematical geometry. Spatial arrangements in the domain under consideration should be combined with the temporal order of events in a way that reflects relationships determined by physical knowledge.

Permissibility criteria and spatiotemporal geometry

The choice of a geometry may have significant consequences in geostatistical analysis. One such important consequence is related to the covariance permissibility criteria in $R^n \times T$ (*i.e.*, the criteria for a function to be a covariance model); similar consequences are valid for other spatiotemporal correlation functions, including the variogram and the generalized covariance models. In fact, the validity of the following postulate will be demonstrated in this section.

POSTULATE 2.10: The permissibility criteria for spatiotemporal correlation models depend on the geometry assumed.

Mathematically, a necessary and sufficient condition for a function to be a permissible covariance model is that it be nonnegative definite. Furthermore, *Bochner's theorem* shows that for a function $c_x(\boldsymbol{h}, \tau)$ to be nonnegative definite it is necessary and sufficient that its spectral density $\tilde{c}_x(\boldsymbol{\kappa}, \omega)$ be a real-valued, integrable, and nonnegative function of the spatial frequency $\boldsymbol{\kappa}$ and the temporal frequency ω (for details, see Christakos and Hristopulos, 1998; Gneiting, 1999). Postulate 2.10 implies that the formulation of Bochner's theorem depends on the coordinate system and the geometric metric used and, therefore, a model that is a permissible covariance for a specific spatiotemporal geometry may not be so for another geometry. As we will see below, this is indeed the case with certain commonly used models.

The Gaussian function is a permissible covariance model for the Euclidean distance (Eq. 2.12). However, as is demonstrated in the following proposition, the same model is not permissible when a non-Euclidean distance is assumed.

PROPOSITION 2.1: The Gaussian function in $R^2 \times T$

$$c_x(\boldsymbol{h},\ \tau) = c_0 \exp(-\boldsymbol{h}^2 - v^2\tau^2) \tag{2.53}$$

where the spatial distance is defined as in Equation 2.13, *i.e.*

$$|\boldsymbol{h}| = |h_1| + |h_2| \tag{2.54}$$

is not a permissible covariance.

Proof: Since Equation 2.53 is a separable space/time function, we can study spatial and temporal parts separately. We first concentrate on the spatial part

$$c_x(\boldsymbol{h}) = \exp(-\boldsymbol{h}^2) \tag{2.55}$$

Let us suppose that $c_x(\boldsymbol{h})$ is a nonnegative definite function. Then, according to Bochner's theorem, there exists a unique Borel measure μ_x on R^2 such that

$$c_x(\boldsymbol{h}) = \int_{R^2} d\mu_x(\boldsymbol{k}) \exp[i\, \boldsymbol{k} \cdot \boldsymbol{h}] \tag{2.56}$$

Since $c_x(\lambda\, \boldsymbol{h}) = \exp[-\lambda^2\, |\boldsymbol{h}|^2]$ for all real λ, the uniqueness of the Fourier transform implies that $\boldsymbol{k} \cdot \boldsymbol{h}$ follows a one-dimensional Gaussian distribution with variance $2\, |\boldsymbol{h}|^2 = \int_{R^2} (\boldsymbol{k} \cdot \boldsymbol{h})^2\, d\mu_x(\boldsymbol{k})$. Hence, $\Phi_h(\boldsymbol{k}) = \boldsymbol{k} \cdot \boldsymbol{h}$ is square integrable with respect to μ_x, *i.e.*, $\Phi_h(\boldsymbol{k}) \in L_2(\mu_x)$. The map $\boldsymbol{h} \rightarrow \frac{1}{\sqrt{2}}\, \Phi_h$ becomes a norm preserving embedding $(R^2,\ |\cdot|)$ into $L_2(\mu_x)$. It then follows that $(R^2,\ |\cdot|)$ should be a Hilbert space—as a subspace of $L_2(\mu_x)$. However, if the norm $|\cdot|$ is defined as the distance (Eq. 2.54), $(R^2,\ |\cdot|)$ is not a Hilbert space, because we cannot define an inner product on R^2 such that $|\boldsymbol{h}|^2 = (\boldsymbol{h},\ \boldsymbol{h})$, which leads to a contradiction. Therefore, we conclude that $c_x(\boldsymbol{h})$ is not a nonnegative definite function. □

In fact, a stronger result can be proven, *i.e.*, that under some general assumptions, the metric in Equation 2.55 must necessarily be Euclidean (Christakos and Papanicolaou, 2000). The validity of Proposition 2.1 may be illustrated by means of a numerical example.

EXAMPLE 2.29: The spectral density of the Gaussian covariance of Equation 2.55 is given by the following expression

$$\tilde{c}_x(k_1,\, k_2) = \int_{-\infty}^{\infty} \int_{-\infty}^{\infty} dh_1\, dh_2\, \exp[-(|h_1| + |h_2|)^2]\; \exp[-i\,(k_1 h_1 + k_2 h_2)] \tag{2.57}$$

The spectral density of Equation 2.57 is plotted in Figure 2.15 and, as shown in this illustration, has negative values at certain regions. Thus, the Gaussian function of Equation 2.53 is not a permissible covariance model for the distance in Equation 2.54. □

The example below shows that, unlike the Gaussian function, the exponential space/time function is a permissible covariance for the spatial distance in Equation 2.54.

EXAMPLE 2.30: Consider the *exponential* function in $R^2 \times T$

$$c_x(\boldsymbol{h},\ \tau) = c_0 \exp(-\,|\boldsymbol{h}| - v\tau) \tag{2.58}$$

where $|\boldsymbol{h}|$ is defined as in Equation 2.54. This is also a separable function, thus we can focus on its spatial component $c_x(\boldsymbol{h}) = \exp(-\,|\boldsymbol{h}|)$. The spectral density of the latter is

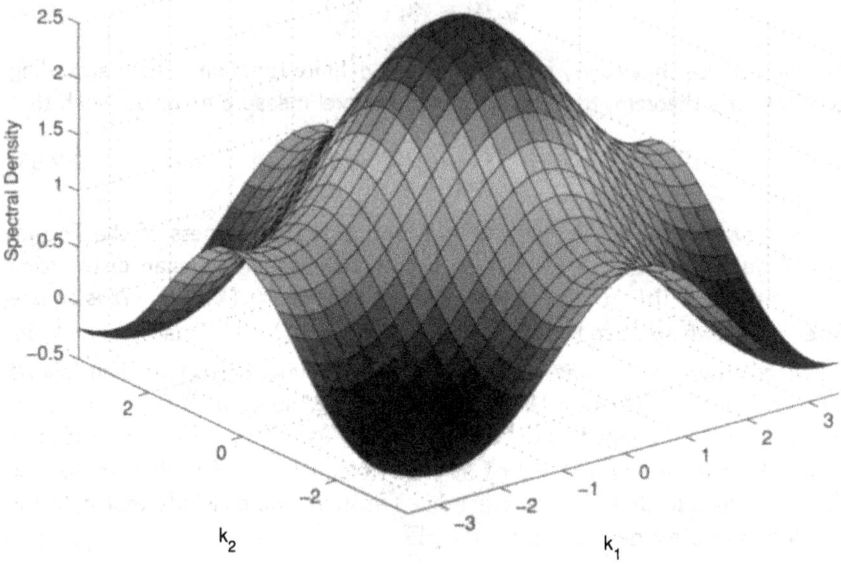

Figure 2.15. Plot of the spectral density of Equation 2.57.

$$\begin{aligned}\tilde{c}_x(k_1, k_2) &= \int_{-\infty}^{\infty}\int_{-\infty}^{\infty} dh_1\, dh_2 \exp[-(|h_1| + |h_2|)]\ \exp[-i\,(k_1h_1 + k_2h_2)] \\ &= 4[(1+k_1^2)(1+k_2^2)]^{-1} \qquad (2.59)\end{aligned}$$

which is nonnegative for all $\boldsymbol{k} \in R^2$. Hence, the exponential function of Equation 2.58 is a permissible covariance for the distance in Equation 2.54. □

As the above analysis shows, the geostatistician must be extra cautious when using a covariance or a variogram model: only those functions that are compatible with the space/time geometry background should be selected.

Effect of spatiotemporal geometry on mapping

There are several ways in which spatiotemporal geometry can affect the outcome of scientific mapping. A kriging map, *e.g.*, is a function of the data set and the correlation function (covariance, variogram, *etc.*) characterizing the variability of the natural variable under consideration. Since the data are functions of the coordinate system used, so is the map obtained from these data. Furthermore, since the correlation function is the necessary input to most mapping techniques, it should be expected that maps produced from these techniques will also be affected by the metrical structure assumed. This line of reasoning leads to the final postulate of this chapter.

POSTULATE 2.11: The geometry assumed in a physical application can have significant effects on the spatiotemporal map produced.

The term "map" in this case may include both spatiotemporal estimation and simulation of natural variables (*e.g.*, kriging estimation and turning bands simulation). Some practical demonstrations of Postulate 2.11 follow.

EXAMPLE 2.31: In Christakos *et al.* (2000a), a two-dimensional field with a constant mean and an exponential covariance $c_x(\boldsymbol{h}) = \exp(-1.5\,|\boldsymbol{h}|)$ $\boldsymbol{h} = (h_1,\ h_2)$ was considered. A metric should in principle be derived based on a physical model of the field. For the sake of illustration, assume that the physically appropriate metric for this field had the non-Euclidean form $|h_1| + |h_2|$. Under these conditions, spatial estimates were generated on the basis of a hard data set $\boldsymbol{\chi}_{hard}$ using a geostatistical kriging technique, thus leading to the map in Figure 2.16a.

Practitioners of geostatistics often favor a theory-free approach that focuses solely on the data set available and ignores physical models. By ignoring, *e.g.*, the underlying physics and using commercial geostatistics software (which allows only for the Euclidean metric $\sqrt{h_1^2 + h_2^2}$ in covariance modeling and kriging), the same data set $\boldsymbol{\chi}_{hard}$ as above results in the map of Figure 2.16b. As was expected, the two maps display considerable differences. The map generated by the standard means (Fig. 2.16b) is based on a convenient but inadequate choice of metric, whereas the correct one (Fig. 2.16a) properly accounts for the physical geometry. □

Most mapping techniques of classical geostatistics have been developed assuming Cartesian coordinate systems (*e.g.*, Deutch and Journel, 1992). However, as is demonstrated in Example 2.32, this system of coordinates may be inadequate for many physical situations and can lead to inaccurate maps.

EXAMPLE 2.32: Figure 2.17 presents variograms and maps of paleo-sea surface conditions from a study by Schäfer-Neth and Stattegger (1998). The variogram and map obtained using Cartesian coordinates are considerably different than those produced using spherical coordinates. While analysis in terms of spherical coordinates provided an adequate representation of spatial variability, the Cartesian coordinates were shown to be inappropriate because they caused a distortion of spatial dependencies in the case of large scales and led to inaccurate predictions. Estimation in terms of spherical coordinates produced maps that were in better agreement with the data and offered a more realistic description of physical reality than maps produced using Cartesian coordinates. □

The implications of producing an inaccurate map—as a result of making the incorrect metric assumption—may be even more serious at subsequent stages of BME analysis. For example, a map of porous geometry that is based on a physically inappropriate metrical structure and that serves as input to the subsurface laws could lead to erroneous predictions of subsurface flow and contaminant transport processes.

(a)

(b)

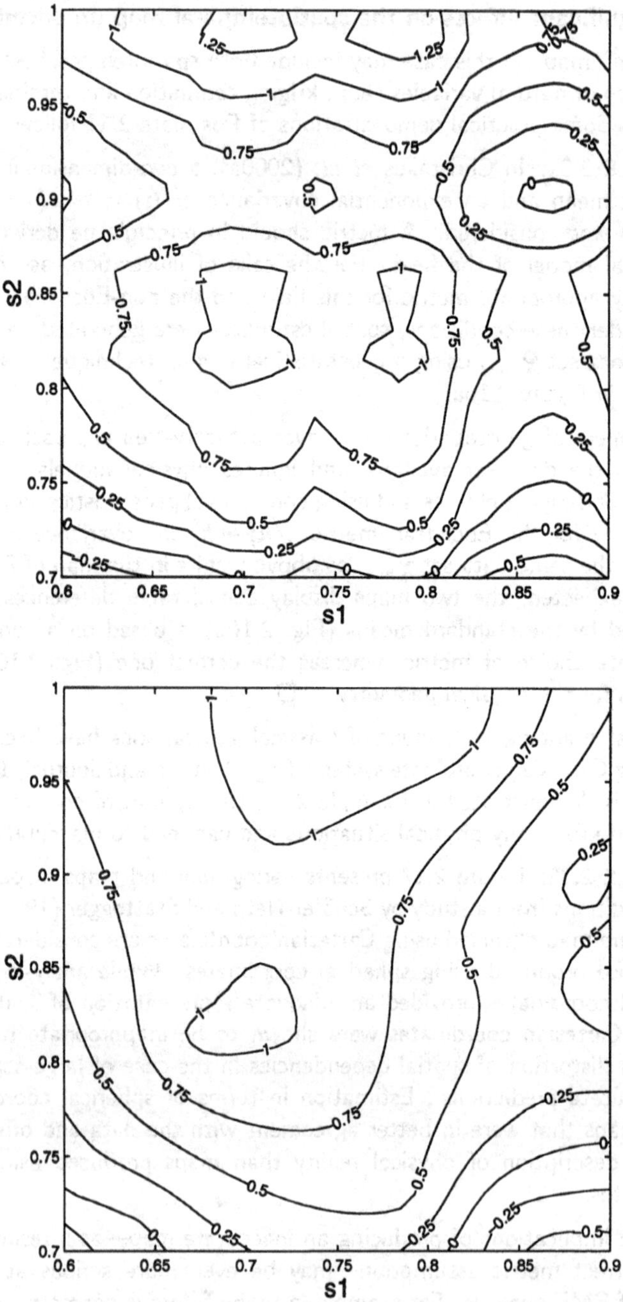

Figure 2.16. Maps obtained (a) using the appropriate non-Euclidean metric and (b) using the inappropriate Euclidean metric.

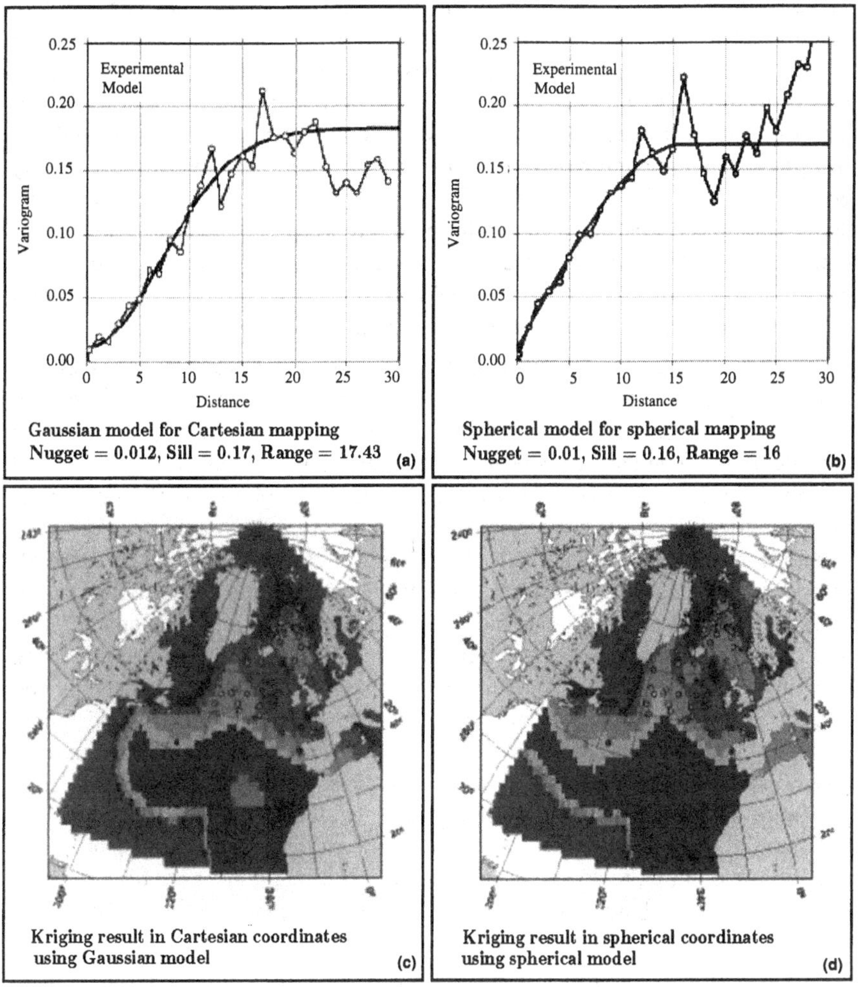

Figure 2.17. Geostatistical study of paleo-sea surface conditions: (a) and (b) are variograms produced using Cartesian *vs.* spherical coordinates (line with open circles = experimental, smooth dark line = model); (c) and (d) are maps produced using Cartesian *vs.* spherical coordinates (after Schäfer-Neth and Stattegger, 1998; reproduced with permission).

Some Final Thoughts

This chapter has emphasized the importance of space/time in the study of natural phenomena. Space/time can be dealt with on several levels of reality associated with different physical insights, considerations of symmetry, and restrictions on empirical understanding. We derive a particular geometry, however, not only empirically but also by theoretical reasoning, because our physical concepts and measurements have a particular meaning. Before attempting to apply any sort of geostatistical approach to the spatiotemporal analysis and mapping of natural variables, a system of coordinates and metric properties needs to be established whereby these variables can be quantified. The selection of the appropriate coordinates and metric properties depends on the physical knowledge relevant to the specific situation, and is not a purely geometric affair. Nature does not allow the natural processes to vary arbitrarily, but imposes constraints in the form of physical laws which, in turn, imply restrictions on the background space/time geometry. Geometrical approaches based on the intrinsic (internal) as well as the extrinsic (external) visualization of the spatiotemporal domain were considered in this chapter.

We have examined several groups of coordinate systems and metrical structures. The coordinate systems allow representations of space/time geometry on the basis of the underlying symmetry of the natural process under consideration. The use of a specific system also depends on the mathematical convenience afforded by the system (*e.g.*, in the case of a natural process that has cylindrical symmetry, the cylindrical coordinate system may be more convenient for mathematical manipulations than a rectangular coordinate system). In addition to coordinate systems, an important issue for geostatistical applications is the introduction of a metric that measures distances in space/time. The definition of an appropriate metric depends on the local properties of space and time (*e.g.*, the space/time curvature), as well as on rules imposed by the specific natural process (*e.g.*, diffusion). The space/time metric is used in formulating parametric models for the covariance function, which are then used in geostatistical mapping studies. In some cases it is possible, based on invariance principles and other physical considerations, to obtain explicit expressions for the metric. If explicit expressions for the metric are not available, it is still possible to obtain the space/time correlation functions for specific processes based on numerical simulations or experimental observations (*e.g.*, this is the case with processes that occur in fractal spaces).

It is possible that future experience will show that our tried and trusted systems and metric structures are inadequate, and that a new, conceptually different spatiotemporal geometry will arise. To properly describe observational evidence one may need to introduce, *e.g.*, a fourth spatial dimension. It must be emphasized, however, that four-dimensional space should not be confused with four-dimensional space/time. Time has some physical properties that space does not have, and *vice versa*. Certain activities and processes that are possible in pure four-dimensional space are not possible in four-dimensional space/time. To give another example, the Riemannian geometry we discussed in this chapter needs to be modified in order to describe consistently the new short-distance physics of superstring theory. Future developments may also show that all physical knowledge can be unified into a single science expressed in terms of space/time geometrical conceptions.

3

PHYSICAL KNOWLEDGE

> ***"All the measurements in the world are not the equivalent of a single theorem that produces a significant advance in our greatest of sciences."*** C.F. Gauss

From the General to the Specific

Modern spatiotemporal geostatistics is a multidisciplinary affair that involves knowledge from various sciences, and is not the province of pure *a priori* mathematics. The *physical knowledge* used in the analysis and mapping of spatiotemporal phenomena may come from a variety of sources. One should include all kinds of valid knowledge that are available at a given moment and can be obtained by the competent scientist using a scientific procedure effectively. In this sense, the availability of physical knowledge is objective. Subjective bias enters when the scientist fails to use the appropriate procedures leading to the valid knowledge, even when such procedures are available.

What we know *about* the physical world depends on us. But it does not necessarily follow that the way things *are* in the world depends on us. Several distinctions can be made between different forms of knowledge and knowing. One distinction is between knowledge that is obtained by the senses and knowledge that is obtained by the mind. Another distinction is between knowledge gained by the direct experience of the knower and knowledge gained through the experience of others. A *knowledge base* is a collection of knowledge sources relevant to the problem at hand to be invoked by a reasoning process aiming at the solution of the problem. For the purposes of modern geostatistics, we shall distinguish between two prime *knowledge bases*:

- The *general* knowledge base $\mathcal{G}$ (*i.e.*, obtained from physical laws and scientific theories, summary statistics, logical principles, *etc.*), and

• The *specificatory* or *case-specific* knowledge base $\mathcal{S}$ (obtained through experience with the specific situation).

Thus, $\mathcal{K} = \mathcal{G} \cup \mathcal{S}$ represents the *total* physical knowledge available.

COMMENT 3.1: *For modern geostatisticians it would be an ostrich-like policy to ignore the significance of these knowledge bases. It has been argued that spatial correlation models (belonging to the $\mathcal{G}$ base) constitute one of the most consequential steps of any geostatistical study, the rest is well-known calculus (Journel, 1989). This being the case, it makes no sense to leave out of the mapping process other important $\mathcal{G}$ and $\mathcal{S}$ knowledge bases (physical laws, local theories, empirical relationships, uncertain observations,* etc.*), and restrict analysis to only the usually limited amount of exact measurements available.*

Among the most significant developments in the frontier of research nowadays is that a vast body of knowledge is becoming available as a result of a new *horizontal* integration among disparate scientific disciplines. The realization on the part of many researchers that the problems they are confronted with are shared by other researchers in disparate fields leads to new highly *interdisciplinary* knowledge bases.

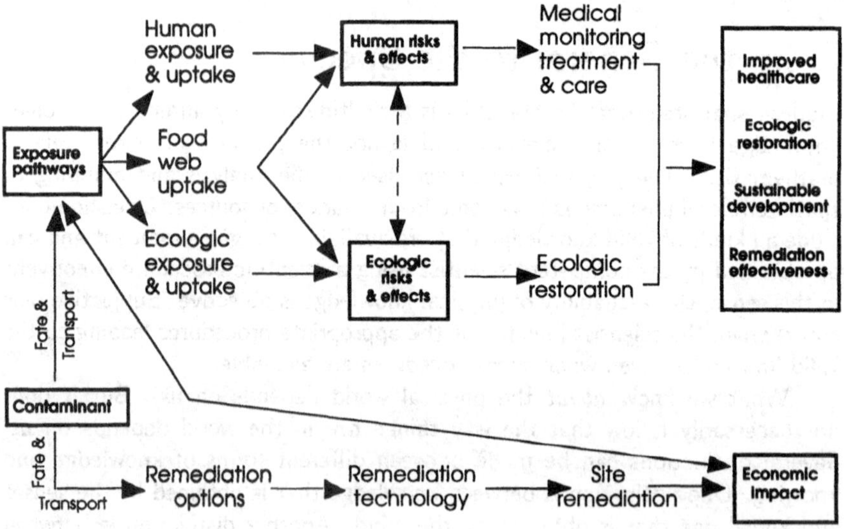

Figure 3.1. An integration framework of subsurface contamination studies.

EXAMPLE 3.1: Figure 3.1 shows an integration framework of a subsurface contamination scenario with its health, ecologic, and economic consequences. Various biomedical and non-biomedical fields are involved in this framework, including biology, toxicology, epidemiology, medicine, Earth sciences, engineering, ecology, economics, and management. The framework emphasizes the need for interdisciplinary research focusing on two crucial areas: (*i.*) unifying scientific disciplines, and (*ii.*) linking contamination with health endpoints for disease prevention and intervention. These endpoints include, *e.g.*, infectious diseases, cancer, birth defects, immune system dysfunction, behavioral abnormalities, decreased fertility, and altered metabolism. □

An important characteristic shared by both the $\mathcal{G}$ and $\mathcal{S}$ knowledge bases is that they should be *reliable*, meaning that the geostatistician can depend upon them in terms of predictability. Indeed, the geostatistician's persisting goal is to understand phenomena to such an extent that the quality of space/time mapping can be consistently improved. It is only through valid, organized physical knowledge bases that truly *scientific* mapping can be achieved. The two knowledge bases are described in considerable detail in the following sections.

The General Knowledge Base

The general knowledge base $\mathcal{G}$ relies on our ability to draw on the experience of previous generations and the discoveries of others in the present. Therefore, $\mathcal{G}$ denotes the background knowledge and the justified beliefs relative to the mapping situation overall, and it may include laws of science and universal statements that make claims about the properties of some aspect of the universe, structured patterns and assumptions, local (phenomenological) theories, previous experience with similar situations independent of any case-specific observations, analytic statements (justified on the basis of relations of concepts and theories), synthetic statements (whose truth or falsehood is a matter of experience), *etc.* The knowledge offered by these statements is considered "general" in the sense that it is vague enough to characterize a large class of fields or situations.

EXAMPLE 3.2: The statements "acids turn litmus red," "the ozone concentration field has specified mean and covariance," and "subsurface flow is governed by Darcy's law" are general statements. Indeed, all acids have the property of turning litmus red, several ozone fields may share the same mean and covariance, and various flow fields satisfy Darcy's law. Such statements thus constitute general knowledge. □

From an epistemic viewpoint, the general knowledge base $\mathcal{G}$ may be divided into two main categories:

• *Analytic* knowledge, which involves abstract quantities, symbols, logical relations, concepts, *etc.* without real counterparts (*e.g.*, "if A or B and not B, then A"). Other analytical statements may have real counterparts, but their validity is a matter of definition and/or logic only (*e.g.*, "soil permeability values are nonnegative," or "all sisters are female").

• *Synthetic* knowledge, which involves statements of fact, laws, and theories. The law of gravitation, *e.g.*, states that "two bodies exert a force between each other which varies inversely as the square of the distance between them, and varies directly as the product of their masses."

A *theory* is an intellectual construct that enquires deeply into the fundamental significance of a phenomenon and has explanatory power. A theory may also include nonobservable quantities. A *law* is a representation of a relation that has been observed consistently in the past and is expected to occur consistently in the future (*e.g.*, Fick's law of mass diffusion, or Fourier's law for heat flow). In this sense, a law is a summary description of numerous facts and is neither fundamental nor explanatory. A law may be derived from a theory (*e.g.*, under certain conditions Darcy's law may be deduced from first principles). *Local* (or *phenomenological*) laws could be obtained by means of well justified approximations or intuitive interpretations of a larger theory. Phenomenological laws may also include local experimental results.

A scientific theory, a physical law, *etc.* may be incorporated directly or indirectly into the BME framework. In the second case, a law may be used to derive a logical construct that expresses general knowledge about certain feature of the natural phenomenon. Such a construct may be, *e.g.*, a *statistical moment* that represents knowledge about spatiotemporal correlations and interactions.

Other forms of general knowledge may be obtained from sources such as literature searches (reports from state and federal agencies, printed material from professional organizations, computerized data bases, *etc.*), and experience with similar physical situations.

A mathematical formulation of the general knowledge base

In natural sciences our data sets, concepts, and theories strongly depend on spatiotemporal correlations and probabilistic dependencies. Therefore, it is useful to express the general knowledge $\mathcal{G}$ mathematically in terms of *stochastic* functions or operations involving the natural variables of interest. Assume that the general knowledge base $\mathcal{G}$ involves a series of funtionals $\mathcal{G}_\alpha$ $(\alpha = 0, 1, \ldots, N_c)$ such that

$$\mathcal{G} : \overline{h_\alpha}(\boldsymbol{p}_{map}) = \mathcal{G}_\alpha\left[\boldsymbol{\chi}_{map}, \boldsymbol{p}_{map}; f_\mathcal{G}\right] \tag{3.1}$$

The right-hand side of Equation 3.1 depends on the $\boldsymbol{\chi}_{map}$ values, the $\boldsymbol{p}_{map}$ coordinates, and the pdf $f_\mathcal{G}$ associated with the general knowledge $\mathcal{G}$. The left-hand side represents a set of stochastic expectations of the natural variables involved. As we shall see later, the $\overline{h_\alpha}$ can be formulated in a number of ways, depending on the general knowledge base available. Equation 3.1 accounts for a variety of sources of general knowledge, including: (*i.*) statistical correlation (or moment) models, (*ii.*) empirical relationships, and (*iii.*) physical laws and scientific theories. Equation 3.1 is, thus, the kind of quantity that makes scientists feel at home and enables them to pull out the tricks of their trades. An important function of modern spatiotemporal geostatistics is to endeavor constantly to refine and improve formulation (Eq. 3.1) so that it may continually and as closely as possible represent the general knowledge available from various sciences. A specific formulation of Equation 3.1 deserves additional attention. In many geostatistical applications, the physical laws, empirical relationships, *etc.* can be transformed into a suitable set of *moment equations* as follows

$$\left.\begin{aligned} \overline{h_\alpha}(\boldsymbol{p}_{map}) &= \overline{g_\alpha(\boldsymbol{x}_{map})}, \quad \alpha = 0, 1, \ldots N_c \\ with \quad \overline{g_\alpha(\boldsymbol{x}_{map})} &= \int d\boldsymbol{\chi}_{map}\, g_\alpha(\boldsymbol{\chi}_{map})\, f_{\mathcal{G}}(\boldsymbol{\chi}_{map};\, \boldsymbol{p}_{map}) \end{aligned}\right\} \tag{3.2}$$

where g_α is a set of known functions of $\boldsymbol{\chi}_{map}$. It is worth mentioning that there are certain mathematical and physical criteria involved in the choice of the g_α. By convention, $g_0 = 1$, and the respective $\overline{g_0(\boldsymbol{x}_{map})} = 1$ is a normalization constant. It is also necessary that the g_α $(\alpha > 0)$ are chosen so that the stochastic expectations $\overline{h_\alpha}$ on the left-hand side of Equation 3.2 can either be calculated directly from field data and experimental surveys or inferred from other sources of knowledge (physical laws, empirical charts, *etc.*). As will become clear from the subsequent examples, the g_α and h_α functions do not necessarily have the same mathematical form.

General knowledge in terms of statistical moments

To clarify certain basic aspects of the preceding formulations, let us discuss a few examples. The general knowledge considered in these examples includes functions characterizing the statistical behavior of the natural system (spatiotemporal means, variograms, ordinary and generalized covariances, multiple-point statistics, *etc.*).

EXAMPLE 3.3: Assume that the means $\overline{x_i}$, the variances $\overline{(x_i - \overline{x_i})^2}$, the third-order (centered) moments $\overline{(x_i - \overline{x_i})^3}$, and the ordinary (centered) covariances $\overline{(x_i - \overline{x_i})\,(x_{i'} - \overline{x_{i'}})}$ of the S/TRF $X(\boldsymbol{p})$ are known at points $\boldsymbol{p}_i$, $i = 1, ..., m, k$. The resulting functions $g_\alpha(\boldsymbol{\chi}_{map})$ of Equation 3.2 are shown in Table 3.1. The total number of g_α functions in this case is $1 + (m+1)(m+6)/2$. Spatiotemporal statistics of higher order, including multiple-point statistics, can be incorporated in a similar fashion. □

The following example requires some knowledge of the theory of S/TRF–ν/μ developed in Christakos (1991b, 1992).

EXAMPLE 3.4: Assume that the generalized spatiotemporal covariances of an S/TRF–1/1 $X(\boldsymbol{p})$ are known between the points $\boldsymbol{p}_i$ $(i = 1, 2, 3, 4, k)$. In view of the S/TRF theory, we let $g(\boldsymbol{\chi}_{map}) = (\chi_k - \frac{1}{4}\sum_{i=1}^{4} \chi_i)^2$, where $\boldsymbol{\chi}_{map} = (\chi_1,\ \chi_2,\ \chi_3,\ \chi_4,\ \chi_k)$ and $g_0 = 1$, as usual. In view of Equation 3.2, the corresponding statistical function is

$$\begin{aligned} \overline{h}(\boldsymbol{p}_{map}) = \overline{g(\boldsymbol{x}_{map})} = c_Q(\boldsymbol{\kappa}_{map}) &= \kappa_x(\boldsymbol{p}_k, \boldsymbol{p}_k) + \tfrac{1}{16}\sum_{i=1}^{4} \kappa_x(\boldsymbol{p}_i, \boldsymbol{p}_i) \\ &+ \tfrac{1}{8}\sum_{i=1}^{4}\sum_{j>i}^{4} \kappa_x(\boldsymbol{p}_i, \boldsymbol{p}_j) - \tfrac{1}{2}\sum_{i=1}^{4} \kappa_x(\boldsymbol{p}_i, \boldsymbol{p}_k) \end{aligned} \tag{3.3}$$

Table 3.1. The g_α functions of Example 3.3.

	α	g_α
	0	$g_0(\boldsymbol{\chi}_{map}) = 1$
$m+1$	1	$g_1(\chi_1) = \chi_1$
	$\vdots$	$\vdots$
	m	$g_m(\chi_m) = \chi_m$
	$m+1$	$g_{m+1}(\chi_k) = \chi_k$
$m+1$	$m+2$	$g_{m+2}(\chi_1, \chi_1) = (\chi_1 - \overline{x_1})^2$
	$\vdots$	$\vdots$
	$2m+1$	$g_{2m+1}(\chi_m, \chi_m) = (\chi_m - \overline{x_m})^2$
	$2m+2$	$g_{2m+2}(\chi_k, \chi_k) = (\chi_k - \overline{x_k})^2$
$m+1$	$2m+3$	$g_{2m+3}(\chi_1, \chi_1) = (\chi_1 - \overline{x_1})^3$
	$\vdots$	$\vdots$
	$3m+2$	$g_{3m+2}(\chi_m, \chi_m) = (\chi_m - \overline{x_m})^3$
	$3m+3$	$g_{3m+3}(\chi_k, \chi_k) = (\chi_k - \overline{x_k})^3$
$\frac{m(m+1)}{2}$	$3m+4$	$g_{3m+4}(\chi_k, \chi_1) = (\chi_k - \overline{x_k})(\chi_1 - \overline{x_1})$
	$\vdots$	$\vdots$
	$\frac{(m+1)(m+6)}{2}$	$g_{\frac{(m+1)(m+6)}{2}}(\chi_k, \chi_m) = (\chi_k - \overline{x_k})(\chi_m - \overline{x_m})$

where κ_x denotes the generalized covariance of $X(\boldsymbol{p})$ and $\boldsymbol{\kappa}_{map}$ is the corresponding generalized covariance matrix of the map. □

The covariances, variograms, and generalized covariances can have *separable* or *nonseparable* forms (see also "Spatiotemporal Covariance and Variogram Models" in Chapter 11, p. 224). Another interesting situation arising in practical applications is that in which the $\overline{g_\alpha}$ statistics contain measurement errors. Such a situation is discussed in the following example.

EXAMPLE 3.5: Assume that the $\overline{g_\alpha}$ statistics at the space/time point $\boldsymbol{p}$ ($\alpha \neq 0$ is the degree of the statistical moment considered) are contaminated by additive zero-mean white Gaussian measurement errors $\nu_\alpha(\boldsymbol{p})$. As a consequence, the actual $\overline{g_\alpha}$ statistics are related to the measured $\hat{\overline{g}}_\alpha$ statistics by the relationship

$$\hat{\overline{g}}_\alpha(\boldsymbol{p}) = \overline{g}_\alpha(\boldsymbol{p}) + \nu_\alpha(\boldsymbol{p}) = \int d\chi\, \chi^\alpha f_{\mathcal{G}}(\chi; \boldsymbol{p}) + \nu_\alpha(\boldsymbol{p}) \qquad (3.4)$$

where $\alpha = 1, \ldots, N_c$. As we shall see in Chapter 5, this is the relationship to be used in BME analysis. □

General knowledge in terms of physical laws

A physical law is a generalization obtained from the scientific study of the facts of observations. This generalization constitutes a very important $\mathcal{G}$ base, because it asserts knowledge of what would be the result of an observational procedure if it were carried out. The spatiotemporal distribution of most natural variables is expressed through physical laws which, thus, constitute important aspects of the mapping process. For the purposes of this book, two prime classes of laws, A and B, will be considered.

> **Class A:** Physical laws that are expressed by means of empirical equations of the general form
>
> $$\mathcal{G}: \Phi(X, \boldsymbol{Y}) = 0 \tag{3.5}$$
>
> where $\Phi(\cdot)$ is an algebraic function, $X(\boldsymbol{p})$ is the primary natural variable, and $\boldsymbol{Y}(\boldsymbol{p}) = \big(Y_1(\boldsymbol{p}), \ldots, Y_k(\boldsymbol{p})\big)$ is a vector of secondary (measurable) natural variables in space/time.

As was already mentioned in the section on mathematical formulation (p. 74), the physical laws should be transformed into an appropriate set of stochastic equations. For the physical laws of Class A, the general knowledge $\mathcal{G}$ involves a set of h_α and g_α functions such that Equation 3.2 can be written

$$\left.\begin{aligned} \overline{h_\alpha} &= \overline{h_\alpha(\boldsymbol{y})} \\ \overline{g_\alpha} &= \iint d\boldsymbol{\chi} d\boldsymbol{\psi}\, g_\alpha(\boldsymbol{\chi}, \boldsymbol{\psi})\, f_{\mathcal{G}}(\boldsymbol{\chi}, \boldsymbol{\psi}) \end{aligned}\right\} \tag{3.6}$$

where the subscripts α account for all space/time points considered. The meaning of the quantities involved in Equation 3.6 is better illustrated with the help of the following example.

EXAMPLE 3.6: Standard penetration resistance X and vertical stress Y for a cohesionless soil are related by an empirical law of the form (Lambe and Whitman, 1969)

$$\mathcal{G}: X(\boldsymbol{p}) - bY(\boldsymbol{p}) - c = 0 \tag{3.7}$$

where X, Y are represented as random fields, and b, c are experimental (deterministic) coefficients. In this case, the functions involved in Equation 3.6 are as follows,

$$\left.\begin{aligned} \overline{h_\alpha} &= b\overline{Y(\boldsymbol{p}_\alpha)} + c \\ \overline{g_\alpha(x_\alpha)} &= \int d\chi_\alpha\, \chi_\alpha\, f_{\mathcal{G}}(\chi_\alpha; \boldsymbol{p}_\alpha) \end{aligned}\right\} \tag{3.8}$$

where α accounts for the space/time point $\boldsymbol{p}_\alpha$ considered; and

$$\left.\begin{aligned} \overline{h_{\alpha'}} &= b\, C_y(\boldsymbol{p}_i, \boldsymbol{p}_j) + c\overline{Y(\boldsymbol{p}_j)} \\ \overline{g_{\alpha'}(x_i, y_j)} &= \iint d\chi_i\, d\psi_j\, \chi_i\, \psi_j\, f_{\mathcal{G}}(\chi_i, \psi_j; \boldsymbol{p}_i, \boldsymbol{p}_j) \end{aligned}\right\} \tag{3.9}$$

where $C_y(\boldsymbol{p}_i, \boldsymbol{p}_j)$ is the (non-centered) covariance of Y between the pair of points $\boldsymbol{p}_i$ and $\boldsymbol{p}_j$ and α' accounts for the pair considered. Note that the $\overline{g_{\alpha'}(x_i, y_j)}$ in Equation 3.9 may also be expressed in terms of the multivariate pdf of the set of mapping points $\boldsymbol{p}_{map}$, *i.e.*,

$$\overline{g_{\alpha'}(x_i, y_j)} = \iint d\boldsymbol{\chi}_{map}\, d\boldsymbol{\psi}_{data}\, \chi_i\, \psi_j\, f_{\mathcal{G}}(\boldsymbol{\chi}_{map}, \boldsymbol{\psi}_{data}; \boldsymbol{p}_{map}) \tag{3.10}$$

where α' accounts for the pair of points $\boldsymbol{p}_i, \boldsymbol{p}_j \in \boldsymbol{p}_{map}$ under consideration. Other functions of χ_i and ψ_j resulting from the physical law (Eq. 3.7) may be expressed in a similar manner, *e.g.*,

$$\overline{g_{\alpha'}(\boldsymbol{x}_{map}, \boldsymbol{y}_{data})} = \iint d\boldsymbol{\chi}_{map}\, d\boldsymbol{\psi}_{data}\, g_{\alpha'}(\boldsymbol{\chi}_{map}, \boldsymbol{\psi}_{data})\, f_{\mathcal{G}}(\boldsymbol{\chi}_{map}, \boldsymbol{\psi}_{data}; \boldsymbol{p}_{map}) \tag{3.11}$$

where $\alpha' = 1, \ldots, N_c$ now accounts for all combinations of points $\boldsymbol{p}_i \in \boldsymbol{p}_{map}$; *etc.* (see also Chapter 9 on multipoint and multivariable BME analysis). □

Class B: Physical laws that can be expressed in terms of the general form

$$\mathcal{G}: D_{\boldsymbol{p}}[X, \boldsymbol{Y}] = \Theta(X, \boldsymbol{Y}) \tag{3.12}$$

where $D_{\boldsymbol{p}}$ is a differential operator in space/time, and Θ is an algebraic function.

Equation 3.12 offers a powerful physical basis for relating $X(\boldsymbol{p})$ with $\boldsymbol{Y}(\boldsymbol{p})$. Two cases may be considered. First, assume that a solution to Equation 3.12 can be obtained so that we can write $X = H(\boldsymbol{Y})$. Then, the representation of Equation 3.6 is valid, where now the possible forms of the h_α and g_α functions above include

$$\left.\begin{aligned} h_\alpha &= H[\boldsymbol{y}], \ = H[\boldsymbol{y}]\, H[\boldsymbol{y}'], \ etc. \\ g_\alpha &= x, \ = x\, x', \ etc. \end{aligned}\right\} \tag{3.13}$$

i.e., the moments $\overline{h_\alpha}$ are expressed simply in terms of the known $\boldsymbol{Y}(\boldsymbol{p})$ statistics. In the case that an explicit solution is difficult to obtain or does not exist, the physical Equation 3.12 can be used in BME analysis by considering the following representation

$$\left.\begin{aligned} \overline{h_\alpha} &= \overline{h_\alpha(\boldsymbol{x}, \boldsymbol{y})} \\ \overline{g_\alpha} &= \iint d\boldsymbol{\chi}\, d\boldsymbol{\psi}\, g_\alpha(\boldsymbol{\chi}, \boldsymbol{\psi}) f_{\mathcal{G}}(\boldsymbol{\chi}, \boldsymbol{\psi}) \end{aligned}\right\} \tag{3.14}$$

where the subscripts α account for all space/time points considered, and the possible forms of the h_α and g_α functions include

$$\left.\begin{aligned} h_\alpha &= D_{\boldsymbol{p}}[x], \ = D_{\boldsymbol{p}}[x]\, D_{\boldsymbol{p}}[x'], \ etc. \\ g_\alpha &= \Theta(x, \boldsymbol{y}), \ = \Theta(x, \boldsymbol{y})\, \Theta(x', \boldsymbol{y}'), \ etc. \end{aligned}\right\} \tag{3.15}$$

The choice of the h_α and g_α functions above is not unique, since there is a number of factors (mathematical and physical) that may influence such a choice. It is possible, *e.g.*, that one may need to consider functions that involve derivatives of the pdf with respect to the space and time coordinates. As it turns out, while the statistics of the random field $X(\boldsymbol{p})$ are implicit in

Equation 3.14, in many cases we do not need to solve for them. The preceding analysis is exhibited most easily by means of examples.

EXAMPLE 3.7: Consider the differential equation representing the temporal variation of a lumped parameter Earth system (Jones, 1997),

$$\mathcal{G}: \frac{dX(t)}{dt} = bX(t) \tag{3.16}$$

where X is represented as a random field, and b is a deterministic parameter. In light of the physical law (Eq. 3.16), we assume that the h_α and g_α functions involved in Equation 3.14 are

$$\left.\begin{aligned} \overline{h_\alpha} &= (\lambda b)^{-1} \frac{d}{dt} \overline{X^\lambda(t_\alpha)} \\ \overline{g_\alpha(x)} &= \int d\chi_\alpha \, \chi_\alpha^\lambda \, f_{\mathcal{G}}(\chi_\alpha;\, t_\alpha), \quad \lambda = 1,\, 2 \end{aligned}\right\} \tag{3.17}$$

where α denotes the time t_α of interest. Other statistics can be taken into account in a similar fashion, *e.g.*,

$$\left.\begin{aligned} \overline{h_{\alpha'}} &= b^{-2} \frac{\partial^2 C_x(t_i,\, t_j)}{\partial t_i \partial t_j} \\ \overline{g_{\alpha'}(x_i, x_j)} &= \iint d\chi_i \, d\chi_j \, \chi_i \chi_j \, f_{\mathcal{G}}(\chi_i, \chi_j;\, t_i,\, t_j) \end{aligned}\right\} \tag{3.18}$$

where C_x is the (non-centered) covariance of X, and α' accounts for the pair of time instants t_i and t_j in consideration. Equations 3.17 and 3.18 are quite appropriate for the geostatistical analysis in the prior stage (see Chapter 5). However, the choice of $g_{\alpha'}$ functions is not unique. Other $g_{\alpha'}$ forms may be used as well. In fact, just as for Class A above, the analysis is easily extended to the whole set of mapping times t_{map}. The $\overline{g_{\alpha'}(x_i, x_j)}$, *e.g.*, may also be generalized in terms of Equation 3.11, where $g_{\alpha'}$ may have any form resulting from the physical law (Eq. 3.16) and $t_{map} = (t_1, \ldots, t_m, t_k)$; *etc.* □

EXAMPLE 3.8: Three-dimensional, steady-state subsurface flow in which the mean hydraulic gradient J is in the s_1 direction could be approximated by the following law (*e.g.*, Dagan, 1989)

$$\nabla^2 X(\boldsymbol{s}) = J \frac{\partial u(\boldsymbol{s})}{\partial s_1} \tag{3.19}$$

where $X(\boldsymbol{s})$ represents the random hydraulic head fluctuation, and $u(\boldsymbol{s})$ denotes the log-conductivity fluctuation with the isotropic covariance $c_u(r) = \sigma_u^2 \exp[-r/\varepsilon]$. Under these conditions, the hydrologic law (Eq. 3.19) leads to

$$\left.\begin{aligned} \overline{h_\alpha} &= \sigma_X^2 = \tfrac{1}{3} \sigma_u^2 J^2 \varepsilon^2 \\ \overline{g_\alpha(x_\alpha)} &= \int d\chi_\alpha \, \chi_\alpha^2 \, f_{\mathcal{G}}(\chi_\alpha;\, s_\alpha) \end{aligned}\right\} \tag{3.20}$$

where the subscript α accounts for the spatial locations considered. □

Spatial covariances, flow moments of higher order, *etc.* may be introduced in a similar fashion.

EXAMPLE 3.9: Under certain assumptions, one-dimensional groundwater flow may be represented by the differential equation (Bear, 1972)

$$\mathcal{G}: \frac{d}{ds}X + Y\,X = 0 \tag{3.21}$$

where X is the random hydraulic gradient and Y is the random hydraulic log-conductivity slope at a spatial location s. Equation 3.21 leads to

$$\left.\begin{aligned} \overline{h_\alpha} &= -\lambda^{-1}\,\overline{\frac{dX^\lambda}{ds_\alpha}} \\ \overline{g_\alpha(x_\alpha, y_\alpha)} &= \iint d\chi_\alpha\, d\psi_\alpha\, \chi_\alpha^\lambda\, \psi_\alpha\, f_{\mathcal{G}}(\chi_\alpha, \psi_\alpha;\ s_\alpha) = \overline{x_\alpha^\lambda y_\alpha}\,,\ \lambda = 1,\, 2 \end{aligned}\right\} \tag{3.22}$$

where $\lambda = 1, 2$ accounts for the moments considered, and the subscript α denotes the location of interest. Equations for moments of higher order derived from the flow law (Eq. 3.21) can be processed in a similar manner and the analysis can be extended to the entire set of locations $\boldsymbol{s}_{map} = (s_1, \ldots, s_m, s_k)$. Suppose, *e.g.*, that we seek to process knowledge about the covariance of X. Then, Equation 3.21 gives rise to the system of equations

$$\left.\begin{aligned} \overline{h_{\alpha'}} &= \frac{\partial^2}{\partial s_i\, \partial s_j}\, C_x(s_i,\, s_j) \\ \overline{g_{\alpha'}(x_i, y_j)} &= \iint d\boldsymbol{\chi}_{map}\, d\boldsymbol{\psi}_{data}\, \chi_i\, \chi_j\, \psi_i\, \psi_j\, f_{\mathcal{G}}(\boldsymbol{\chi}_{map}, \boldsymbol{\psi}_{data};\ \boldsymbol{s}_{map}) \end{aligned}\right\} \tag{3.23}$$

where the subscript α' accounts for the pair of locations s_i and s_j under consideration. □

COMMENT 3.2: *In some situations, the $\overline{h_\alpha}$ functions may be formulated indirectly: the stochastic moment equations associated with the physical equations are derived and the $\overline{h_\alpha}$ functions are chosen so that they express the statistical moments appearing in these equations (mean, variance, covariance,* etc.*). See also Chapter 5.*

In many applications, it may be preferable to transfer a problem expressed in terms of Class B physical laws into a Class A problem. This may be done in the following manner.

> **Class B→Class A:** A Class B problem can be transferred to a Class A problem by discretizing the differential equation representing the physical law, thus obtaining an algebraic equation of the form given in Equation 3.5, *i.e.*,

$$\mathcal{G}: D_p[X, Y] = \Theta(X, Y) \xrightarrow{\textit{Discretization}} \Phi(X, Y) = 0 \tag{3.24}$$

The function Φ may have the form of a finite difference of any order, *etc.* One such case is discussed in the following example.

EXAMPLE 3.10: The partial differential equation (pde) governing subsurface flow in two dimensions (Bear, 1972) is

$$\mathcal{G}: \nabla \cdot [Y(\boldsymbol{p}) \nabla X(\boldsymbol{p})] = S \frac{\partial X(\boldsymbol{p})}{\partial t} \tag{3.25}$$

where $\boldsymbol{p} = (s_1, s_2, t)$, X is the random hydraulic head, Y is the random conductivity field, and S is the storativity. The flow equation offers a physical basis for relating the stochastic moments of hydraulic head and hydraulic conductivity. For the flow law (Eq. 3.25), a possible discretization is as follows

$$\begin{aligned} \mathcal{G}: & \Delta s_1^{-2} \left[\psi_{i+1/2,j,k}(\chi_{i+1,j,k} - \chi_{i,j,k}) - \psi_{i-1/2,j,k}(\chi_{i,j,k} - \chi_{i-1,j,k})\right] \\ & + \Delta s_2^{-2} \left[\psi_{i,j+1/2,k}(\chi_{i,j+1,k} - \chi_{i,j,k}) - \psi_{i,j-1/2,k}(\chi_{i,j,k} - \chi_{i,j-1,k})\right] \\ & = S \Delta t^{-1}(\chi_{i,j,k+1} - \chi_{i,j,k}) \end{aligned} \tag{3.26}$$

where $\chi_{i,j,k}$ and $\psi_{i,j,k}$ are, respectively, the head and conductivity values at the space/time grid node (i, j, k) associated with (s_1, s_2, t); Δs_1 and Δs_2 are spatial discretization steps along the s_1 and s_2 directions, respectively; and Δt is the time step. Equation 3.26 is an algebraic equation of the form given in Equation 3.5 and, hence, we can proceed as in the case of the Class A problems above. In a realistic study, discretizations (Eq. 3.26) at various sets of points are considered. □

COMMENT 3.3: *As we shall see in Chapter 5 ("General knowledge in the form of physical laws," p. 109), the fact that in many cases we do not need to solve the differential equations for the space/time moments of $X(\boldsymbol{p})$—which are implicit in these equations—has a special significance in the mathematical formulation of BME. In certain applications, the calculation of the $\overline{g_\alpha}$ statistics is based on a theory that may include correlation modeling (Deutsch and Journel, 1992), stochastic pde's, and entropy maximization (Christakos, 1992). Specific observational data sets may be involved in an indirect way,* e.g., *in the estimation of the parameters of the theoretical model. This way is justified by means of an empirical process that moves from a set of particulars to general knowledge.*

Some other forms of general knowledge

Depending on the application of interest, several other sorts of general knowledge may be incorporated into the BME analysis. In most real-world problems, the development of useful general knowledge bases involves a constant balancing of theory and practicality.

In some cases a geostatistician may feel comfortable considering the univariate pdf of the natural variable as part of the available general knowledge.

There are several ways to incorporate such knowledge into BME analysis. Assuming that the univariate pdf is the same for all points in space/time, a simple approach is to consider a transformation of the original random field into a Gaussian field with known mean and covariance functions, which brings us back to the earlier section on general knowledge in terms of statistical moments (p. 75).

In some other circumstances, while the multivariate pdf $f_{\mathcal{G}}$ is unknown, an equation that describes the evolution of a lower level pdf may be derived from physical or mathematical considerations (*e.g.*, a diffusion-type equation). In order to incorporate this knowledge into BME analysis, one may select a set of g_α functions so that the stochastic expectations $\overline{g_\alpha}$ with respect to $f_{\mathcal{G}}$ are calculable (*e.g.*, with the help of the evolution equation of $f_{\mathcal{G}}$). The statistical functions $\overline{g_\alpha}$ are now the general knowledge, *i.e.*, the original situation again has been converted to that of the earlier section (p. 75). It may be instructive to demonstrate the approach in terms of a very simple example.

EXAMPLE 3.11: Suppose that the prior pdf $f_{\mathcal{G}}$ satisfies the equation

$$\frac{\partial}{\partial t} f_{\mathcal{G}}(\chi;\, t) = \beta\, f_{\mathcal{G}}(\chi;\, t) \tag{3.27}$$

$\chi \in [a,\, b]$, with initial condition $f_{\mathcal{G}}(\chi;\, 0) = \delta(\chi)$. We choose $g_0 = 1$ (as usual) and let $g_1(\chi)$ be a known function of χ only, with $g_1(0) = c_0$. We can then write

$$\frac{d}{dt}\overline{g_1}(t) = \frac{d}{dt}\int_a^b d\chi\, g_1(\chi)\, f_{\mathcal{G}}(\chi;\, t) = \beta \int_a^b d\chi\, g_1(\chi)\, f_{\mathcal{G}}(\chi;\, t) = \beta\, \overline{g_1}(t) \tag{3.28}$$

In light of the initial condition, Equation 3.28 gives $\overline{g_1}(t) = c_0 \exp[\beta t]$. □

Finally, in some applications, useful physical knowledge may be expressed in the form of *integral equations*. This is the case, *e.g.*, of perturbation approximations (ordinary or diagrammatic) to groundwater flow and solute transport problems (for a detailed study of such perturbations see Christakos and Hristopulos, 1998).

The Specificatory Knowledge Base

The *specificatory* or *case-specific* knowledge base $\mathcal{S}$ is knowledge about the specific situation. For example, the singular statements, "the litmus paper that Mr. Nicolaou gave me turned red when immersed in the liquid" and "the distribution of the West Lyons field porosity data is shown in Figure 7.4," constitute specificatory knowledge. Unlike general knowledge, the specificatory knowledge base $\mathcal{S}$ refers to a particular occurrence or state of affairs at a particular location and at a particular time. The $\mathcal{S}$ includes both external or demonstrative evidence (actual measurements, perceptual or data of sense, *etc.*) and internal or inductive evidence (inferring one thing from another thing, empirical propositions, expertise with the particular situation, *etc.*).

The bulk of specificatory knowledge $\mathcal{S}$ usually consists of data sets representing natural variables. The entities of a data set may be media samples (geologic, chemical, environmental, *etc.*), readings of measurement devices, well-log profiles, weather station reports, topographic features, census data, state registries, *etc.* A single data set may contain information on several natural variables. Depending on the scientific discipline, numerous data types may be considered. These data types are of varying quality and accessibility, which may impose restrictions on the sort of data analysis and processing techniques to be employed.

EXAMPLE 3.12: In geological sciences, main data types (Swan and Sandilands, 1995) include: (a) ratio scale data (ordinary measurements, such as length or weight, which are usually the best-quality data); (b) interval scale data (temperature in °C or °F, *etc.*); (c) closed data (such as percentages or ppm); (d) directional data (*e.g.*, angles); (e) discrete data (counts of objects, *etc.*); and (f) nominal or categorical data (lists of minerals or fossils, *etc.*). □

It is axiomatic in science that the most difficult of all problems is that of asking the *right* questions. These questions have a direct effect on decisions about what sort of data is to be gathered and which data gathering procedures are to be implemented. For purposes of modern spatiotemporal geostatistics, the data sets $\boldsymbol{\chi}_{data}$ available as specificatory knowledge are usually divided into two main caregories

$$\mathcal{S}: \boldsymbol{\chi}_{data} = (\boldsymbol{\chi}_{hard}, \boldsymbol{\chi}_{soft}) = (\chi_1, \ldots, \chi_m) \tag{3.29}$$

where $\boldsymbol{\chi}_{hard}$ denotes *hard* data (*i.e.*, accurate measurements obtained from real-time observation devices, computational algorithms, or simulation processes); and $\boldsymbol{\chi}_{soft}$ denotes *soft* data (uncertain observations expressed in terms of interval values, probability statements, empirical charts, assessments by experts, *etc.*).

Each member of the data set $\boldsymbol{\chi}_{data}$ is an *operationally* defined measurement, *i.e.*, a measurement obtained by following a specified procedure of action and involving the appropriate instruments. It is imperative that the competent geostatistician understand clearly the essential features of the various data gathering techniques in terms of their strengths or weaknesses. There is no doubt that the outcome of the geostatistical analysis depends on the sophistication of the instrumentation available. Advanced instrumentation of space/time measurement can produce new maps, thereby improving the geostatistical analysis. Certainly, theoretical understanding and intuition are important factors in geostatistical efforts, not only to acquire knowledge but also to apprehend correctly the physical and logical relations that exist between a body of knowledge and the specific phenomenon under consideration.

POSTULATE 3.1: Operationally defined measurements are vital components of the epistemic observer–observed interplay and, thus, can have

a significant effect on the design and development of a geostatistical method.

An important consequence of Postulate 3.1 is the well-known support effect of geostatistics, which is described in the following example.

EXAMPLE 3.13: In many geostatistical applications, the sizes of the operationally defined measurement units (samples) of a data set are different than the size of the mapping domain of interest (*e.g.*, a block). This difference leads to the so-called *support effect*, which, if not properly taken into consideration, can lead to serious mapping errors. The epistemic nature of the BME model allows it to be formulated in a way that accounts for the support effect rigorously and efficiently (see "The support effect" in Chapter 9, p. 168). □

Specificatory knowledge in the form of either hard or soft data may be related to the boundary and initial conditions of a phenomenon; these conditions are usually complicated, reflecting the complexity of the real world. The construction of high-quality specificatory knowledge bases may entail considerable investments in time and effort.

Let us now consider separately the two groups of data sets presented in Equation 3.29 above, and attempt to throw some more light on some of their most significant features.

Specificatory knowledge in terms of hard data

A scientist or an engineer embarked on any empirical enquiry implements certain well-established scientific procedures in order to ascertain in a finite time the knowledge relevant to his/her enquiry. In many of these procedures, theory-based instruments form the conditions for and are the mediators of a significant part of scientific knowledge. In this context, the hard data vector $\boldsymbol{\chi}_{hard}$ represents sets of measurements obtained with the help of the instruments which, for all practical purposes, are considered *accurate*. In real-world situations, the latter could mean either of the following two things:

(*i.*) there is a high degree of confidence that the data obtained are not contaminated by measurement errors, operator biases, computational blunders, *etc.*; or,

(*ii.*) any existing errors are of the sort that can be eliminated effectively by carefully repeating the operations, successively refining the experimental techniques, using experience from earlier experiments and theoretical predictions, *etc.* (see Bevington and Robinson, 1992, for example).

Since two essential attributes of geostatistical analysis are objectivity and accuracy, the growth of the field should be in a way related to the development of objectifying instruments which generate accurate hard data. In much of the following analysis we will suppose that the hard data existing at a set of m_h ($< m$) points are represented as follows

$$S\colon \boldsymbol{\chi}_{hard} = (\chi_1, \ldots, \chi_{m_h}) \tag{3.30}$$

The knowledge base $\mathcal{S}$ in Equation 3.30 includes single-valued measurements $\chi_i\,(i = 1, \ldots, m_h)$ in space/time. Data of the form in Equation 3.30 usually constitute the first act of a space/time analysis, and have been employed traditionally by classical geostatistics and spatial statistics techniques (*e.g.*, Agterberg, 1974; Davis, 1986; Cressie, 1991; Kitanidis, 1997). In practice, hard data may include measurement sets, meteorological surveys, remote-sensing observations, census data, *etc.*; and they are available on regular space/time grids, lattices, arbitrary sampling networks, *etc.*

COMMENT 3.4: *In most geostatistical applications it is presupposed that the natural phenomenon under investigation has not been modified by the experimental procedures leading to the data set (Eq. 3.30). If, however, one is dealing with a situation in which the experiments modify certain features of the natural variable, this effect should be taken into account by modern geostatistical analysis.*

Specificatory knowledge in terms of soft data

As we discussed in previous sections, observations presuppose a theoretical or conceptual framework. Insofar as some of these theories are incomplete and uncertain, the guidance they offer as to what kind of observations should be made and in what manner could be incomplete and misleading (important factors may be overlooked, *etc.*); see Shafer and Pearl (1990), for example. An empirical feature of human knowledge is that it does not only rely on sets of hard data and pure facts. It also includes incomplete or qualitative data linked to experts' opinions, experience, intuition, *etc.* In the case of such qualitative data, the *impossibility theorem* (Arrow and Raynaud, 1986) shows that reconciling experts' opinions may imply a considerable amount of uncertainty. The result of all this is the generation of an amount of uncertainty about the observed variables.

In situations such as the above, the observation statements take the form of *soft* data, which are assumed to be available at the remaining $m_s = m - m_h$ points, *i.e.*,

$$\mathcal{S}: \boldsymbol{\chi}_{soft} = (\chi_{m_h+1}, \ldots, \chi_m) \tag{3.31}$$

Soft data may represent varying levels of understanding of uncertain observations leading to the direct calculation of the *probabilities* or their indirect estimation from accumulated experience. In fact, depending on the situation, several types of soft data may be available to the geostatistician. A strategy for evaluating the soft data types available in a particular situation would be based on criteria such as consistency, completeness, and relevance to stated objectives. A few specific examples are considered next.

EXAMPLE 3.14: The $\boldsymbol{\chi}_{soft}$ is often expressed in terms of *intervals* I_i of possible values of the χ_i $(i = m_h + 1, \ldots, m)$, *i.e.*,

$$\mathcal{S}: \boldsymbol{\chi}_{soft} \in \boldsymbol{I}, \;\; \chi_i \in I_i = [l_i, u_i] \text{ for } i = m_h + 1, \ldots, m \tag{3.32}$$

where $\boldsymbol{I} = (I_{m_h+1}, \ldots, I_m)$ is the domain of $\boldsymbol{\chi}_{soft}$; l_i (u_i) are the lower (upper) bounds of the intervals I_i. □

EXAMPLE 3.15: Other forms of commonly used soft data have a *probabilistic* character, such as

$$\mathcal{S}: \boldsymbol{\chi}_{soft} \in \boldsymbol{I},\ P_{\mathcal{S}}[\boldsymbol{x}_{soft} \leq \boldsymbol{\zeta}] = F_{\mathcal{S}}(\boldsymbol{\zeta}) \tag{3.33}$$

where $F_{\mathcal{S}}$ is the cumulative distribution function (cdf) obtained from $\mathcal{S}$; and

$$\mathcal{S}: \boldsymbol{\chi}_{soft} \in \boldsymbol{I},\ P_{\mathcal{S}}[h(\boldsymbol{x}_{soft}) \leq \zeta] = F_{\mathcal{S}}(\zeta;\ h) \tag{3.34}$$

where the function $h(\cdot)$ represents an empirical chart, a model, a justified belief, *etc.* (for instance, a hydrologist may be uncertain about the accuracy of individual hydraulic conductivities obtained at two different wells, but he/she can justifiably assign probability values to their differences). Equation 3.32 may be derived as a special case of Equation 3.33 for a properly selected cdf (*i.e.*, the χ_i values lie with probability one within known intervals I_i). Soft data can also be expressed in terms of *interval probabilities*, *i.e.*,

$$\mathcal{S}: \boldsymbol{\chi}_{soft} \in \boldsymbol{I},\ P_{\mathcal{S}}[\boldsymbol{x}_{soft} \in \boldsymbol{I}] = p_{\mathcal{S}}(\boldsymbol{I}) \tag{3.35}$$

where $\boldsymbol{I} = \{I_i;\ i = m_h + 1, \ldots, m\}$, $I_i = [l_i,\ u_i]$. □

EXAMPLE 3.16: In some cases the knowledge base $\mathcal{S}$ may involve *probabilistic logic* relationships, such as

$$\mathcal{S}: \chi_i,\ \chi_j \in \boldsymbol{\chi}_{soft} \quad s.t. \begin{cases} P_{\mathcal{S}}[\chi_i \wedge \chi_j] = p_{ij} \\ P_{\mathcal{S}}[\chi_i \rightarrow \chi_j] = p'_{ij}, \\ etc. \end{cases} \tag{3.36}$$

where the symbols "$\wedge$" and "$\rightarrow$" denote conjunction and material conditional, respectively (see also the discussion of conditional probability and the probability of conditionals in Chapter 4, beginning on p. 98). □

EXAMPLE 3.17: There are cases of practical importance in which soft data are available not only at the data points $\boldsymbol{p}_{soft}$ but also at the estimation point $\boldsymbol{p}_k$ itself. Such a case is one in which the specificatory knowledge has the form

$$\mathcal{S} = \mathcal{S}_0 \cup \mathcal{S}_1 : \begin{cases} P_{\mathcal{S}}[\boldsymbol{x}_{soft} \leq \boldsymbol{\zeta},\ x_k \leq \zeta_k] = F_{\mathcal{S}}(\boldsymbol{\chi}_{soft},\ \chi_k) \\ \mathcal{S}_0 : \boldsymbol{\chi}_{soft} \in \boldsymbol{I},\ \mathcal{S}_1 : \ \chi_k \in I_k \end{cases} \tag{3.37}$$

See also Chapter 6. □

In practice, there exist several methods for *encoding* the soft probability functions mentioned above ($F_{\mathcal{S}}$, $p_{\mathcal{S}}$, *etc.*). Some of these methods depend on the physical situation considered and the background of the experts involved.

EXAMPLE 3.18: Given hard measurements χ_{hard} of a natural variable X at points p_{hard}, a technique (*e.g.*, polynomial fitting or model simulation) can be used to derive new values at the estimation points p_k. The new X-values, which are uncertain, can be used to develop soft data at these points, such as probability functions having these values as means. □

COMMENT 3.5: *Frequently used* encoding *methods rely on a series of questions to establish points on the soft probability functions* (e.g., *Morgan and Henrion, 1990). Asking that the experts provide sound scientific justification and reasons for and against judgments can improve the quality of encoding. In the case of probabilistic logic, the encoding methods usually involve linear or nonlinear programming techniques* (e.g., *Chandru and Hooker, 1999).*

Finally, in the case that there is not enough knowledge to yield a probability law, intervals for probabilities may be introduced, *e.g.*,

$$\mathcal{S}: P_S[\boldsymbol{x}_{soft} \in \boldsymbol{I}] \in [a, b] \tag{3.38}$$

where $0 \leq a, b \leq 1$. In addition to the probabilistic formulation, soft data may be available in the form of *fuzzy* statements (*e.g.*, "the temperature is high" or "the range of excessive contamination is short"). These kinds of statements as well as fuzzy sets result from the experience one has of reality and the way one constructs and organizes that knowledge. Fuzzy statements and sets can be processed by approximate reasoning (*i.e.*, fuzzy logic), and then defuzzified (*e.g.*, converted into definite values; Tanaka, 1997). Soft data may also be derived from fuzzy information by means of α-cuts (Klir and Yuan, 1995). Generally, the soft data obtained from fuzzy information may have forms similar to the ones mentioned above (interval data, *etc.*).

Summa Theologica

If the laborious effort of data gathering and processing is going to be valuable, it must be combined with careful preliminary planning and a conceptual groundwork identifying what one is looking for, what it is likely to look like, and how to find it. As a consequence, general knowledge (in the form of scientific theories, laws, conceptual relationships, *etc.*) can play an important role in the acquisition and integration of specificatory knowledge (in the form of hard and soft data). Depending on the underlying theory, the same data may be classified according to a variety of categories. In other words, what we already know influences what we are going to observe.

Specificatory knowledge may become available from a variety of sources. In many situations in applied sciences, the uncertain knowledge expressed in terms of soft data is of vital importance. By limiting oneself to knowledge that is considered certain beyond doubt, one minimizes the risk of some errors associated with soft data, but at the same time the risk of missing out on what might be a very important and subtle source of knowledge is maximized. As a matter of fact, the importance of uncertain (soft) data was recognized centuries ago. Saint Thomas Aquinas, following Aristotle, argued (see, *e.g.*, Schumacher, 1977) that

> The uncertain knowledge that may be obtained of the highest things is more desirable than the most certain knowledge obtained of lesser things.
>
> *Summa Theologica*

In light of the above considerations, modern spatiotemporal geostatistics may be viewed as a field of concepts and methods whose boundary conditions are the available knowledge bases. Both knowledge bases $\mathcal{G}$ (general) and $\mathcal{S}$ (specificatory) will be used in the BME theoretical construct leading to the spatiotemporal map of the phenomenon. These two knowledge bases must mesh in coherent interaction with the new information provided by the map in order to provide us with an *explanatory* rationale for the phenomenon of interest.

The modern geostatistics paradigm requires not only that the mathematical model or technique chosen be the best possible, but also that the processing of the various forms of knowledge be achieved by means of logically plausible rules and the updated knowledge be derived from coherent inferences. These *epistemic* requirements are discussed in the following chapter.

By being able to incorporate physical knowledge (about structural connectivity, laws, mechanistic models, *etc.*), BME may move one step ahead of empirical (or statistical) methods. Indeed, unlike empirical mapping techniques that describe an existing set of data and are only *locally* predictive (*i.e.*, interpolation is possible only within the range of the available data), BME is able to integrate physical knowledge (in the form of scientific laws, empirical relations, *etc.*) and, thus, it has *explanatory* and *global* prediction features (*i.e.*, extrapolation is possible beyond the range of observations). This is important, if geostatistics is to be considered a respectable scientific discipline.

In "Sources of physical knowledge" (Chapter 1, p. 20), strong emphasis was laid on the argument that, as an *applied* scientific discipline, geostatistics is intended to produce marketable products, capitalizing on the stores of basic knowledge that have accumulated thus far in a richly productive century. Surely, the meaning of the term "basic knowledge" in the definition of applied science above goes far beyond observational facts and includes several other $\mathcal{G}$ and $\mathcal{S}$ bases of physical knowledge. Hence, in the following chapters we will be concerned with the development of a group of modern spatiotemporal geostatistics models which have the epistemic and technical capacity to account for these knowledge bases in a rigorous and systematic fashion.

4

THE EPISTEMIC PARADIGM

"Whenever one lights upon more exact proofs, then we must be grateful to the discoverer, but for the present we must state what seems plausible." Aristotle (*De Caelo*, ca. 330 B.C.)

Acquisition and Processing of Physical Knowledge

To think intelligently, geostatisticians combine empirical reasoning with positive thinking. As they ponder over the insights their findings are giving them into "objective reality," they discover that the issue is not merely how to deal with data but also how to interpret and integrate them into the process of understanding and prediction. In a sense, this expands the study domain to include the *observer* (geostatistician) as well as the *observed* (natural processes). The meaning of such an expansion is that geostatisticians—through the inescapable demands of their own subjects—are forced to become epistemologists, just as pure mathematicians have been forced to become logicians. Before proceeding any further with epistemic analysis, let us formulate a general spatiotemporal mapping problem of interest in the natural sciences (see also Fig. 4.1):

> **The spatiotemporal mapping problem:** Consider a natural variable $X(\boldsymbol{p})$ characterized by a set of general knowledge functions as in Chapter 3, Equation 3.2 (p. 75), and a set of specificatory data represented by Equation 3.29 later in that chapter (p. 83). We seek an S/TRF estimator $\hat{X}(\boldsymbol{p})$ that provides estimates of the actual (but unknown) $X(\boldsymbol{p})$ values at an arbitrary set of space/time points $\boldsymbol{p}_{k_\ell}$ $(\ell = 1, \ldots, \rho)$.

While *single-point* analysis deals with one estimate $\hat{\chi}_k$ at a time, *multi-point* analysis is concerned with estimates $\hat{\boldsymbol{\chi}}_k = (\hat{\chi}_{k_1}, \ldots, \hat{\chi}_{k_\rho})$ of $\boldsymbol{\chi}_k = (\chi_{k_1}, \ldots, \chi_{k_\rho})$ at several points $\boldsymbol{p}_{k_\ell}$ $(\ell = 1, \ldots, \rho)$ simultaneously. In most

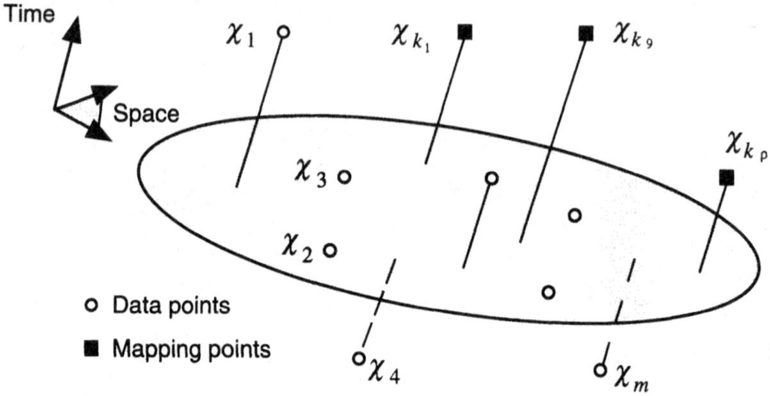

Figure 4.1. An illustration of the space/time mapping problem.

applications, the mapping points lie on the nodes of a space/time grid. Technically, one may distinguish between spatial, temporal, and spatiotemporal maps, depending upon whether the intention is to capture a single instantaneous snapshot (a picture), a sequence of successive snapshots at a single geographical location (a temporal profile), or a video sequence of successive spatial pictures (a movie).

COMMENT 4.1: *In some mapping problems, while we seek to map the natural variable $X(\boldsymbol{p})$, the available physical knowledge is about another variable $Y(\boldsymbol{p})$ or about a set of variables $\boldsymbol{Y} = (Y_1(\boldsymbol{p}), \ldots, Y_N(\boldsymbol{p}))$. These variables may be related,* e.g., *by a physical law, a theory, or an empirical relationship (as we discussed in Chapter 3). Hard and/or soft data may also be available for some of these variables. This is the case of the so-called* multivariable *(or* vector*) mapping problem, which is examined in detail in Chapter 9. At the moment, without loss of generality, we focus on the* scalar *mapping problem in which only one natural variable $X(\boldsymbol{p})$ is involved.*

As a product of scientific reasoning, a spatiotemporal map should be derived by means of a sound *epistemic paradigm* or *framework*. Understanding epistemology can enlighten considerably our mathematical investigations for the best mapping approach possible. In the following sections we will first discuss some general characteristics of the epistemic investigations of modern spatiotemporal geostatistics, and then we will focus our attention to a particular method, namely, the *Bayesian maximum entropy (BME)* analysis.

Epistemic Geostatistics and the BME Analysis

One uses logical reasoning to obtain answers to questions related to the acquisition, meaning, and processing of knowledge. Epistemically, the BME method of reasoning is stochastic and not the natural process. In light of this concept, what should be the important epistemic features of spatiotemporal mapping?

The paradigm considered in this book is based on considerations that most philosophers of science will find acceptable. Along with the dominant tradition (*e.g.*, Goldman, 1986), we regard it as an evaluative or normative paradigm, not a purely descriptive one. This is an epistemic geostatistics paradigm which distinguishes between three main stages of knowledge acquisition, interpretation, and processing as follows:

1. The *prior* stage. Spatiotemporal analysis and mapping does not work in an intellectual vacuum. Instead it always starts with a basic set of assumptions and the general knowledge base $\mathcal{G}$.

2. The *meta-prior* stage. The specificatory knowledge base $\mathcal{S}$ is considered, including hard and soft data.

3. The *integration* or *posterior* stage. Information from (1.) and (2.) is processed by means of logical rules to produce the required spatiotemporal map.

Each one of the above stages will be discussed in considerable detail in the following chapters. The epistemic paradigm includes elements from both the empiricist and the rationalist traditions (just as scientific reasoning combines empiricist and rationalist approaches in an inseparable method). Within the context of this paradigm, knowledge comes from a synthesis of concepts and experience. The distinction between the prior and meta-prior stages is not meant so much in a temporal sense. Rather the real meaning is that the boundary line between the prior and meta-prior stages of the mapping paradigm coincides with the boundary line between general and specificatory knowledge. However, while the distinction between general and specificatory knowledge is a matter of logic, the distinction between the prior and meta-prior stages is an epistemic issue. At the prior stage, we process knowledge that has been characterized as general on the basis of some logical arguments (Chapter 3, "The General Knowledge Base," p. 73) and at the meta-prior stage we incorporate knowledge which—again on the basis of logical arguments—is considered specificatory (Chapter 3, "The Specificatory Knowledge Base," p. 82). The use of the terms prior and meta-prior is natural, in this connection. Prior knowledge comes epistemically—though not necessarily genetically—before observational experience. Meta-prior knowledge is born pragmatically from experience and may change with use.

COMMENT 4.2: *In the view of many researchers, Socrates was the first to clearly point out the distinction between the prior and meta-prior stages (*i.e.*, the distinction between what one knows before experience with the specific situation and what one learns from the case-specific data). Indeed, several references to this distinction can be found in Plato's famous* Dialogues *(von Foerster, 1962; Vlastos, 1971). In his discussion with Menon,* e.g., *Socrates makes the point that education is in many cases not a transfer of knowledge from teacher to student, but an awakening of the awareness of knowledge already possessed by the student. Without this prior knowledge, Socrates argues, experience cannot be gathered (at the meta-prior stage).*

A simple yet illuminating (and, perhaps, entertaining) example may help fix certain ideas about the aforementioned epistemic stages.

EXAMPLE 4.1: Imagine Nature as a furnished house. At the prior stage our general knowledge $\mathcal{G}$ is that Nature bought the furniture from KMart (but we have not seen any specific articles). On the basis of $\mathcal{G}$, we may derive some conclusions about the total value of the furniture (*e.g.*, no article is worth more than $200). At the meta-prior stage we inspect certain articles and we obtain their actual values (case-specific knowledge $\mathcal{S}$), which improves our prior estimate of the total value of the furniture. □

Epistemically important qualities of the mapping paradigm are those that can be determined before the event (*i.e.*, before the phenomenon predicted by the map actually takes place). With this in mind, we want maps that carry as much *information* as possible, that are well-supported with evidence, and that have a high *probability* of being correct (rather than *certain* correctness, which is an aspect that cannot be guaranteed before the event).

Reflecting upon such an epistemic paradigm, the BME mapping approach was proposed about a decade ago (Christakos, 1990, 1991a, 1992). The crux of BME is that the spatiotemporal analysis and mapping of natural phenomena should be both *informative* and *cogent*. Due to the natural variations and uncertainties involved in the description of such phenomena, both of these requirements involve probabilities, but they are conditional probabilities relative to the different knowledge bases considered at each stage. This double epistemic goal of BME is summarized by the following postulate.

POSTULATE 4.1: BME aims at informativeness (in terms of prior information relative to the general knowledge $\mathcal{G}$) as well as cogency (in terms of posterior probability relative to the specificatory knowledge $\mathcal{S}$).

The two epistemic ideals of Postulate 4.1 are at the conceptual heart of BME. Let us, therefore, examine them in more detail. [Multipoint mapping will be considered in this section; the single-point mapping can always be derived as a special case.]

Prior stage

At the *prior stage*, the probability function considered is relative to the general knowledge $\mathcal{G}$, *i.e.*,

$$\text{Prob}_{\mathcal{G}}[\boldsymbol{\chi}_{map}] = p \in [0,1] \tag{4.1}$$

which means "the probability of the map $\boldsymbol{\chi}_{map} = (\boldsymbol{\chi}_{data}, \boldsymbol{\chi}_k)$ given the general knowledge base $\mathcal{G}$ is p." Another way of expressing the meaning of Equation 4.1 is by saying that probability judgments about $\boldsymbol{\chi}_{map}$ are relative to knowledge $\mathcal{G}$. Furthermore, an interpretation can be given in terms of the complementarity idea of Chapter 2 (p. 58): Given $\mathcal{G}$, the probability function (Eq. 4.1) provides a measure of the relative quantity of random field realizations in which $\boldsymbol{\chi}_{map}$ occurs over all possible realizations.

EXAMPLE 4.2: A well-known situation of general knowledge processing is *unconditional simulation* producing various field realizations on the basis of its mean and covariance. While useful in the characterization of spatiotemporal variability, unconditional simulation is of limited value for prediction purposes. □

The informativeness of Postulate 4.1 implies information maximization at the prior stage, which is also conditioned to the available general knowledge $\mathcal{G}$. This stage assumes an inverse relation between information and probability: The more informative an assessment about a mapping situation is, the less probable it is to occur. This expresses a standard epistemic rule, namely, the more vague and general a theory is, the more alternatives it includes (it is, hence, more probable) and the less informative it is. Conversely, the more alternatives a theory excludes, the more informative (less probable) it is. From a Popperian standpoint: "The more a theory forbids, the more it tells us." So, while the statement "Que será será" ("what will be, will be") is an absolutely safe prediction model, it provides no information at all. Let us pause and discuss another example.

EXAMPLE 4.3: A weather forecast theory A predicts that tomorrow it will either snow, or rain, or be cloudy (but not rain), or be sunny. Another theory B predicts that tomorrow it will either rain or it will be cloudy (but not rain). A is a very general theory that includes several possible alternatives. Hence, while it has a high probability (Prob_A) of turning out to be true, it is not a particularly informative theory (for it is incapable of discriminating among alternatives). Theory B, on the other hand, includes only two alternatives and, thus, the probability of being true is $\text{Prob}_B < \text{Prob}_A$. Since, however, it is capable of reducing the alternatives to only two, theory B is more informative than A. □

An informative scientific theory, therefore, is a prohibition: it forbids certain things to happen. In quantitative terms, the inverse relation between information and probability can be expressed as $\text{Info}_{\mathcal{G}}[\boldsymbol{\chi}_{map}] = \{\text{Prob}_{\mathcal{G}}[\boldsymbol{\chi}_{map}]\}^{-1}$, *i.e.*, the information about the actual mapping situation provided by $\mathcal{G}$ is inversely proportional to the probability model constructed on the basis of $\mathcal{G}$. Given that, for technical reasons, probabilities can be very small, it is often more convenient to work with logarithms so that information is mathematically defined by

$$\text{Info}_{\mathcal{G}}[\boldsymbol{\chi}_{map}] = \ell og\{\text{Prob}_{\mathcal{G}}[\boldsymbol{\chi}_{map}]\}^{-1} = -\ell og\{\text{Prob}_{\mathcal{G}}[\boldsymbol{\chi}_{map}]\} \tag{4.2}$$

According to the first epistemic ideal (informativeness) of Postulate 4.1, Equation 4.2 should be maximized—in a stochastic sense—subject to the available general knowledge $\mathcal{G}$. (In fact, this epistemic ideal may be viewed as a stochastic version of a general rule of scientific reasoning, often referred to as *ampliative* reasoning: One should use all but no more knowledge than is available.) The detailed mathematical analysis of this ideal is presented in Chapter 5.

COMMENT 4.3: *Being philosophical in nature, the concept of information is associated with more than one interpretation (Aczel and Daroczy, 1975; Jumarie, 1990; Baldi and Brunak, 1998; Ebanks* et al., *1998). Indeed, the various meanings of information encountered in scientific investigations reflect the different aspects of this fundamental concept (syntactic, semantic, pragmatic,* etc. *aspects). In scientific practice, it is often more appropriate to emphasize the epistemic features of information. Furthermore, many researchers approach information the same way they approach other important concepts like energy and gravity (these concepts lead to powerful scientific theories and models that have important applications, but their existence cannot be proven ontologically from first principles). As it turns out, while the concept of information is philosophical in nature, we are nevertheless able to investigate its intriguing mathematical structure and obtain very useful results for our scientific applications.*

Meta-prior stage

At the *meta-prior* stage we collect and organize specificatory knowledge $\mathcal{S}$ in appropriate quantitative forms that can be explicitly incorporated into the BME formulation. $\mathcal{S}$ includes case-specific empirical evidence concerning, *e.g.*, the existence of certain physical factors associated with the natural variable under consideration. In real-world applications, we may be dealing with various bodies of case-specific knowledge (Chapter 3, p. 82*ff*). The quality and quantity of the hard and soft data collected is a matter of experimental and computational investigations, which can provide with precious feedback the theoretical processes involved in BME analysis.

There could also be several causes for the existence of constraints on the knowledge to be considered at this stage, like data storage limitations, costs, and communication restrictions. In many cases, the existence of such constraints may raise the issue of deciding "what to know."

COMMENT 4.4: *A note may be relevant here. Just as any method of scientific reasoning and prediction, the BME method relies on the continuing communication and cooperation between theoreticians, on the one side, and experimentalists, on the other. Indeed, good scientists on both sides will always find ways to communicate, and it is their cooperation that has characterized some of the brightest moments of scientific endeavor. As a matter of fact, it is now widely recognized that the potentially greatest obstacle to scientific research is not any artificial schism between theoreticians and experimentalists, but rather the absence of inspiration, illumination, constructive criticism, and critical thinking, which have always been the forces that move scientific thought and action. There is nothing inspiring or even vaguely satisfying in the flood of undigested cookbook recipes, fashionable clichés, and quick fixes that serve to obfuscate fundamental questions of substance, value, and purpose. Instead, approaches are needed that are capable of integrating scientific and technological progress with philosophical ideals and humanistic values.*

Integration or posterior stage

At the *integration* or *posterior* stage, the new probability function is relative to the total knowledge $\mathcal{K}$, *i.e.*,

$$\text{Prob}_{\mathcal{K}}[\boldsymbol{\chi}_k] = p' \in [0, 1] \tag{4.3}$$

Equation 4.3 means that "the probability of a map $\boldsymbol{\chi}_k$ given the total knowledge base $\mathcal{K} = \mathcal{G} \cup \mathcal{S}$ is p'." Equation 4.3 offers a measure of the credibility or assertibility of the proposition "if $\mathcal{K}$, then $\boldsymbol{\chi}_k$," *i.e.*, it asserts a logical relation between knowledge $\mathcal{K}$ and the map $\boldsymbol{\chi}_k$. Given $\mathcal{K}$, the probability function (Eq. 4.3) may provide a measure of the relative quantity of random field realizations in which $\boldsymbol{\chi}_k$ occurs over all possible realizations. Note that while at the prior stage the probability (Eq. 4.1) refers to the whole domain (including data and estimation points), *i.e.*, $\boldsymbol{p}_{map} = (\boldsymbol{p}_{data}, \boldsymbol{p}_k)$, the probability (Eq. 4.3) of the posterior stage includes only estimation points $\boldsymbol{p}_k$.

EXAMPLE 4.4: A well-known situation of specificatory knowledge processing in geostatistics is *conditional simulation* which—by incorporating a set of measurements—is of much greater predictive value than unconditional simulation discussed in Example 4.2 above. □

The probability functions (Eqs. 4.1 and 4.3) assume a connection between mapping predictions and the available knowledge. In other words, the probability is epistemic, supported by empirical data and related to inductive evidence. While we seek posterior predictions that are highly probable, we nevertheless want them to achieve this probability on the basis of total knowledge and not on general knowledge alone. As already mentioned, in the mapping context the specificatory knowledge $\mathcal{S}$ refers to the $\boldsymbol{\chi}_{data}$, which for this reason is sometimes denoted as $\boldsymbol{\chi}_{data}(\mathcal{S})$.

The analysis above has left us with a final issue to be considered within the epistemic framework of modern spatiotemporal geostatistics; namely, how should we process knowledge of the prior and meta-prior stages to the integration stage. There are various ways to do this. A particularly efficient way is by means of the *knowledge processing rule* suggested by the following fundamental proposition.

PROPOSITION 4.1: The posterior mapping probability (Eq. 4.3) is related to the prior mapping probability (Eq. 4.1) by means of the relationship

$$\text{Prob}_{\mathcal{K}}[\boldsymbol{\chi}_k] = A^{-1}\,\text{Prob}_{\mathcal{G}}[\boldsymbol{\chi}_{map}(\mathcal{S})] \tag{4.4}$$

where $A = \text{Prob}_{\mathcal{G}}[\boldsymbol{\chi}_{data}(\mathcal{S})]$.

Proof : It is valid that

$$\text{Prob}_{\mathcal{K}}[\boldsymbol{\chi}_{data}(\mathcal{S})] = 1 \tag{4.5}$$

which expresses the obvious fact that given that the data vector $\boldsymbol{\chi}_{data}(S)$ has been observed—as the subscript $\mathcal{K}$ in Equation 4.5 indicates—its probability is one. On the other hand, at the prior stage it is usually valid that $\text{Prob}_{\mathcal{G}}[\boldsymbol{\chi}_{data}(S)] < 1$. Furthermore, we have the following relation between conditional probabilities

$$\text{Prob}_{\mathcal{K}}[\boldsymbol{\chi}_k|\boldsymbol{\chi}_{data}(S)] = \text{Prob}_{\mathcal{G}}[\boldsymbol{\chi}_k|\boldsymbol{\chi}_{data}(S)] \tag{4.6}$$

Equation 4.6 simply means that the evidential impact of $\boldsymbol{\chi}_{data}(S)$ on $\boldsymbol{\chi}_k$ is already fully assessed in assigning the prior conditional probability and, hence, the fact that $\boldsymbol{\chi}_{data}(S)$ actually occurred is no reason to change this assessment. [In other words, Equation 4.6 means that the probability we assign to $\boldsymbol{\chi}_k$ *assuming* that $\boldsymbol{\chi}_{data}(S)$ will turn out to be true is equal to the probability we would assign to $\boldsymbol{\chi}_k$ on learning that $\boldsymbol{\chi}_{data}(S)$ *indeed* turned out to be true.]

By the definition of conditional probability

$$\text{Prob}_{\mathcal{K}}[\boldsymbol{\chi}_k|\boldsymbol{\chi}_{data}(S)] = \frac{\text{Prob}_{\mathcal{K}}[\boldsymbol{\chi}_k \text{ and } \boldsymbol{\chi}_{data}(S)]}{\text{Prob}_{\mathcal{K}}[\boldsymbol{\chi}_{data}(S)]} \tag{4.7}$$

which in light of Equation 4.5 gives

$$\text{Prob}_{\mathcal{K}}[\boldsymbol{\chi}_k|\boldsymbol{\chi}_{data}(S)] = \text{Prob}_{\mathcal{K}}[\boldsymbol{\chi}_k \text{ and } \boldsymbol{\chi}_{data}(S)] = \text{Prob}_{\mathcal{K}}[\boldsymbol{\chi}_k] \tag{4.8}$$

Taking Equation 4.6 into account, Equation 4.8 yields the *Bayesian conditionalization principle*

$$\text{Prob}_{\mathcal{K}}[\boldsymbol{\chi}_k] = \text{Prob}_{\mathcal{G}}[\boldsymbol{\chi}_k|\boldsymbol{\chi}_{data}(S)] \tag{4.9}$$

which is the desired result. [This principle, in fact, was responsible for the first letter in the acronym "BME," originally used to name the spatiotemporal mapping approach; Christakos, 1990.] In light of Equation 4.9, the mathematical form of $\text{Prob}_{\mathcal{K}}$ is obtained from that of $\text{Prob}_{\mathcal{G}}$ by fixing $\boldsymbol{\chi}_{data}(S)$ as the conditioning factor. □

COMMENT 4.5: *Equation 4.4 may be given the following interpretation in terms of the random field complementarity idea that was discussed in Chapter 2 (p. 58): If $\boldsymbol{\chi}_{data}(S)$ does indeed entail $\boldsymbol{\chi}_k$, then $\boldsymbol{\chi}_{map}(S)$ occurs in every X-realization in which $\boldsymbol{\chi}_{data}(S)$ occurs. But if $\boldsymbol{\chi}_k$ is neither entailed by $\boldsymbol{\chi}_{data}(S)$ nor inconsistent with it, then $\boldsymbol{\chi}_{map}(S)$ occurs only in some of the possible X-realizations in which $\boldsymbol{\chi}_{data}(S)$ occurs, (*i.e., *those realizations in which $\boldsymbol{\chi}_k$ also happens to occur). Therefore, we may take the ratio of the quantity of random field realizations in which $\boldsymbol{\chi}_{map}(S)$ occurs to the quantity of realizations in which $\boldsymbol{\chi}_{data}(S)$ occurs as determining the extent to which $\boldsymbol{\chi}_{data}(S)$ entails $\boldsymbol{\chi}_k$ and, thus, defining the probability of $\boldsymbol{\chi}_k$ given $\boldsymbol{\chi}_{data}(S)$, namely* $\text{Prob}_{\mathcal{G}}[\boldsymbol{\chi}_k|\boldsymbol{\chi}_{data}(S)]$. *See also the discussion in the section on conditional probability on p. 98.*

The cogency requirement of Postulate 4.1 may seek a map $\boldsymbol{\chi}_k$ that maximizes probability $\text{Prob}_{\mathcal{K}}[\boldsymbol{\chi}_k]$, *i.e.*, the *mode* of the probability law. Other representations of the cogency requirement may lead to a map $\boldsymbol{\chi}_k$ that is the *median* of $\text{Prob}_{\mathcal{K}}[\boldsymbol{\chi}_k]$, the *conditional mean, etc.* Equation 4.4 relates this new probability function at the integration stage with the probability function at the prior stage. The $\text{Prob}_{\mathcal{K}}[\boldsymbol{\chi}_k]$ is a monadic function of $\boldsymbol{\chi}_k$ (because its value depends on just one issue—whether $\boldsymbol{\chi}_k$ occurs or not). But the conditional probability $\text{Prob}_{\mathcal{G}}[\boldsymbol{\chi}_k|\boldsymbol{\chi}_{data}(\mathcal{S})]$ is a dyadic function of $\boldsymbol{\chi}_k$ and $\boldsymbol{\chi}_{data}(\mathcal{S})$ because its value depends on the two issues that it relates.

There are various notational ways of representing the effects of knowledge bases $\mathcal{G}$ and $\mathcal{K}$ on the probability functions considered in each one of the mapping stages. In order to familiarize the reader with the conditional probabilities used in BME mapping, in the following example we choose to reexamine Proposition 4.1 above using a slightly different notation.

EXAMPLE 4.5: Let us denote the prior probability function relative to the general knowledge $\mathcal{G}$ as the conditional probability

$$\text{Prob}[\boldsymbol{\chi}_{map}|\mathcal{G}] = p \in [0, 1] \tag{4.10}$$

Consider the probability function $\text{Prob}'[\cdot]$, such that $\text{Prob}'[\mathcal{G}] = 1$. This implies that

$$\text{Prob}'[\boldsymbol{\chi}_k|\mathcal{G}] = \frac{\text{Prob}'[\boldsymbol{\chi}_k \text{ and } \mathcal{G}]}{\text{Prob}'[\mathcal{G}]} = \text{Prob}'[\boldsymbol{\chi}_k] \tag{4.11}$$

Furthermore, given that $\text{Prob}'[\boldsymbol{\chi}_k|\mathcal{G}] = \text{Prob}[\boldsymbol{\chi}_k|\mathcal{G}]$, we get

$$\text{Prob}'[\boldsymbol{\chi}_k] = \text{Prob}[\boldsymbol{\chi}_k|\mathcal{G}] \tag{4.12}$$

At the posterior stage we define the probability function $\text{Prob}''[\cdot]$ such that $\text{Prob}''[\boldsymbol{\chi}_{data}(\mathcal{S})] = 1$. This means that $\text{Prob}''[\boldsymbol{\chi}_k|\boldsymbol{\chi}_{data}(\mathcal{S})] = \text{Prob}''[\boldsymbol{\chi}_k \text{ and } \boldsymbol{\chi}_{data}(\mathcal{S})] = \text{Prob}''[\boldsymbol{\chi}_k]$; and since $\text{Prob}''[\boldsymbol{\chi}_k|\boldsymbol{\chi}_{data}(\mathcal{S})] = \text{Prob}'[\boldsymbol{\chi}_k|\boldsymbol{\chi}_{data}(\mathcal{S})]$, we get the following expression

$$\text{Prob}''[\boldsymbol{\chi}_k] = \text{Prob}'[\boldsymbol{\chi}_k|\boldsymbol{\chi}_{data}(\mathcal{S})] \tag{4.13}$$

Furthermore, using Equation 4.12, we find

$$\text{Prob}'[\boldsymbol{\chi}_k|\boldsymbol{\chi}_{data}(\mathcal{S})] = \frac{\text{Prob}'[\boldsymbol{\chi}_k \text{ and } \boldsymbol{\chi}_{data}(\mathcal{S})]}{\text{Prob}'[\boldsymbol{\chi}_{data}(\mathcal{S})]} = \frac{\text{Prob}[\boldsymbol{\chi}_k \text{ and } \boldsymbol{\chi}_{data}(\mathcal{S})|\mathcal{G}]}{\text{Prob}[\boldsymbol{\chi}_{data}(\mathcal{S})|\mathcal{G}]} \tag{4.14}$$

$$= \frac{\text{Prob}[\boldsymbol{\chi}_k \text{ and } \boldsymbol{\chi}_{data}(\mathcal{S}) \text{ and } \mathcal{G}]}{\text{Prob}[\boldsymbol{\chi}_{data}(\mathcal{S}) \text{ and } \mathcal{G}]} = \text{Prob}[\boldsymbol{\chi}_k|\boldsymbol{\chi}_{data}(\mathcal{S}) \text{ and } \mathcal{G}] \tag{4.15}$$

or

$$\text{Prob}''[\boldsymbol{\chi}_k] = \text{Prob}[\boldsymbol{\chi}_k|\boldsymbol{\chi}_{data}(\mathcal{S}) \text{ and } \mathcal{G}] = \text{Prob}[\boldsymbol{\chi}_k|\mathcal{K}] \tag{4.16}$$

which relates the probability function Prob $[\cdot]$ of the prior stage with the probability function Prob$''$ $[\cdot]$ of the posterior stage. Finally, one may notice that

$$\left.\begin{aligned} \text{Prob}[\cdot|\mathcal{G}] &= \text{Prob}_{\mathcal{G}}[\cdot] \\ \text{Prob}'[\cdot|\mathcal{S}] &= \text{Prob}_{\mathcal{G}}[\cdot|\mathcal{S}] \\ \text{Prob}''[\cdot] &= \text{Prob}_{\mathcal{K}}[\cdot] \end{aligned}\right\} \tag{4.17}$$

which relates the probability functions used in the approach of Proposition 4.1 with the probability functions of this example. □

Conditional Probability of a Spatiotemporal Map and its Relation to the Probability of Conditionals

The concept of "cause and effect" is of paramount importance in scientific investigations. Even if the indetermination of modern physics may show that not every natural process has a cause (certainly not a deterministic one), most phenomena can, indeed, have causes. Certainly there exist various sorts of *causation*, including deterministic causation in which the causes are necessary and sufficient for their effects (see, *e.g.*, Mellor, 1995) and also probabilistic or stochastic causation, which includes causes that raise the chances of their effects.

The sufficiency and necessity of causes and their effects is usually expressed by *conditionals*. Within the framework of modern geostatistics we can distinguish between *truth-functional* and *non-truth-functional* conditionals related to spatiotemporal maps. Truth-functional map conditionals include *material* and *strict* conditionals. The most important among the non-truth-functional conditionals are *physical* and *logical* map conditionals. We will now discuss these concepts in more detail.

Material and strict map conditionals

From the mathematical logic viewpoint, the *material* conditional of a spatiotemporal map is a structure of the form "if $\boldsymbol{\chi}_{data}(\mathcal{S})$ occurs, then $\boldsymbol{\chi}_k$ occurs," or "$\boldsymbol{\chi}_{data}(\mathcal{S})$ implies $\boldsymbol{\chi}_k$," for short, $\boldsymbol{\chi}_{data}(\mathcal{S}) \rightarrow \boldsymbol{\chi}_k$. Material map conditionals are logical structures based on purely *truth-functional* concepts (conjunction $\wedge$, negation $\neg$, disjunction $\vee$, truth tables, *etc.*; see for example Burris, 1998), *i.e.*, they express a truth-functional relation. A material map conditional is equivalent to the statement "it is not the case that $\boldsymbol{\chi}_{data}(\mathcal{S})$ and not $\boldsymbol{\chi}_k$," in short, $\neg(\boldsymbol{\chi}_{data}(\mathcal{S}) \wedge \neg\boldsymbol{\chi}_k)$. Indeed, this is how the word "implies" is used in mathematics. Another sort of related conditional is the *strong* map conditional, denoted by $\chi_{data}(\mathcal{S}) \mapsto \chi_k$, which is, by definition, true if and only if the $\chi_{data}(\mathcal{S}) \rightarrow \chi_k$ is necessarily true.

In terms of the complementarity concept (discussed in Chapter 2, p. 59), one can say that a material conditional holds between $\chi_{data}(\mathcal{S})$ and χ_k in all possible random field realizations when $\chi_{data}(\mathcal{S})$ and $\neg\chi_k$ is not the case. In other words, material conditionality holds in every possible random field realization except in those possible realizations in which the case is $\chi_{data}(\mathcal{S})$ and $\neg\chi_k$.

The conditional probability of a map is not necessarily the probability of its material conditional. Whether these two are or are not equal may depend on the *knowledge base* considered. In particular, the probability of truth of the material map conditional "$\chi_{data}(\mathcal{S}) \rightarrow \chi_k$," *i.e.*,

$$\mathrm{Prob}_{\mathcal{G}}[\chi_{data}(\mathcal{S}) \rightarrow \chi_k] = \mathrm{Prob}_{\mathcal{G}}[\neg(\chi_{data}(\mathcal{S}) \wedge \neg\chi_k)] \tag{4.18}$$

is not necessarily equal to $\mathrm{Prob}_{\mathcal{G}}[\chi_k|\chi_{data}(\mathcal{S})]$. $\mathrm{Prob}_{\mathcal{G}}[\chi_k|\chi_{data}(\mathcal{S})]$, in other words, is the probability that χ_k occurs given that $\chi_{data}(\mathcal{S})$ does, whether or not $\chi_{data}(\mathcal{S})$ gave rise to χ_k, and whether or not there is any (probabilistic) subjunctive connection between χ_k and $\chi_{data}(\mathcal{S})$. It can be shown that the above map probabilities are related by

$$\mathrm{Prob}_{\mathcal{G}}[\chi_{data}(\mathcal{S}) \rightarrow \chi_k] = \mathrm{Prob}_{\mathcal{G}}[\neg\chi_{data}(\mathcal{S})] + \mathrm{Prob}_{\mathcal{G}}[\chi_{data}(\mathcal{S})]\,\mathrm{Prob}_{\mathcal{G}}[\chi_k|\chi_{data}(\mathcal{S})] \tag{4.19}$$

which implies that $\mathrm{Prob}_{\mathcal{G}}[\chi_k|\chi_{data}(\mathcal{S})] \leq \mathrm{Prob}_{\mathcal{G}}[\chi_{data}(\mathcal{S}) \rightarrow \chi_k]$. However, in light of Equation 4.5 it is valid that

$$\mathrm{Prob}_{\mathcal{K}}[\chi_{data}(\mathcal{S}) \rightarrow \chi_k] = \mathrm{Prob}_{\mathcal{K}}[\chi_k|\chi_{data}(\mathcal{S})] = \mathrm{Prob}_{\mathcal{K}}[\chi_k] \tag{4.20}$$

In other words, the meaning of Equation 4.20 is that, given that $\chi_{data}(\mathcal{S})$ has indeed been observed, the probability of the material map conditional is equal to the conditional probability of the map.

An interesting representation of the analysis above is obtained in the context of map *truth tables*. It is instructive to illustrate this representation by means of the following example.

EXAMPLE 4.6: Consider the map truth tables shown in Figure 4.2. Given $\mathcal{G}$, the $\mathrm{Prob}_{\mathcal{G}}$ of the material map conditional $\chi_{data}(\mathcal{S}) \rightarrow \chi_k$ includes all three realizations $(\chi_{data}(\mathcal{S}), \chi_k)$, $(\neg\chi_{data}(\mathcal{S}), \chi_k)$, and $(\neg\chi_{data}(\mathcal{S}), \neg\chi_k)$ of the truth table in Figure 4.2a. On the other hand, the $\mathcal{K}$-based $\mathrm{Prob}_{\mathcal{K}}$ includes only the realization $(\chi_{data}(\mathcal{S}), \chi_k)$ of Figure 4.2b (*i.e.*, from the probabilistic logic viewpoint, if we select a world at random according to $\mathrm{Prob}_{\mathcal{K}}$, the latter gives the probability that we selected a world in which both $\chi_{data}(\mathcal{S})$ and χ_k occur). Note that nothing is written into the conditional map probability $\chi_k|\chi_{data}(\mathcal{S})$ column when $\neg\chi_{data}(\mathcal{S})$ is considered (Fig. 4.2a). □

(a)

$\chi_{data}(S)$	χ_k	$\chi_{data}(S) \rightarrow \chi_k$	$\chi_k \vert \chi_{data}(S)$
T	T	T	T
T	F	F	F
F	T	T	...
F	F	T	...

(b)

$\chi_{data}(S)$	χ_k	$\chi_{data}(S) \rightarrow \chi_k$	$\chi_k \vert \chi_{data}(S)$
T	T	T	T
T	F	F	F

Figure 4.2. Truth tables of spatiotemporal mapping: (a) given $\mathcal{G}$; (b) given $\mathcal{K} = \mathcal{G} \cup \mathcal{S}$.

Other map conditionals

Material map conditionals are expressed in terms of truth-functional concepts, which implies that the connective "if, then" is used to express the (truth-functional) concept of material conditionality and this concept alone. As already mentioned, there exist other important uses of conditionals which are not based on purely truth-functional concepts, but which express more than a mere truth-functional relation. These non-truth-functional map conditionals may assert stronger (physical or logical) connections between $\chi_{data}(S)$ and χ_k than the purely truth-functional relation of material conditionality. In the case of a *logical* map conditional, the antecedent logically implies the consequent. A *physical* map conditional is one in which the antecedent physically implies the consequent. In this sense, and since in the mapping case the $\chi_{data}(S)$ and χ_k represent the same natural variable, it is likely that the material conditional $\chi_{data}(S) \rightarrow \chi_k$ expresses a physical connection.

It is a result worth mentioning that, in many cases, no mapping relationship expressed as any sort of conditional (physical, logical, *etc.*) is valid unless the same relationship viewed as a material conditional is valid. In other words, if the relation "$\chi_{data}(S)$ implies χ_k" is not valid when considered as a material conditional, then it is also not valid when considered as a physical or a logical conditional. As a consequence, the following postulate seems justified.

POSTULATE 4.2: Even though most of the conditionals we encounter in natural sciences are physical or logical conditionals, the material conditionals can play a basic role in modern geostatistics.

The preceding discussion remains valid in the case of two or more natural variables, as illustrated with the help of the following example.

EXAMPLE 4.7: Consider the multivariable (vector) mapping case briefly discussed in Comment 4.1 (p. 90). Assume, for simplicity, that we are dealing with two natural variables $X(\boldsymbol{p})$ and $Y(\boldsymbol{p})$, in which case the material conditional has the form

$$\boldsymbol{\chi}_{data}(\mathcal{S}) \rightarrow \boldsymbol{\psi}_k \tag{4.21}$$

Then, if $X(\boldsymbol{p})$ physically or logically implies $Y(\boldsymbol{p})$, the connection "if, then" of the material conditional could be considered as causally or deductively valid. This has important consequences in pollution monitoring and control, and in environmental health studies involving cause-effect analysis in which $\boldsymbol{\chi}$ may represent environmental exposure and $\boldsymbol{\psi}$ denotes the resulting health effect map (see Chapter 9, "Associations between environmental exposure and health effect," on p. 183). □

We conclude our discussion of map conditionals by noticing that the examination of spatiotemporal mapping in the light of the map conditionals introduced in these sections deserves to be studied in more depth by modern geostatisticians.

The BME Net

As the domain of geostatistics keeps expanding in search of new concepts and applications, a return to the foundations will be necessary because each of the two processes nourishes the other. The epistemic component of BME analysis is concerned with the acquisition, modification, integration, and processing of knowledge by scientific reasoning and experience. Hence, like most epistemologies, BME incorporates a varying degree of commitment to both rationalism and empiricism. At the prior stage, *e.g.*, BME emphasizes the importance of scientific reasoning, physical theories, and laws in advancing knowledge. By comparison, the meta-prior and integration stages require good knowledge based on evidence derived from observations, experience, *etc.* Conditional probabilities make explicit the changes in the probabilities of maps in light of physical knowledge. This makes conditional probabilities especially relevant to logic and to epistemology in general.

The epistemic method offers a higher set of standards for appraising the quality of a mapping process based on scientific theories and empirical facts, and for adjudicating between them. In short, the better the underlying epistemic method, the more rational a mapping process is deemed to be. BME analysis leads us to study the nature of the reasoning frame, and so be forewarned of its impress on the physical knowledge to be processed by the frame. From a scientific reasoning point of view, one may argue that the aim of the mapping paradigm in this chapter is to constrain induction. This is the purpose of the general knowledge $\mathcal{G}$-constraints on information maximization at the prior stage, as well as the specificatory knowledge $\mathcal{S}$-constraints on probability maximization at the integration stage. Constraining can avoid generating innumerable fruitless maps in the search for useful generalizations.

In practice, in some cases, the same knowledge base may be considered either at the prior stage or at the meta-prior stage, which is an issue of continuing debate among science methodologists. Various possibilities may exist, depending on the situation, but one thing is certain: We always have prior knowledge, but at different times we treat different knowledge as prior. This situation allows a certain amount of flexibility in the application of the BME approach and will be revisited later in this book.

COMMENT 4.6: *A fascinating case of prior* vs. *posterior knowledge was introduced by the discovery of non-Euclidean geometries (Chapter 2). From a logical and mathematical standpoint there is no* a priori *means of deciding which kind of geometry does in fact represent the space/time relations between natural variables in a specific situation. Thus, it is necessary to appeal to specificatory evidence (experimentation, empirical investigation,* etc.*) to find out whether the question of geometry can be settled* a posteriori.

In a sense, the BME model introduces a sort of *net* that is built (*i.*) to "catch" certain knowledge bases of application-specific interest, and (*ii.*) to satisfy a set of rules of logic. Part (*i.*) of the BME net is concerned with the external representation of a scientific problem in terms of the physical knowledge bases. On the other hand, the main focus of part (*ii.*) is on how to order these knowledge bases internally and the relations among them. Therefore, the twofold goal of the BME net is to acquire various knowledge bases *and* to order these bases in an appropriate manner so that when taken all together they form a realistic picture of the phenomenon of interest. As a consequence of the twofold goal, map interpretations based on certain knowledge bases and the examination of the properties of the BME net possess a high level of security. (However, map interpretations based on the same knowledge bases but independent of the properties of the BME net should not be considered secure.) Metaphorically speaking, BME's net somehow resembles the fisherman's net; by examining the net (properties of the specific BME model used), the fisherman (the geostatistician) feels quite secure about the kind of fish he can catch (the map interpretation he can obtain). The sea in which the modern geostatistician often throws his net, however, is not any becalmed Sargasso sea, but one whipped by the winds of uncertainty, governed by complicated physical phenomena, and teetering on the edge of anomalies and counter-examples.

This chapter has provided a general conceptual account of physical knowledge acquisition and processing rules. The derivation of analytically tractable mathematical formulations of these rules which hold true in a variety of real-world applications is the topic of the following chapters.

5

MATHEMATICAL FORMULATION OF THE BME METHOD

"Any empirical science in its normal healthy development begins with a more purely inductive emphasis ... and then comes to maturity with deductively formulated theory in which formal logic and mathematics play a most significant part." F.S.C. Northrop

A Pragmatic Framework of the Mapping Problem

The intention of this book is to contribute to a perspective of geostatistics which is, to a large extent, epistemic. A salient point of the previous chapter is that the proposed epistemic paradigm places modern spatiotemporal geostatistics in a pragmatic framework, the central theses of which are that a spatiotemporal approach should:

- be *context-dependent* (being guided by the physical knowledge bases),
- satisfy *logically plausible* rules, and
- always be *relevant to the goals* of the specific study.

From the BME mapping viewpoint, the issue is not merely how to deal with data, but also how to interpret and integrate them into the understanding process which, as already mentioned, implies that the study domain is expanded to include the observer (the geostatistician) as well as the observed (a physical phenomenon).

Consider the spatiotemporal mapping problem described in Chapter 4 (p. 89). Because in many situations the basic problem is the lack of a sufficient number of hard data, bringing diverse sources of general and specificatory knowledge to bear on mapping can have an especially large payoff. Below we will follow the steps of the BME epistemic paradigm discussed in Chapter 4 (p. 91) in order to obtain a mathematical solution to the mapping problem.

This paradigm suggested certain essential directions for studying space/time variability and producing a map that can be formalized in both the continuum and the discrete domains. In the following theoretical analysis, the continuum mapping formalization is presented for reasons of mathematical convenience and generality, but its discrete version may be used in the implementation of BME by means of computer algorithms.

The Prior Stage

Each stage of the BME analysis processes physical knowledge. We do not do scientific reasoning in a void. Before we reason from a specificatory data set to a particular map, we already have some general knowledge about the distribution of the natural variable or the phenomenon been mapped. This general knowledge $\mathcal{G}$ is the result of earlier instances of scientific reasoning, as well as background beliefs relative to the situation overall (Chapter 3, p. 73, "The General Knowledge Base").

Map information measures in light of general knowledge

Let $f_{\mathcal{G}}(\boldsymbol{\chi}_{map})$ be the multivariate pdf model associated with the general knowledge $\mathcal{G}$, before any specificatory knowledge $\mathcal{S}$ (*e.g.*, hard and/or soft data) has been taken into consideration. As we saw in Chapter 3, the general knowledge $\mathcal{G}$ considered at the prior stage may be expressed mathematically in terms of a series of functions g_α which represent known statistics of the S/TRF $X(\boldsymbol{p})$. The g_α's can be associated with various types of knowledge about $X(\boldsymbol{p})$. Examples were given in Chapter 3.

At the prior stage, the knowledge contained in the pdf about the random vector $\boldsymbol{x}_{map}$ can be expressed mathematically in terms of *information measures*. Generally, the more probable a model of $\boldsymbol{x}_{map}$ is, the more alternatives it allows; but, it is also less informative. Conversely, the more informative the model is, the more alternatives it excludes. These standard epistemic rules imply the inverse relationship between prior information and probability, and have already been discussed in the previous chapter. There are various information measures satisfying this inverse relation. One such measure is suggested by the following postulate.

POSTULATE 5.1: Given the general knowledge base $\mathcal{G}$, the information contained in the map $\boldsymbol{x}_{map}$ will be expressed as follows

$$\mathrm{Info}_{\mathcal{G}}[\boldsymbol{x}_{map}] = -\ell og\ f_{\mathcal{G}}(\boldsymbol{\chi}_{map}) \tag{5.1}$$

which is sometimes referred to as the Shannon information measure (see also Eq. 4.2, p. 93).

The epistemic rules mentioned above have, thus, a quantitative side that can be expressed by means of the information measure (Eq. 5.1). Postulate 5.1 also represents the uncertainty regarding the random vector $\boldsymbol{x}_{map}$, which is characterized by the pdf and provides the formal ground for the epistemic rule: The higher the probability, the lower the uncertainty about $\boldsymbol{x}_{map}$ and the lesser the amount of information provided by the pdf about $\boldsymbol{x}_{map}$. The expected information is then given naturally by

$$\overline{\text{Info}_{\mathcal{G}}[\boldsymbol{x}_{map}]} = -\int d\boldsymbol{\chi}_{map}\, f_{\mathcal{G}}(\boldsymbol{\chi}_{map})\, \ell og\, f_{\mathcal{G}}(\boldsymbol{\chi}_{map}) \tag{5.2}$$

where all the components of the vector $\boldsymbol{\chi}_{map}$ are assumed to be integrated in the above. In other words, on the basis of the available general knowledge $\mathcal{G}$, we construct a probability model of the mapping situation (how we do this is discussed in the following section). The amount of information about the natural map provided by $\mathcal{G}$ and carried in the model is expressed by Equation 5.2, which is called the *entropy* function. In fact, the possible interpretation of Equation 5.2 in terms of entropy was responsible for the third letter in the acronym "BME," used originally to name the spatiotemporal mapping approach (Christakos, 1990). Intuitively, the entropy (Eq. 5.2) of a pdf is its degree of diffuseness, so that the more concentrated it is, the smaller its entropy. Various bases can be used for the logarithm in Equations 5.1 and 5.2; in most applications the natural basis e is assumed. Equation 5.2 is the continuum entropy. In the discrete-domain formulation, the integral should be replaced simply by a summation (the latter is used in the computer implementation of BME). In a different setting than the space/time mapping problem considered above, entropy-based analyses have been used to study a variety of problems, including the inverse problem of groundwater hydrology (Woodbury and Ulrych, 1998).

COMMENT 5.1: *Note that while the continuum entropic definition (Eq. 5.2) as a measure of uncertainty is the subject of some theoretical debate, nevertheless, it is fully justified from a practical point of view (*i.e.*, by the physical arguments which support it and the results it yields;* e.g., *Jumarie, 1990). In certain cases group-invariance requirements may make it more appropriate to replace $\ell og\, f_{\mathcal{G}}$ in the right-hand side of Equation 5.2 with $\ell og\,(f_{\mathcal{G}} f_0^{-1})$, where f_0 is a noninformative prior pdf (*i.e.*, one that is form-invariant under the transform groups that leave the physical laws invariant; Jaynes, 1983). Usually, an $\boldsymbol{\chi}_{map}$ parameterization is sought for which the corresponding f_0 is a natural constant. Moreover, Eddington (1959, p. 133) notices that in real-world applications involving natural phenomena: "Probability is always relative to knowledge (actual or presumed) and there is no* a priori *probability of things in a metaphysical sense,* i.e., *a probability relative to complete ignorance." In any case, the theoretical issues associated with the representation of complete ignorance (noninformativeness) are not of practical concern in BME mapping applications. Even if it is the case* *that f_0 cannot be determined in a convenient way for a specific mapping application, the discrete-domain formalization of BME, which does not involve a noninformative prior, can be used.*

General knowledge-based map pdf

In BME analysis, the shape of the pdf $f_{\mathcal{G}}$ at the prior stage is determined on the basis of the following postulate.

POSTULATE 5.2: The prior ($\mathcal{G}$-based) pdf $f_{\mathcal{G}}$ of the map is obtained by maximizing the expected prior information (Eq. 5.2) subject to the physical constraints introduced by Equation 3.2 (p. 75).

The prior pdf suggested by Postulate 5.2 is associated with a certain knowledge base and a method of using it. This $\mathcal{G}$-based pdf will be used in subsequent BME stages. The application of Postulate 5.2 requires the introduction of the mathematical method of *Lagrange* multipliers (the basic concepts of the method may be found in Ewing, 1969).

Lagrange multipliers method: The maximum value of the integral

$$\mathcal{M} = \int d\boldsymbol{\chi}\,\Phi[\boldsymbol{\chi}, f_{\mathcal{G}}(\boldsymbol{\chi})] \tag{5.3}$$

with respect to the function $f_{\mathcal{G}}$ and subject to the conditions

$$\overline{h_\alpha} = \int d\boldsymbol{\chi}\,\varphi_\alpha[\boldsymbol{\chi}, f_{\mathcal{G}}(\boldsymbol{\chi})], \quad \alpha = 0, 1, \ldots, N_c \tag{5.4}$$

can be found from the Euler–Lagrange equation

$$\frac{\partial \Phi}{\partial f_{\mathcal{G}}} + \sum\nolimits_{\alpha=0}^{N_c} \mu_\alpha \frac{\partial \varphi_\alpha}{\partial f_{\mathcal{G}}} = 0 \tag{5.5}$$

where μ_α are Lagrange multipliers. The μ_α are determined by introducing the function $f_{\mathcal{G}}$ found from Equation 5.5 into the constraints (Eq. 5.4).

In the case of BME analysis, $\Phi[\boldsymbol{\chi}, f_{\mathcal{G}}(\boldsymbol{\chi})] = -f_{\mathcal{G}}(\boldsymbol{\chi})\,\ell og\, f_{\mathcal{G}}(\boldsymbol{\chi})$ and $\varphi_\alpha[\boldsymbol{\chi}, f_{\mathcal{G}}(\boldsymbol{\chi})] = g_\alpha(\boldsymbol{\chi})\, f_{\mathcal{G}}(\boldsymbol{\chi})$. The shape of the prior pdf $f_{\mathcal{G}}(\boldsymbol{\chi}_{map})$ is derived by maximizing Equation 5.2 with respect to $f_{\mathcal{G}}(\boldsymbol{\chi}_{map})$, subject to the constraints imposed by general knowledge $\mathcal{G}$. This is a straightforward exercise of the Lagrange multipliers method, leading to the following proposition.

PROPOSITION 5.1: The mathematical implementation of Postulate 5.2 produces the prior pdf

$$f_{\mathcal{G}}(\boldsymbol{\chi}_{map};\, \boldsymbol{p}_{map}) = Z^{-1} \exp\left\{ \mathcal{Y}_{\mathcal{G}}[\boldsymbol{\chi}_{map};\, \boldsymbol{p}_{map}] \right\} \tag{5.6}$$

where

$$\mathcal{Y}_{G}[\boldsymbol{\chi}_{map};\, \boldsymbol{p}_{map}] = \sum_{\alpha=1}^{N_c} \mu_\alpha(\boldsymbol{p}_{map})\, g_\alpha(\boldsymbol{\chi}_{map}) \tag{5.7}$$

and Z is the partition function given by

$$\log Z = -\mu_0 \tag{5.8}$$

In view of Proposition 5.1, the general knowledge equations (Eq. 3.2) are written as follows

$$\overline{h_\alpha}(\boldsymbol{p}_{map}) = \int d\boldsymbol{\chi}_{map}\, g_\alpha(\boldsymbol{\chi}_{map}) \exp\{\mu_0 + \mathcal{Y}_{G}[\boldsymbol{\chi}_{map};\, \boldsymbol{p}_{map}]\},$$
$$\alpha = 0, 1, \ldots, N_c \tag{5.9}$$

Equation 5.9 is part of the *basic BME system* of equations that we will develop below for our spatiotemporal mapping purposes. The solution of the system of $N_c + 1$ equations (Eq. 5.9) determines the Lagrange multipliers μ_α.

COMMENT 5.2: *Normalization of the pdf leads to the following expression for the partition function*

$$Z = \int d\boldsymbol{\chi}_{map} \exp\{\mathcal{Y}_{G}[\boldsymbol{\chi}_{map};\, \boldsymbol{p}_{map}]\} \tag{5.10}$$

Then, the general knowledge constraints (Eq. 5.9) are also written as

$$\overline{h_\alpha}(\boldsymbol{p}_{map}) = \frac{\partial \log Z}{\partial \mu_\alpha}, \quad \alpha = 1, \ldots, N_c \tag{5.11}$$

Hence, the functional $\log Z$ *can be viewed as the generator of the general knowledge equations. The solution of the system of equations (Eqs. 5.10 and 5.11) also determines the Lagrange multipliers* μ_α. *Note that in this system the Lagrange multiplier* μ_0 *has been replaced by* Z, *and the number of constraint equations is* N_c; *the constraint for* $\alpha = 0$ *has been replaced by Equation 5.10.*

General knowledge in the form of random field statistics (including multiple-point statistics)

As we saw in a previous chapter, very often in geostatistical applications the available knowledge has the form of *random field statistics* of any order in space/time. In the language of geostatistics, the latter term includes: space/time moments of any order $(1, \ldots, \lambda)$ involving up to two points at a time; and space/time moments involving more than two points at a time—also called *multiple-point* statistics.

This sort of general knowledge is associated with a distinct class of $\mathcal{Y}_{G}$-operators, which is best illustrated by means of examples. The first example deals with the cases in which the prior pdf $f_{\mathcal{G}}$ of the map is sought from the general knowledge base $\mathcal{G} =$ {space/time statistical moments of orders $1, \ldots, \lambda$} and the base $\mathcal{G} =$ {multiple-point statistics of orders $\lambda_1, \ldots, \lambda_v$}.

EXAMPLE 5.1: Assume that the g_α functions at a point $\boldsymbol{p}$ include the normalization constraint and the spatiotemporal statistical moments, as follows

$$\left.\begin{aligned} \overline{h_0}(\boldsymbol{p}) &= \overline{g_0(x)} = 1, \\ \overline{h_\alpha}(\boldsymbol{p}) &= \overline{g_\alpha(x)} = \overline{x^\alpha}(\boldsymbol{p}) = \int d\chi\, \chi^\alpha f_{\mathcal{G}}(\chi;\, \boldsymbol{p}) \end{aligned}\right\} \tag{5.12}$$

where $\alpha = 1, \ldots, \lambda$ accounts for the orders of the corresponding statistical moments in space/time. Then, the $\mathcal{Y}_{\mathcal{G}}$-operator in Equation 5.7 is given by

$$\mathcal{Y}_{\mathcal{G}}[\chi;\, \boldsymbol{p}] = \sum\nolimits_{\alpha=1}^{\lambda} \mu_\alpha(\boldsymbol{p})\, \chi^\alpha \tag{5.13}$$

where the Lagrange multipliers are found by substituting Equation 5.13 into Equation 5.12 and solving for the μ_α's at point $\boldsymbol{p}$ for all moment orders considered (in this case, $N_c = \lambda$). In addition to the solution in Equation 5.13, other approaches could also be used to determine $\mathcal{Y}_{\mathcal{G}}$ (see below section, "Possible modifications and generalizations of the prior stage," on p. 118). The analysis is easily extended to more complicated situations. We could, *e.g.*, involve the whole set of mapping points $\boldsymbol{p}_{map}$, which leads to

$$\mathcal{Y}_{\mathcal{G}}[\boldsymbol{\chi}_{map};\, \boldsymbol{p}_{map}] = \sum\nolimits_{\alpha=1}^{m,k} \sum\nolimits_{\zeta=1}^{\lambda} \mu_{\alpha,\zeta}(\boldsymbol{p}_\alpha)\, \chi_\alpha^\zeta$$

or we could assume a series of more elaborate g_α functions such as,

$$\left.\begin{aligned} \overline{h_0}(\boldsymbol{p}_{map}) &= \overline{g_0(\boldsymbol{x}_{map})} = 1, \\ \overline{h_\alpha}(\boldsymbol{p}_{map}) &= \overline{g_\alpha(\boldsymbol{x}_{map})} = \int d\boldsymbol{\chi}_{map}\, g_\alpha(\boldsymbol{\chi}_{map})\, f_{\mathcal{G}}(\boldsymbol{\chi}_{map};\, \boldsymbol{p}_{map}),\ \alpha = 1, \ldots, m, k \end{aligned}\right\}$$

in which case $\mathcal{Y}_{\mathcal{G}}$ is generally given by Equation 5.7; *etc.* Note that the above BME equation includes the case of multiple-point statistics, as well. Indeed, as we saw in previous sections, the g_α functions may have the form $g_\alpha(\boldsymbol{\chi}_{map}) = \chi_1^{\lambda_1} \chi_2^{\lambda_2} \ldots \chi_v^{\lambda_v}$, which implies the *multiple-point* statistics

$$\overline{g_\alpha(\boldsymbol{x}_{map})} = \overline{x_1^{\lambda_1}\, x_2^{\lambda_2} \ldots x_v^{\lambda_v}} \tag{5.14}$$

of orders λ_j ($j = 1, 2, \ldots, v$); *etc.* The BME approach will account for this kind of statistical knowledge rigorously and efficiently. □

COMMENT 5.3: *The epistemic significance of the situation studied in the above example (Eqs. 5.12 and 5.13) can best be appreciated if we look at the familiar way in which a scientific theory fails to be constrained or determined by evidence (Quine, 1970; Friedman, 1983). Lower level (or inductive) indetermination (Chapter 1, p. 16) consists in the fact that we* *possess only a limited amount of evidence (expressed by the moments of order $\alpha \leq \lambda$). Thus, depending on the unknown higher order moments (*i.e.*, for $\alpha > \lambda$), different $\mathcal{Y}_{\mathcal{G}}$'s might have been obtained.*

An interesting circumstance arises when the calculation of the statistical moments includes some measurement error, as was discussed in Chapter 3 (Example 3.5, p. 76). This case is revisited in the example below.

EXAMPLE 5.2: The g_α functions at point $\boldsymbol{p}$ include the normalization constraint and the moments in Equation 3.4 (p. 76). Then, the constraints to be considered in Equation 5.6 are obtained using the chi-square statistic, *i.e.*,

$$\left.\begin{array}{l} \hat{\bar{g}}_0(\boldsymbol{p}) = 1, \\ \sum_{\alpha=1}^{N_c} (\sigma_{\nu,\alpha}^2)^{-1} \left[\int_\ell^u d\chi\, \chi^\alpha f_{\mathcal{G}}(\chi;\, \boldsymbol{p}) - \hat{\bar{g}}_\alpha(\boldsymbol{p})\right]^2 = N_c \end{array}\right\}$$

where $\sigma_{\nu,\alpha}^2(\boldsymbol{p})$ is the variance of $\nu_\alpha(\boldsymbol{p})$. The $\mathcal{Y}_{\mathcal{G}}$-operator in Equation 5.6 is given by Equation 5.13, where λ is now replaced by N_c, *i.e.*, $\mathcal{Y}_{\mathcal{G}}(\chi;\, \boldsymbol{p}) = \sum_{\alpha=1}^{N_c} \mu_\alpha(\boldsymbol{p})\, \chi^\alpha$. In this case the general knowledge constraints yield

$$\left.\begin{array}{l} \int_\ell^u d\chi\, \chi^\alpha \exp\left[\sum_{\alpha=0}^{N_c} \mu_\alpha\, \chi^\alpha\right] = \hat{\bar{g}}_\alpha - \frac{1}{2}\sigma_{\nu,\alpha}^2\, \mu_\alpha\, \eta^{-1}, \quad \alpha = 1, ..., N_c\,; \\ \sum_{\alpha=1}^{N_c} (\sigma_{\nu,\alpha}^2)^{-1} \left\{ \int_\ell^u d\chi\, \chi^\alpha \exp\left[\sum_{\alpha=0}^{N_c} \mu_\alpha\, \chi^\alpha\right] - \hat{\bar{g}}_\alpha \right\}^2 = N_c\,; \\ \int_\ell^u d\chi \exp\left[\sum_{\alpha=0}^{N_c} \mu_\alpha\, \chi^\alpha\right] = 1 \end{array}\right\} \tag{5.15}$$

The Lagrange multipliers μ_0, μ_α ($\alpha = 1, \ldots, N_c$) and η are found by solving the system of $N_c + 1$ equations (Eq. 5.15). In most cases of practical interest, these calculations are done numerically. □

General knowledge in the form of physical laws

As is well known, empirical or statistical mapping techniques describe an existing set of data and are only *locally* predictive (*i.e.*, interpolation is possible only within the data range). The situation is different with BME analysis, which is expressed by the following postulate (see also Chapter 3, p. 88).

POSTULATE 5.3: BME incorporates physical laws into space/time mapping and, thus, it can have global prediction features (*i.e.*, extrapolation is possible beyond the range of observations).

Next we will discuss how physical laws of various forms are incorporated into BME analysis. Since these laws constitute general knowledge $\mathcal{G}$, they must be analyzed at the prior stage of the BME approach. At this stage, one essentially seeks to transform the physical law into a set of moment equations and then proceeds as before [*i.e.*, formulates the prior pdf (Eq. 5.6) associated with the moment equations]. More specifically, the main steps of the approach are as follows:

(a) *Transformation*: Obtain the stochastic moment equations (Eq. 3.1 or 3.2) associated with the physical law. The way the transformation is made depends on the form of the physical law.

(b) *Formulation*: Derive the form of the prior pdf (Eq. 5.6) in view of the stochastic moment equations of the previous step.

(c) *Solution*: Insert the pdf (Eq. 5.6) into the stochastic equations and solve for the Lagrange multipliers. These multipliers are then substituted back into Equation 5.6 to obtain the final form of the prior pdf.

The BME analysis of physical laws above deserves some additional comments (an outline of the analysis is given in Fig. 5.1). In the case of physical laws represented by *algebraic* equations (see also Class A in Chapter 3, p. 77), the moment equations in the transformation step (a) are of the form given in Equation 3.6 (p. 77); for physical laws expressed in terms of *differential* equations (ordinary or partial; see Class B in Chapter 3, p. 78), the moment equations are of the form given in Chapter 3, Equations 3.6 and 3.13 or Equations 3.14 and 3.15 (p. 77–78).

The formulation step (b) depends on whether or not the moment equations in step (a) can be solved explicitly for the mean, covariance, *etc.* of the random field of interest $X(\boldsymbol{p})$. Two possible methods are proposed for handling the situation (Fig. 5.1):

> **Method A:** In many cases explicit solutions of the moment equations are intractable or the moments of $X(\boldsymbol{p})$ are not known. In these cases we incorporate directly the moment equations into BME analysis using the stochastic physical law representations discussed in Chapter 3.
>
> **Method B:** In some cases the physical law is such that the moment equations can be solved. The solutions may be exact or they may involve some approximations in terms of perturbation expansions, diagrammatic analysis, *etc.* (Christakos *et al.*, 1999). In these cases, the BME equations are essentially reduced to the set in Equation 5.9.

COMMENT 5.4: *Note that an interesting advantage of Method A over Method B is that the moment equations are rigorously taken into consideration without solving them for the specified moments (which is the case with Method B); see also Comment 3.3, p. 81. This avoids approximations involved in the experimental calculation of the moments (e.g., Stein, 1999) and eliminates the well-known circular problem of geostatistics: the data are first used to calculate the mean, variogram,* etc. *(which are inserted into the estimation system to obtain the kriging weights), and then the same*

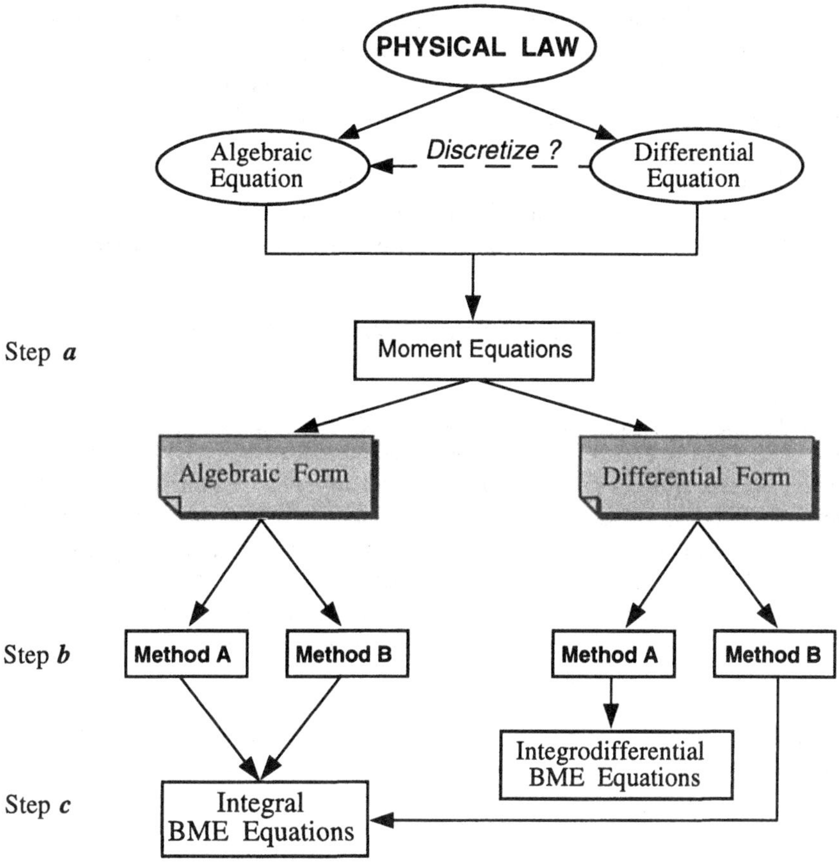

Figure 5.1. BME analysis of physical laws.

data are used again in terms of a weighted summation to derive the kriging estimates (see also Comment 7.2, p. 143).

The solution step (c) is concerned with the solution of a system of equations for the Lagrange multipliers. These equations are obtained by substituting the appropriate pdf formula (Eq. 5.6) into the moment equations of step (a). While in the case of algebraic laws these are integral equations for the Lagrange multipliers, in the case of differential laws they have the form of integrodifferential equations.

When physical laws in the form of differential equations are available, another approach (Fig. 5.1) is to *discretize* the differential as shown in Chapter 3 (Eq. 3.24, p. 80), thus obtaining an algebraic (difference) equation of the form given in Equation 3.5 (p. 77). This approach, which is particularly useful in cases involving complex partial differential equations, has already been discussed in Chapter 3 in the section, "General knowledge in terms of physical laws," beginning on p. 76; see also Example 3.10, p. 81.

The conceptual and technical issues introduced in the preceding discussion are best clarified with the help of a few simple examples. Complex situations can be studied as extensions of the simple, well-understood solutions to these examples. Frequent references to Chapter 3 are made in the following examples.

EXAMPLE 5.3: Consider the empirical law of Example 3.6 (p. 77) which relates standard penetration resistance X and vertical stress Y for a cohesionless soil. In light of constraints (Eqs. 3.8 and 3.9), a $\mathcal{G}$-based prior pdf is given by

$$f_{\mathcal{G}}(\boldsymbol{\chi}_{map}, \boldsymbol{\psi}_{data}; \boldsymbol{p}_{map}) = \exp\left[\mu_0 + \mathcal{Y}_{\mathcal{G}}(\boldsymbol{\chi}_{map}, \boldsymbol{\psi}_{data}; \boldsymbol{p}_{map})\right] \qquad (5.16)$$

where $\mathcal{Y}_{\mathcal{G}}(\boldsymbol{\chi}_{map}, \boldsymbol{\psi}_{data}; \boldsymbol{p}_{map}) = \sum_{\alpha} \mu_{\alpha}(\boldsymbol{p}_{\alpha})\, \chi_{\alpha} + \sum_{\alpha, \alpha'} \mu_{\alpha\alpha'}(\boldsymbol{p}_{\alpha}, \boldsymbol{p}_{\alpha'})\, \chi_{\alpha}\, \psi_{\alpha'}$; the subscripts α and α' account for all points considered. Other constraints involving the multivariate pdf may also be imposed. Equation 5.16 is substituted into the BME equations (Eqs. 3.8 and 3.9) which together with the normalization constraint can be solved for the Lagrange multipliers μ_0, μ_{α} and $\mu_{\alpha\alpha'}$. This process will give us the $\mathcal{G}$-based prior pdf $f_{\mathcal{G}}$. Note that it is not necessary to calculate the cross covariance between the penetration resistance and the vertical stress which is, nevertheless, taken into consideration implicitly in the BME equations above.

EXAMPLE 5.4: Let us study the physical law expressed by Equation 3.16 (p. 79). From the set of equations in Equation 3.17 we obtain the system of moment equations below

$$(\lambda\, b)^{-1} \frac{d}{dt_{\alpha}} \overline{X^{\lambda}(t_{\alpha})} = \int d\chi_{\alpha}\, \chi_{\alpha}^{\lambda}\, f_{\mathcal{G}}(\chi_{\alpha};\, t_{\alpha}) = \overline{x_{\alpha}^{\lambda}}, \quad \lambda = 1, 2 \qquad (5.17)$$

where the subscript $\alpha = 1, \ldots, m, k$ accounts for all times t_{α} of interest. In light of Equation 5.17 and taking $\overline{h_0}(t_{\alpha}) = \overline{g_0}(t_{\alpha}) = 1$ into consideration, the $\mathcal{Y}_{\mathcal{G}}$-operator in Equation 5.7 is given by

$$\mathcal{Y}_{\mathcal{G}}[\boldsymbol{\chi}_{map};\, \boldsymbol{t}_{map}] = \sum\nolimits_{\alpha=1}^{m,k} \sum\nolimits_{\lambda=1}^{2} \mu_{\alpha,\lambda}(t_{\alpha})\, \chi_{\alpha}^{\lambda} \qquad (5.18)$$

From Equation 5.17 we can write in terms of the pdf

$$\int d\boldsymbol{\chi}_{map}\, \chi_{\alpha}^{\lambda} \frac{\partial f_{\mathcal{G}}(\boldsymbol{\chi}_{map};\, \boldsymbol{t}_{map})}{\partial t_{\alpha}} = \lambda\, b \int d\boldsymbol{\chi}_{map}\, \chi_{\alpha}^{\lambda}\, f_{\mathcal{G}}(\boldsymbol{\chi}_{map};\, \boldsymbol{t}_{map}), \quad \lambda = 1, 2 \qquad (5.19)$$

with $f_{\mathcal{G}}(\boldsymbol{\chi}_{map};\, \boldsymbol{t}_{map}) = \exp\left[\mu_0 + \mathcal{Y}_{\mathcal{G}}\right]$. By substituting Equation 5.18 into Equation 5.19, we get

$$\left.\begin{aligned}&\int d\boldsymbol{\chi}_{map}\,\chi_\alpha^\lambda\left[\tfrac{d\mu_0}{dt_\alpha}+\textstyle\sum_{\beta=1}^{2}\tfrac{d\mu_{\alpha,\beta}(t_\alpha)}{dt_\alpha}\,\chi_\alpha^\beta\right]\exp\Big[\mu_0+\\&\quad\textstyle\sum_{\alpha'=1}^{m,k}\sum_{\lambda=1}^{2}\mu_{\alpha',\lambda}(t_{\alpha'})\,\chi_{\alpha'}^\lambda\Big]\\&=\lambda b\int d\boldsymbol{\chi}_{map}\,\chi_\alpha^\lambda\exp\left[\mu_0+\textstyle\sum_{\alpha'=1}^{m,k}\sum_{\lambda=1}^{2}\mu_{\alpha',\lambda}(t_{\alpha'})\,\chi_{\alpha'}^\lambda\right],\\&\lambda=1,\,2\qquad\alpha=1,\ldots,m,k\end{aligned}\right\}\tag{5.20}$$

which together with the normalization equation

$$\int d\boldsymbol{\chi}_{map}\,\exp\left[\mu_0+\sum_{\alpha=1}^{m,k}\sum_{\lambda=1}^{2}\mu_{\alpha,\lambda}(t_\alpha)\,\chi_\alpha^\lambda\right]=1\tag{5.21}$$

must be solved with respect to μ_0, $\mu_{\alpha,\lambda}(t_\alpha)$, $\alpha=1,...,m,k$ and $\lambda=1,\,2$. In this formulation, while the mean $\overline{X(t)}$ may not be known explicitly, it can be considered to be known implicitly as the solution of Equation 5.17. For illustrative purposes, let us consider the simple case in which $\alpha=k$ so that $\chi_\alpha=\chi_k$ and $\mu_{\alpha,\lambda}(t_\alpha)=\mu_\lambda(t_k)$, $\lambda=1,\,2$. After some analytical manipulations, Equations 5.20 and 5.21 give

$$\left.\begin{aligned}\mu_1(t_k)&=\mu_1(0)\exp\left[-bt_k\right]\\\mu_2(t_k)&=\mu_2(0)\exp\left[-2bt_k\right]\end{aligned}\right\}\tag{5.22}$$

where $\mu_\lambda(0)$, $\lambda=1,\,2$ are the initial conditions. These conditions are related to the mean and the variance of $X(t)$ at $t=0$, namely $\mu_1(0)=\overline{X(0)}\big/\sigma_x^2(0)$ and $\mu_2(0)=-1\big/\left[2\sigma_x^2(0)\right]$. By substituting Equation 5.22 into Equation 5.21, the μ_0 is found. The set of equations in Equation 3.18 (p. 79) can also be included in the analysis in a similar fashion. Indeed, the $\mathcal{Y}_G$-operator is given in this case by

$$\mathcal{Y}_G[\chi_i,\,\chi_j;\,t_i,\,t_j]=\sum_{\lambda=1}^{2}\mu_\lambda(t_i)\,\chi_i^\lambda+\mu_2(t_i,\,t_j)\chi_i\,\chi_j\tag{5.23}$$

which is substituted into Equation 3.18, leading to

$$\begin{aligned}&\iint d\chi_i\,d\chi_j\,\chi_i\,\chi_j\,\frac{\partial^2}{\partial t_i\partial t_j}f_G(\chi_i\,,\chi_j;\,t_i,\,t_j)=\\&b^2\iint d\chi_i\,d\chi_j\,\chi_i\,\chi_j\,f_G(\chi_i\,,\chi_j;\,t_i,\,t_j)\end{aligned}\tag{5.24}$$

or

$$\begin{aligned}&\iint d\chi_i\,d\chi_j\,\chi_i\,\chi_j\,\frac{\partial^2}{\partial t_i\partial t_j}\exp\left[\sum_{\lambda=0}^{2}\mu_\lambda(t_i)\,\chi_i^\lambda+\mu_2(t_i,\,t_j)\chi_i\,\chi_j\right]=\\&b^2\iint d\chi_i\,d\chi_j\,\chi_i\,\chi_j\,\exp\left[\sum_{\lambda=0}^{2}\mu_\lambda(t_i)\,\chi_i^\lambda+\mu_2(t_i,\,t_j)\chi_i\,\chi_j\right]\end{aligned}\tag{5.25}$$

etc. As before, the BME analysis can easily be extended to include all the mapping points $\boldsymbol{t}_{map}$. □

COMMENT 5.5: *We can proceed further with the mathematical formulations above in a number of ways. Consider, e.g., Equations 5.20 and 5.21 at a specific time t_α. Differentiating Equation 5.21 with respect to time, we find*

$$\frac{d\mu_0(t_\alpha)}{dt_\alpha} = -\sum_{\lambda=1}^{2} \overline{g_\lambda}(t_\alpha) \frac{d\mu_\lambda(t_\alpha)}{dt_\alpha} \tag{5.26}$$

where $\alpha = 1, \ldots, m, k$ *and* $\overline{g_\lambda}(t_\alpha) = \int d\chi_\alpha\, \chi_\alpha^\lambda \exp\left[\sum_{\zeta=0}^{2} \mu_\zeta(t_\alpha)\, \chi_\alpha^\zeta\right]$. *Similarly, Equation 5.20 gives the following system of equations*

$$\sum_{\beta=0}^{2} \overline{g_{\beta+\lambda}}(t_\alpha) \frac{d}{dt_\alpha} \mu_\beta(t_\alpha) = \lambda\, b\overline{g_\lambda}(t_\alpha), \quad \lambda = 1, 2 \tag{5.27}$$

As before, this system of equations must be solved—usually numerically—with respect to $\mu_\beta(t_\alpha)$, where $\beta = 0, 1, 2$.

The numerical investigations of the following example can serve to illustrate how the mathematical analysis of Example 5.4 above can be used to obtain interesting plots of the $\mathcal{G}$-based prior pdf, the moments, *etc.*

EXAMPLE 5.5: Typically, the first-order moment constraints of the physical law (Eq. 3.16, p. 79) are expressed by means of the integrodifferential equation

$$\int d\chi_\alpha\, \chi_\alpha^\lambda \frac{\partial f_\mathcal{G}(\chi_\alpha;\, t_\alpha)}{\partial t_\alpha} = (\lambda\, b) \int d\chi_\alpha\, \chi_\alpha^\lambda\, f_\mathcal{G}(\chi_\alpha;\, t_\alpha), \quad \lambda = 1, 2 \tag{5.28}$$

where, as usual, the subscript α accounts for all times t_α of interest. Covariances will similarly be taken into account by

$$\int d\chi_\alpha \int d\chi_{\alpha'} \frac{\partial^2 f_\mathcal{G}(\chi_\alpha, \chi_{\alpha'};\, t_\alpha, t_{\alpha'})}{\partial t_\alpha\, \partial t_{\alpha'}} =$$
$$b^2 \int d\chi_\alpha \int d\chi_{\alpha'}\, \chi_\alpha\, \chi_{\alpha'}\, f_\mathcal{G}(\chi_\alpha, \chi_{\alpha'};\, t_\alpha, t_{\alpha'}) \tag{5.29}$$

where the subscripts α and α' account for all pairs of time instants considered (note that the analysis can easily be generalized to *multipoint* statistics). In Christakos *et al.* (1999), for simplicity, we used the symbols t, χ, and μ_λ instead of t_α, χ_α, and $\mu_{\alpha,\lambda}$, respectively, and in order to obtain some numerical results, we let $b = -0.01$ and assumed that the pdf of the random variable $X(t=0) = X_0$ is described by $f_\mathcal{G}(\chi;\, 0) = \exp\left[\mu_0(0) + \mu_1(0) b\chi + \mu_2(0) b\chi^2\right]$, with $\mu_0(0) = -90.40\, b^{-1}$, $\mu_1(0) = 19.54\, b^{-1}$, and $\mu_2(0) = -1.06\, b^{-1}$; these values correspond to an initial mean $\overline{X_0} = 9.20$ and an initial variance $\sigma_0^2 = 0.47$ at $t = 0$. At later times $t > 0$, the evolution of $X(t)$ is governed by the physical

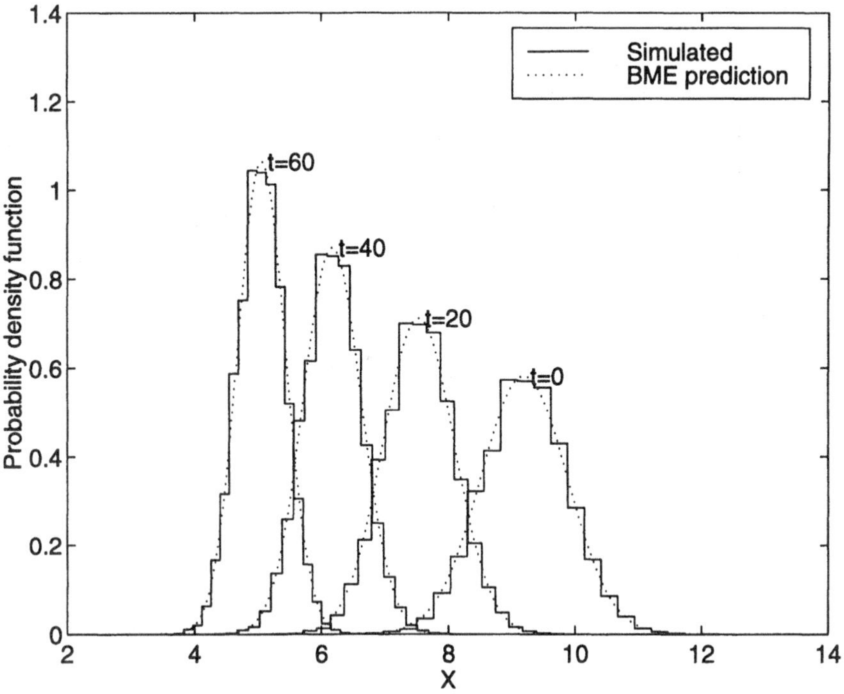

Figure 5.2. Simulated pdf (plain line) and BME pdf (dotted line).

law (Eq. 3.16, p. 79). Using the $f_G(\chi;\ 0)$ above and Equation 3.16 (p. 79) with $b = -0.01$, we generated 10,000 $X(t)$-realizations at each one of the following time instants: $t = 0,\ 20,\ 40,$ and 60. These simulated $X(t)$-values were used to plot the temporal evolution of the corresponding prior pdf (Fig. 5.2), which was then compared to the pdf obtained from BME analysis as follows. First we solved Equations 5.28 and 5.29 for the Lagrange multipliers, which were thus expressed as functions of time. Using these multipliers, we calculated the BME pdf at various times, which are also plotted in Figure 5.2 for comparison.

Notice that the temporal evolution of the BME pdf is in very good agreement with the temporal evolution of the simulated pdf. In this case, the BME implementation does not require one to solve explicitly the physical law (Eq. 3.16) for the moments of $X(t)$. For comparison purposes, however, in Figure 5.3 the moments

$$\left.\begin{aligned} m_{x,k}(t) &= \overline{X^k(t)}\Big/\ \overline{X^k(0)},\ k = 1,\ldots,4 \\ c_{x,k}(t) &= \overline{\left[X(t) - \overline{X(t)}\right]^k},\ k = 2,\ldots,4 \end{aligned}\right\}$$

were calculated from the BME pdf and compared with the simulated moments. Once more, the comparison demonstrates the close agreement between the simulated and the BME moments. □

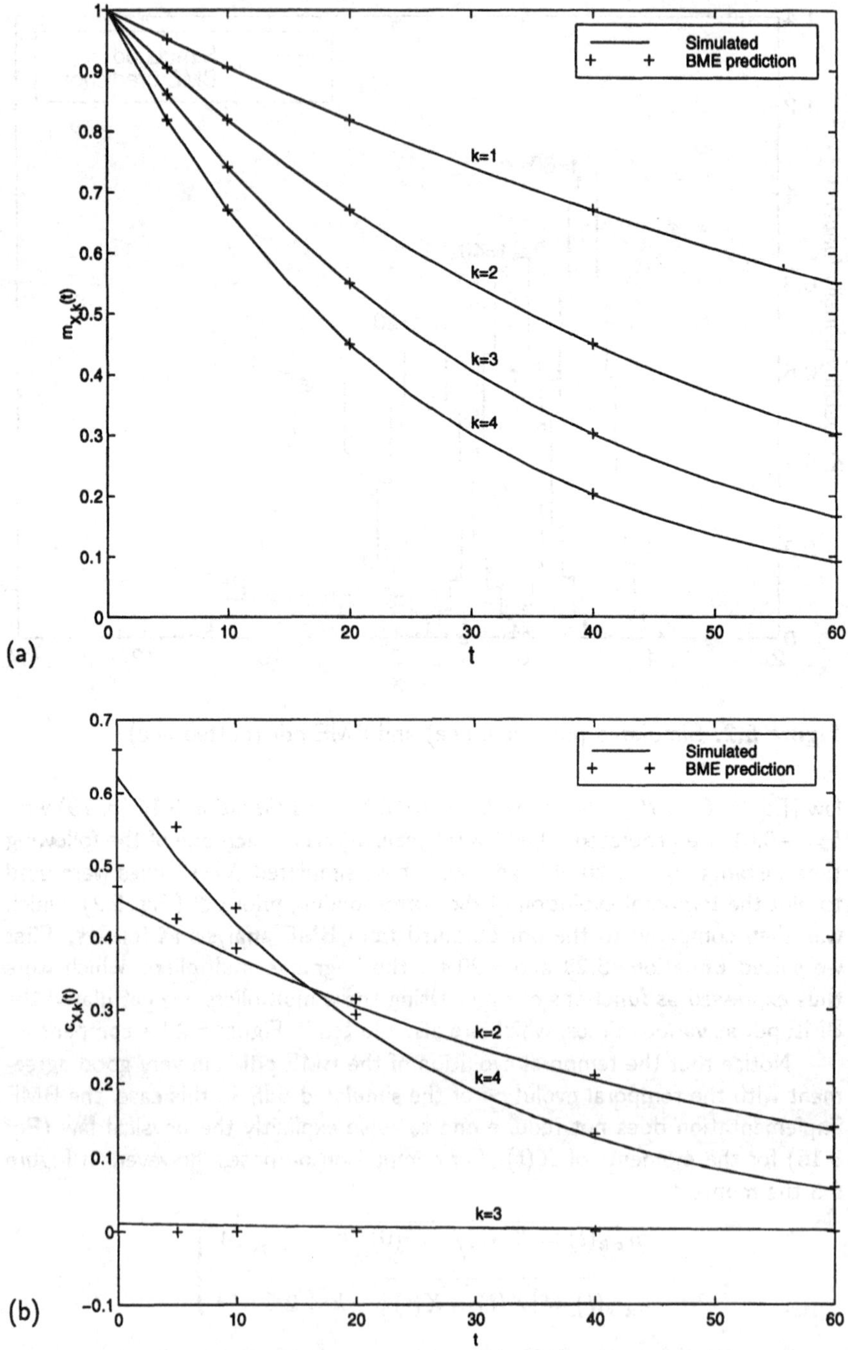

Figure 5.3. (a) Normalized non-centered moments $m_{x,k}(t)$, $(k = 1, \ldots, 4)$; (b) centered moments $c_{x,k}(t)$, $(k = 2, \ldots, 4)$ calculated using the simulated values (plain line), and the BME pdf (crosses).

We continue our discussion with a situation in which the coefficient of the differential equation representing the physical law is a spatial random field (*i.e.*, a function of space). This situation possesses some interesting features that can also be taken into consideration by the BME analysis.

EXAMPLE 5.6: Consider the groundwater flow law expressed by Equation 3.21 (p. 80). From Equation 3.22 we obtain the corresponding system of stochastic moment equations as follows

$$-\lambda^{-1}\frac{d\overline{X^\lambda}}{ds_\alpha} = \int\int d\chi_\alpha\, d\psi_\alpha\, \chi_\alpha^\lambda\, \psi_\alpha\, f_{\mathcal{G}}(\chi_\alpha, \psi_\alpha;\, s_\alpha) = \overline{x_\alpha^\lambda\, y_\alpha}, \quad \lambda = 1, 2 \tag{5.30}$$

or in terms of the pdf

$$-\int\int d\chi_\alpha\, d\psi_\alpha\, \chi_\alpha^\lambda\, \frac{\partial f_{\mathcal{G}}(\chi_\alpha, \psi_\alpha;\, s_\alpha)}{\partial s_\alpha} = \lambda \int\int d\chi_\alpha\, d\psi_\alpha\, \chi_\alpha^\lambda\, \psi_\alpha\, f_{\mathcal{G}}(\chi_\alpha, \psi_\alpha;\, s_\alpha) \tag{5.31}$$

where $\lambda = 1, 2$, and the subscript α denotes the location of interest. In light of Equation 5.30, and taking the normalization equation $\overline{h_0}(s_\alpha) = \overline{g_0}(s_\alpha) = 1$ and the moment equations $\overline{h_\kappa}(s_\alpha) = \overline{Y^\kappa}(s_\alpha)$ into consideration ($\kappa = 1, \ldots, K$ are the orders of the available Y-moments), the $\mathcal{Y}_{\mathcal{G}}$-operator in Equation 5.7 is

$$\mathcal{Y}_{\mathcal{G}}[\chi_\alpha, \psi_\alpha;\, s_\alpha] = \sum\nolimits_{\lambda=1}^{2} \mu_\lambda(s_\alpha)\, \chi_\alpha^\lambda\, \psi_\alpha + \sum\nolimits_{\kappa=1}^{K} \mu_\kappa(s_\alpha)\, \psi_\alpha^\kappa \tag{5.32}$$

Following the same procedure as before, Equation 5.32 is substituted into Equation 5.31 which, together with the normalization and the moment equations, should be solved with respect to the μ_λ's and μ_κ's. Other moment equations arising from the flow law (Eq. 3.21) can also be taken into consideration. Taking the covariance given in Equation 3.23 (p. 80) into account, *e.g.*, leads to the pdf equation

$$\begin{aligned} &\int\int d\boldsymbol{\chi}_{map}\, d\boldsymbol{\psi}_{data}\, \chi_i\, \chi_j\, \frac{\partial^2 f_{\mathcal{G}}(\boldsymbol{\chi}_{map}, \boldsymbol{\psi}_{data};\, \boldsymbol{s}_{map})}{\partial s_i\, \partial s_j} = \\ &\int\int d\boldsymbol{\chi}_{map}\, d\boldsymbol{\psi}_{data}\, \chi_i\, \chi_j\, \psi_i\, \psi_j\, f_{\mathcal{G}}(\boldsymbol{\chi}_{map}, \boldsymbol{\psi}_{data};\, \boldsymbol{s}_{map}) \end{aligned} \tag{5.33}$$

where, as usual, $i, j = 1, \ldots, m, k$; *etc.* □

COMMENT 5.6: *An ME formulation of a stochastic Itô-type equation for purely temporal processes is discussed,* e.g., *in Trebicki and Sobczyk (1996). However, this one-dimensional formulation is restricted to a specific equation involving a univariate pdf at each time and does not account for other physical knowledge sources, whereas in BME mapping one is concerned with a multivariate pdf at several space/time points that integrates general knowledge as well as hard and soft data. In these more complicated physical situations it is many times preferable to first discretize the partial differential equation, and then proceed with BME analysis.*

In view of the preceding analysis, an essential difference between classical and modern geostatistics could be described as follows: Unlike classical geostatistics which essentially capitalized on the techniques of spatial statistics, the emphasis of modern spatiotemporal geostatistics is on the powerful theories and laws of natural sciences. The implication of this difference is that modern geostatisticians, rather than being misled by the numerous statistical models and hypotheses compatible with the data set available, should be able to weed out the physically incorrect models and hypotheses quickly, and get on to the next problem with a great deal more confidence about the maps and conclusions they have drawn. This important feature of modern geostatistics is usually called *strong inference* in scientific reasoning theory.

EXAMPLE 5.7: The strong inference situation described above is a familiar one in science. Molecular biology, *e.g.*, was able to make spectacular progress in a rather short period of time. As Platt (1964) explained in an article published in the journal *Science*, this extraordinary success was due to the fact that, unlike traditional biology which was relying primarily on taxonomy (*i.e.*, collecting, describing, and tabulating observational facts), molecular biology capitalized on the powerful theories of chemistry and the mathematical modeling tools of theoretical physics. □

The BME approach aims at contributing to the continuing dialogue between scientific theories and experimental results. In this sense, the BME space/time maps could offer valuable evidence that a theory may need to be revised or reassessed.

Possible modifications and generalizations of the prior stage

The epistemic framework of modern spatiotemporal geostatistics is very general, thus allowing various modifications of the BME approach. Let us briefly discuss a few possibilities.

The φ_α functions of Equation 5.4 were assumed to be of the form $\varphi_\alpha[\boldsymbol{\chi}, f_{\mathcal{G}}(\boldsymbol{\chi})] = g_\alpha(\boldsymbol{\chi})\, f_{\mathcal{G}}(\boldsymbol{\chi})$. Other φ_α forms may also be considered, thus broadening the range of applicability of the BME analysis (the issue was already raised in Chapter 3, "A mathematical formulation of the general knowledge base," p. 74). In cases where the general knowledge consists of statistical moments (Example 5.1 above), several methods can be used in place of entropy maximization, including *series truncation* and *orthogonal polynomial expansion* (Christakos, 1992).

When the univariate prior pdf is available (considered to be the same at all points in space/time), it may be assumed that a transformation of the original random field into a Gaussian one exists. Using the derived statistics of the Gaussian field, the approach of Example 5.1 can then be applied to define the $\mathcal{Y}_{\mathcal{G}}$-operator. Finally, the operator associated with the original field is obtained by the inverse transformation (Bogaert *et al.*, 1999).

It may be possible that, while the multivariate pdf $f_{\mathcal{G}}$ is unknown, an equation that describes the evolution of a lower level pdf can be derived from physical or mathematical considerations (see "Some other forms of general knowledge," in Chapter 3, p. 81). A set of g_α functions could then be properly selected so that the stochastic expectations $\overline{g_\alpha}$ with respect to $f_{\mathcal{G}}$ can be calculated. Under these conditions, the problem has essentially been reduced to that described above in the section "General knowledge in the form of random field statistics" (p. 107). Let us consider the following rather simple example.

EXAMPLE 5.8: In Chapter 3, in the case of Example 3.11 (p. 82), the operator $\mathcal{Y}_{\mathcal{G}}$ is simply given by $\mathcal{Y}_{\mathcal{G}}[\chi;\ t] = \mu_1(t) g_1(\chi)$, where $\mu_1(t)$ is the solution of the equation

$$c_0 \exp[\beta t] \int_a^b d\chi \exp[\mu_1(t) g_1(\chi)] = \int_a^b d\chi\, g_1(\chi) \exp[\mu_1(t) g_1(\chi)] \quad (5.34)$$

Depending on the form of the function $g_1(\cdot)$, this equation may be solved analytically or numerically. □

In its current formulation, BME uses the information measure (Eq. 5.1) and the expected information or entropy (Eq. 5.2). But it may be worth examining situations in which one could use other measures available in the literature (*e.g.*, Aczel and Daroczy, 1975; Ebanks *et al.*, 1998). New measures could also be proposed by modern geostatisticians, as long as they are physically meaningful and epistemically sound. In such a case, entropy may not be part of the mapping formulation, and the letter "E" in the acronym "BME" should be replaced by another letter denoting the new information measure.

The Meta-Prior Stage

In real-world applications the knowledge bases are continuously refined due to further interaction with the environment. At the meta-prior stage we collect and organize specificatory knowledge $\mathcal{S}$ in appropriate quantitative forms that can be explicitly incorporated into the BME formulation. In many applications a variety of case-specific data are available, which usually makes the analysis at the meta-prior stage not a trivial task. The quality and quantity of the hard and soft data collected is a matter of experimental and/or computational investigations. BME's concern is that accurate spatiotemporal maps be drawn from good-quality data and other sources of knowledge by means of acceptable rules of reasoning, which can be adequately quantified in terms of the analytical and computational tools available.

As we already saw in Chapter 4 ("Epistemic Geostatistics and the BME Analysis," p. 90), the distinction between prior and meta-prior knowledge is an important epistemic issue. Whereas at the prior stage general knowledge $\mathcal{G}$ was introduced, at the meta-prior stage the specificatory knowledge $\mathcal{S}$ is considered, including hard and soft data. Certain of these kinds of data have been discussed in Chapter 3 (p. 73, "The General Knowledge Base" and p. 82, "The

Specificatory Knowledge Base"). Approaches leading to different arrangements of the knowledge bases incorporated in the prior and meta-prior stages can be considered, a fact that adds to the flexibility of the BME analysis. Computational aspects can also play a role in decisions about the physical knowledge to be used at each stage. In some cases, physical knowledge-based constraints may arise for which pdf maximizing of the expected information (Eq. 5.2) is difficult to obtain (see, *e.g.*, Shimony, 1985). It is then possible that these constraints can be put in a form that is easily incorporated at the meta-prior stage.

The Integration or Posterior Stage

In the prior stage, the operator $\mathcal{Y}_{\mathcal{G}}$ processed general knowledge $\mathcal{G}$. The $\mathcal{G}$ knowledge as well as the specificatory knowledge $\mathcal{S}$ considered in the meta-prior stage are incorporated into the mapping process by means of a new operator introduced by the following postulate.

POSTULATE 5.4: At the integration stage, the updated pdf of the map should be expressed in terms of an operator $\mathcal{Y}_{\mathcal{S}}$ which integrates the general knowledge-based operator $\mathcal{Y}_{\mathcal{G}}$ with the specificatory knowledge $\mathcal{S}$ considered at the meta-prior stage, *i.e.*,

$$f_{\mathcal{K}}(\boldsymbol{\chi}_k) = A^{-1}\,\mathcal{Y}_{\mathcal{S}}[\mathcal{Y}_{\mathcal{G}},\ \mathcal{S},\ \boldsymbol{\chi}_{map};\ \boldsymbol{p}_k] \tag{5.35}$$

where $\mathcal{K} = \mathcal{G} \cup \mathcal{S}$, and A is a normalization parameter.

At this point, Equation 5.35 presents an abstract form of $\mathcal{Y}_{\mathcal{S}}$, which indicates its dependence on the general knowledge operator $\mathcal{Y}_{\mathcal{G}}$, the specificatory knowledge $\mathcal{S}$, and the mapping vector $\boldsymbol{\chi}_{map} = (\boldsymbol{\chi}_{hard}, \boldsymbol{\chi}_{soft}, \boldsymbol{\chi}_k)$; the operator $\mathcal{Y}_{\mathcal{S}}$ is associated with the mapping point $\boldsymbol{p}_k$. Conceptually, Postulate 5.4 is a direct consequence of Proposition 4.1 (p. 95), in which case the operator $\mathcal{Y}_{\mathcal{S}}$ and the normalization parameter A account for the probability functions $f_{\mathcal{G}}[\boldsymbol{\chi}_{map}(\mathcal{S})]$ and $f_{\mathcal{G}}[\boldsymbol{\chi}_{data}(\mathcal{S})]$, respectively. The $\boldsymbol{\chi}_k | \boldsymbol{\chi}_{data}(\mathcal{S})$ stands for the possible values $\boldsymbol{\chi}_k$ of the map in the context specified by $\boldsymbol{\chi}_{data}(\mathcal{S})$ and $\mathcal{G}$. Probability models such as Equation 5.35 should always be understood to apply in the appropriate context, which defines the current state of physical knowledge. Postulate 5.4 suggests a *natural synthesis* of the general knowledge of the prior stage and the specificatory knowledge of the meta-prior stage. In the *multipoint* case (see Chapter 4, p. 89), Postulate 5.4 is concerned with the multivariate pdf $f_{\mathcal{K}}(\boldsymbol{\chi}_k)$, $\boldsymbol{\chi}_k = (\chi_{k_1}, \ldots, \chi_{k_\rho})$. In the *single-point* case, on the other hand, we are dealing with several univariate pdf's $f_{\mathcal{K}}(\chi_{k_i})$, $i = 1, \ldots, \rho$ (see Fig. 5.4).

Certain $\mathcal{Y}_{\mathcal{G}}$-operators have been introduced earlier in this chapter (p. 106). In order, then, to obtain an explicit analytical form for $\mathcal{Y}_{\mathcal{S}}$, the form of the specificatory knowledge $\mathcal{S}$ must first be described explicitly (in addition to the hard facts, a geostatistician may find it useful to look to the judgments of those

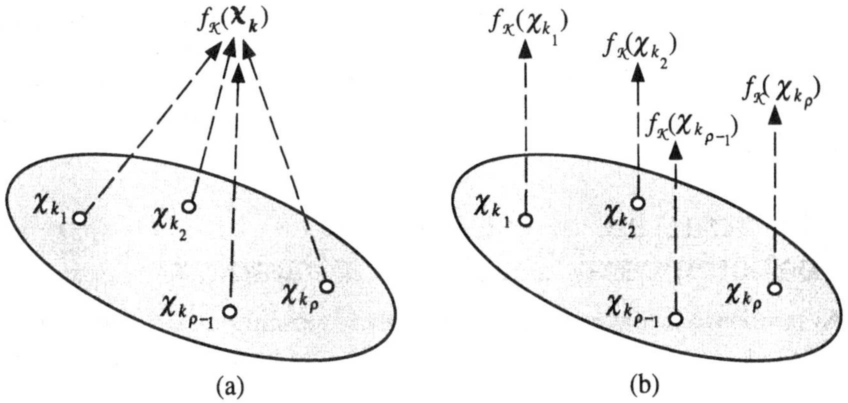

Figure 5.4. Posterior pdf associated with: (a) multipoint mapping at all space/time points simultaneously; and (b) single-point mapping at a succession of points independently.

who have the right expertise and experience; see Chapter 3). Given $\mathcal{Y}_G$ and $\mathcal{S}$, a variety of $\mathcal{Y}_S$-operators can be developed which express knowledge related to actual measurements, uncertain evidence, probability distributions, functional relationships, *etc.* Indeed, several examples of $\mathcal{Y}_S$-operators are discussed in Chapter 6.

On the basis of Equation 5.35, the effects of different prior $\mathcal{Y}_G$-operators can be assessed by comparing the corresponding posterior pdf related to single-point *vs.* multipoint situations. Finally, in Chapter 7, we will discuss various practical applications in which the general and specificatory knowledge operators are combined to provide informative spatiotemporal maps.

COMMENT 5.7: *The importance of notation in modern spatiotemporal geostatistics cannot be overemphasized. A well-chosen notation suggests the right operations and liberates the mind from pointless distractions, while an ill-chosen symbolism may be a hindrance to reasoning. For notational convenience, some of the symbols* $\mathcal{Y}_G$, $\mathcal{S}$, $\boldsymbol{\chi}_{map}$, *and* $\mathbf{p}_k$ *that appear in Equation 5.35 will be dropped occasionally in the following chapters. In some cases,* e.g., *we will write* $\mathcal{Y}_S[\boldsymbol{\chi}_{map}; \mathbf{p}_k]$ *to denote that the operator involves the values of a natural variable* X, *as opposed to* $\mathcal{Y}_S[\boldsymbol{\chi}_{map}, \boldsymbol{\psi}_{data}; \mathbf{p}_k]$ *which involves two natural variables* X *and* Y *(this is the multivariable or vector case; see also Chapter 9). Furthermore, when its meaning is obvious from the context, the operator will be denoted simply as* $\mathcal{Y}_S$.

In concluding this section, the following point is worth mentioning. In light of the developments so far, the *posterior BME information* associated with a spatiotemporal map can be defined naturally as

$$\text{Info}_{\mathcal{K}}[\boldsymbol{\chi}_k] = \log A - \log \mathcal{Y}_S \tag{5.36}$$

where the A parameter and the $\mathcal{Y}_S$-operator have been defined above. There is an interesting intuitive implication of Equation 5.36: The observation of the data set χ_{data} lowers the number of possible $X(\boldsymbol{p}_k)$ realizations. Equations 5.35 and 5.36 could be viewed as *knowledge processing rules*. Indeed, just as certain rules tell us how to measure distances or how to weigh objects, these equations tell us how to update our evaluation of a situation given new knowledge.

The Structure of the Modern Spatiotemporal Geostatistics Paradigm

As mentioned in preceding chapters, classical geostatistics is based on pure induction (Chapter 1, p. 11) which interprets the relationship between data and mathematical analysis as a linear one, *i.e.*,

$$\textit{Data} \rightarrow \textit{Mathematical Fitting} \rightarrow \textit{Prediction and Testing} \tag{5.37}$$

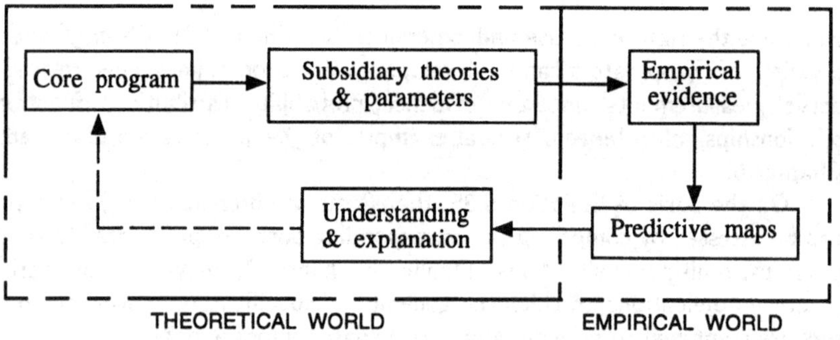

Figure 5.5. The structure of the modern geostatistical paradigm.

On the basis of our discussion so far, we can conclude that the modern spatiotemporal geostatistics paradigm proposed by BME is hypothetico–deductive rather than linear, involving the following parts (Fig. 5.5):

(a) The *core program*, which includes the S/TRF concepts and the epistemic ideals.

(b) *Subsidiary theories* provided by the general knowledge base and *subsidiary parameters* (various classes of correlation models, space/time metrics, *etc.*).

(c) *Empirical evidence* provided by the specificatory knowledge base.

(d) *Predictive* maps.

(e) *Understanding* and *explanation*.

Part (a) functions in a *Kuhnian* programmatic way: it provides modern geostatisticians with a particular way of acquiring, interpreting, and processing knowledge. Within this framework, in part (b), subsidiary theories and parameters can be generated that specify how the core program works in practice.

The subsidiary theories and parameters are combined with empirical evidence in part (c), and then interpreted, tested, and evaluated in detail in parts (d) and (e). The testing and evaluation may involve (but are not limited by) the *Popperian* falsification principle, which attempts to define the applicability limits of the models (*e.g.*, the areas in which the models are falsified, or they make incorrect predictions).

While parts (a), (b), and (e) belong to the theoretical world, parts (c) and (d) belong to the empirical world. These two worlds are linked through a feedback process, which allows modern geostatisticians to learn from their own mistakes. Indeed, if the predictions are not verified or an appropriate level of understanding and explanation is not achieved, the implication is that there must be unknown influences which are relevant and which should be sought at the ontological level of part (b) or there is a lack of sufficient data in part (c). In rare cases one may need to modify the core epistemology of part (a).

The core program aims at drawing attention to salient features of the observational world and its relation to the understanding process of the observer (the geostatistician). The program involves a strong formal part (mathematical concepts and tools, *etc.*). Depending on the results obtained (accuracy of the predictive pattern, explanatory content of the maps, *etc.*), a subsidiary theory may be modified or replaced by another, if necessary. This was the meaning, *e.g.*, of Comment 1.4 (p. 21; *e.g.*, information measures other than entropy may be considered).

The Two Legs on Which the BME Equations Stand

To summarize, what we have demonstrated is that the BME approach of modern spatiotemporal geostatistics forces the researcher to determine explicitly the physical knowledge bases that are objectively available, and to develop logically plausible rules for knowledge integration and processing. All of this is epistemically incorporated in the mapping process; nothing is swept under the carpet. As a consequence, the conclusions that can be reached by the BME method have a security that is denied to conclusions reached merely by empirical procedures.

We have also established that there are two fundamental operators involved in the mathematical formulation of the BME approach:

(*i.*) the $\mathcal{Y}_{\mathcal{G}}$-operator that incorporates general knowledge $\mathcal{G}$, and

(*ii.*) the $\mathcal{Y}_{\mathcal{S}}$-operator that incorporates specificatory knowledge $\mathcal{S}$.

These two operators are, indeed, the two *legs* on which the BME equations stand. Much of the following chapters are essentially mathematical results in terms of these two operators. The mathematics that we will use, however, are usually much simpler than what most geostatisticians are capable of handling. In fact, the most important problems are conceptual rather than mathematical.

COMMENT 5.8: *As already mentioned, the characterization "Bayesian" in the acronym BME denotes the fact that Bayesian conditionalization is used in the theoretical analysis of the integration stage of the approach (see,* e.g.*, Proposition 4.1 of Chapter 4, p. 95). In the eyes of some statisticians, this analysis may not fit in the orthodox Bayesian framework. This is hardly surprising. As Wang (1993; p. 158) notices: "There are at least 46,656 varieties of Bayesians." This statement is, obviously, a hyperbole that serves to stress the fact that there are, indeed, various kinds of Bayesian approaches, including the orthodox, the subjective, and the epistemic Bayesianisms. On the other hand, the characterization "Maximum Entropy" in the acronym BME is due to the fact that Equation 5.2 has the mathematical form of the entropy function used by Boltzmann and Jaynes (in thermodynamics and statistical mechanics;* e.g.*, Boltzmann, 1964 [1896–98], Jaynes, 1983), by Shannon (in the description, storage, and transmission of messages; Shannon, 1948), and by many others (see,* e.g.*, Ebanks* et al.*, 1998). In all the above cases, however, the entropy functions were developed in different scientific contexts than the spatiotemporal mapping situation considered in Equation 5.2. As a matter of fact, it is not uncommon in scientific investigations to start from different origins and to end up with similar mathematical formulations of otherwise different physical situations. Entropy is a case in point. In 1896, the term was introduced by Boltzmann in the kinetic theory of gases in an effort to measure disorder by means of the probabilities of molecular arrangements. In 1948, while working on communication engineering problems, Shannon derived a working definition of syntactic information which, when translated into mathematical symbols, was identical to the Boltzmann entropy function. Certainly, from a physical interpretation point of view, the two entropies were drastically different. In light of these examples, we conclude that the mathematical form of entropy arises in various scientific disciplines, in which, however, it has very different physical interpretations.*

By way of a summary, the problem of how to integrate the $\mathcal{G}$-base (scientific theories, physical laws, *etc.*) with the $\mathcal{S}$-base (case-specific data, empirical evidence, *etc.*) when constructing a space/time map has generated considerable controversy in geostatistics and spatial statistics. The BME approach suggests a solution to this problem by using first the $\mathcal{Y}_{\mathcal{G}}$-operator to provide an initial specification of the probability model of the map in terms of the $\mathcal{G}$-base, and then the $\mathcal{Y}_{\mathcal{S}}$-operator to extend or refine the starting model leading to a form that better represents the $\mathcal{S}$-base.

6

ANALYTICAL EXPRESSIONS OF THE POSTERIOR OPERATOR

"It is theory which decides what we can observe." A. Einstein

Specificatory Knowledge and Single-Point Mapping

Underlying the fundamental equation of Chapter 5 (Eq. 5.35; p. 120) is a positive feedback that takes place between (*i.*) meta-prior specificatory knowledge (*e.g.*, in the form of empirical evidence specific to the problem considered), and (*ii.*) the prior probability law obtained from the processing of general knowledge (*e.g.*, in the form of physical laws, scientific theories, engineering relationships, and statistical models whose validity has already been confirmed and reaches beyond the specific problem). Despite its theoretical elegance, as it stands Equation 5.35 may not offer too much to the practitioner of modern spatiotemporal geostatistics. Its physical meaning and powerful features emerge when particular cases of the prior and the posterior (or integration) operators $\mathcal{Y}_G$ and $\mathcal{Y}_S$, respectively, are worked out.

These two operators represent the geostatistician's state of knowledge of the phenomenon under study. Analytical forms of the $\mathcal{Y}_G$-operator have already been considered in the previous chapter. These forms are associated with a number of important general knowledge sources encountered by the practicioner in a variety of scientific disciplines. The mathematical formulation of the $\mathcal{Y}_S$-operator depends upon the form in which the specificatory knowledge base $\mathcal{S}$ may become available. Analytical expressions of the $\mathcal{Y}_S$-operator which account for some of the most common sources of specificatory knowledge will be developed in this chapter. Other, more complex situations can be studied building on these well-understood analytical $\mathcal{Y}_S$ expressions. The presentation of the analytical results, with their full technical beauty, is combined with an attempt to communicate across the various fields of natural science. In many cases, it is still true that "one month on the computer can easily save a day of paper and pencil research..."; hence, modern geostatisticians should not avoid using their analytical skills as often as possible. In this chapter, single-point mapping situations will be considered. Multipoint analysis will be the subject of later chapters.

Posterior Operators for Interval and Probabilistic Soft Data

The following proposition considers the common case in which interval soft data $\boldsymbol{\chi}_{soft}$ are considered at the meta-prior stage.

PROPOSITION 6.1: Assume that the specificatory knowledge $\mathcal{S}$ consists of the hard data (Eq. 3.30, p. 84) and the interval (soft) data (Eq. 3.32, p. 85). Then, the posterior operator is given by

$$\mathcal{Y}_{\mathcal{S}}[\boldsymbol{\chi}_{map}; \boldsymbol{p}_k] = Z^{-1} \int_I d\boldsymbol{\chi}_{soft} \exp\{\mathcal{Y}_{\mathcal{G}}[\boldsymbol{\chi}_{map}; \boldsymbol{p}_{map}]\} \tag{6.1}$$

where I denotes the domain of $\boldsymbol{\chi}_{soft}$ and Z is the partition function.

Proof : By definition

$$f_{\mathcal{K}}(\chi_k) = f_{\mathcal{G}}[\chi_k \mid \boldsymbol{\chi}_{data}(\mathcal{S})] = \frac{\partial}{\partial \chi_k} F_{\mathcal{G}}[\chi_k \mid \boldsymbol{\chi}_{data}(\mathcal{S})]$$

$$= \frac{\partial}{\partial \chi_k} F_{\mathcal{G}}(\chi_k, \boldsymbol{\chi}_{hard}, \boldsymbol{\chi}_{soft} \in \boldsymbol{I})[F_{\mathcal{G}}(\boldsymbol{\chi}_{hard}, \boldsymbol{\chi}_{soft} \in \boldsymbol{I})]^{-1}$$

$$= [\int_I d\boldsymbol{\chi}_{soft}\, f_{\mathcal{G}}(\boldsymbol{\chi}_{data})]^{-1} \frac{\partial}{\partial \chi_k} [\int_{-\infty}^{\chi_k} d\chi'_k \int_I d\boldsymbol{\chi}_{soft}\, f_{\mathcal{G}}(\chi'_k, \boldsymbol{\chi}_{hard}, \boldsymbol{\chi}_{soft})]$$

$$= [\int_I d\boldsymbol{\chi}_{soft}\, f_{\mathcal{G}}(\boldsymbol{\chi}_{data})]^{-1} \int_I d\boldsymbol{\chi}_{soft}\, f_{\mathcal{G}}(\boldsymbol{\chi}_{map}) \tag{6.2}$$

Since according to Equation 5.6 (p. 106) $f_{\mathcal{G}} = Z^{-1} \exp[\mathcal{Y}_{\mathcal{G}}]$, Equation 6.2 implies that the posterior operator is given by Equation 6.1. Furthermore, from Equation 5.35 (p. 120), the posterior pdf is given by

$$f_{\mathcal{K}}(\chi_k) = A^{-1} \int_I d\boldsymbol{\chi}_{soft}\, f_{\mathcal{G}}(\boldsymbol{\chi}_{map})$$

where $A = \int_I d\boldsymbol{\chi}_{soft}\, f_{\mathcal{G}}(\boldsymbol{\chi}_{data})$ is the normalization parameter. □

Equation 6.1 refers to the updated knowledge after the data (hard and soft) of the meta-prior stage have been incorporated into the analysis. The

$\boldsymbol{I}$-domain corresponds to the subset of data points $\boldsymbol{p}_i$ at which interval (soft) data rather than actual observations are available.

EXAMPLE 6.1: Let $\boldsymbol{\chi}_{soft} = (\chi_4,\ \chi_5,\ \chi_6)$, where $\chi_i \in I_i = [l_i,\ u_i]$, $i = 4,\ 5,\ 6$. Equation 6.1 gives $\mathcal{Y}_S = Z^{-1} \int_{\ell_4}^{u_4} d\chi_4 \int_{\ell_5}^{u_5} d\chi_5 \int_{\ell_6}^{u_6} d\chi_6 \exp[\mathcal{Y}_G]$, and $\boldsymbol{I} = \{I_i\,;\ i = 4,\ 5,\ 6\}$. □

Below we examine a few more cases of $\mathcal{Y}_S$-operators. Unlike Proposition 6.1 in which the probability space of the prior stage was the same as that of the meta-prior stage (conditioning knowledge was known with certainty), in the following proposition the probability space of the soft data at the meta-prior stage is different than the probability space of the prior stage (the conditioning knowledge is uncertain, and the existence of a new probability space will be denoted by the subscript $\mathcal{S}$).

PROPOSITION 6.2: Assume that the specificatory knowledge $\mathcal{S}$ consists of the hard data (Eq. 3.30, p. 84) and the probabilistic (soft) data (Eq. 3.33, p. 86). Then, the posterior operator is given by

$$\mathcal{Y}_S[\boldsymbol{\chi}_{map};\ \boldsymbol{p}_k] = Z^{-1} \int_{\boldsymbol{I}} dF_S(\boldsymbol{\chi}_{soft}) \exp\{\mathcal{Y}_G[\boldsymbol{\chi}_{map};\ \boldsymbol{p}_{map}]\} \tag{6.3}$$

where Z is the partition function.

Proof: Let $\boldsymbol{I} = (I_{m_h+1}, \ldots, I_m)$ be the domain of the soft data vector $\boldsymbol{\chi}_{soft}$ such that $\chi_i \in I_i$ $(i = m_h + 1, \ldots, m)$. The notation $\boldsymbol{x}_{soft} \in \boldsymbol{I}$ denotes an event with respect to the prior probability law $P_G[\boldsymbol{x}_{soft} \in \boldsymbol{I}]$, *i.e.*, before $\mathcal{S}$ is taken into consideration. The notation $\boldsymbol{x}_{soft} \in \boldsymbol{I}(\mathcal{S})$, on the other hand, denotes an event with respect to the new probability law $P_G[\boldsymbol{x}_{soft} \in \boldsymbol{I}(\mathcal{S})]$, *i.e.*, after acquiring $\mathcal{S}$. Let us define

$$f_{\mathcal{K}}(\chi_k) = f_G(\chi_k | \boldsymbol{\chi}_{data}(\mathcal{S})) = \frac{\partial}{\partial \chi_k} F_G(\chi_k | \boldsymbol{\chi}_{data}(\mathcal{S}))$$

where $\boldsymbol{\chi}_{data}(\mathcal{S}) : \boldsymbol{\chi}_{hard}$ and $\boldsymbol{\chi}_{soft} \in \boldsymbol{I}(\mathcal{S})$, see Equation 3.33, so that

$$\begin{aligned} f_{\mathcal{K}}(\chi_k) &= [F_G(\boldsymbol{\chi}_{hard},\ \boldsymbol{\chi}_{soft} \in \boldsymbol{I}(\mathcal{S}))]^{-1} \frac{\partial}{\partial \chi_k} F_G(\chi_k,\ \boldsymbol{\chi}_{hard},\ \boldsymbol{\chi}_{soft} \in \boldsymbol{I}(\mathcal{S})) \\ &= [F_G(\boldsymbol{\chi}_{hard},\ \boldsymbol{\chi}_{soft} \in \boldsymbol{I},\ \boldsymbol{\chi}_{soft} \in \boldsymbol{I}(\mathcal{S}))]^{-1} \\ &\qquad \frac{\partial}{\partial \chi_k} F_G(\chi_k,\ \boldsymbol{\chi}_{hard},\ \boldsymbol{\chi}_{soft} \in \boldsymbol{I},\ \boldsymbol{\chi}_{soft} \in \boldsymbol{I}(\mathcal{S})) \end{aligned} \tag{6.4}$$

Herein, for simplicity, the $\boldsymbol{\chi}_{soft} \in \boldsymbol{I}$ and $\boldsymbol{\chi}_{soft} \in \boldsymbol{I}(\mathcal{S})$ within the probability laws will simply be replaced by $\boldsymbol{I}$ and $\boldsymbol{I}(\mathcal{S})$. Each interval I_i of $\boldsymbol{I}$ is partitioned into a number of mutually exclusive and exhaustive sub-intervals $j_{i,u_i} \subset I_i$, $u_i = 1, \ldots, N_i$. By choosing one j_{i,u_i} from each I_i, we can define all possible combinations of sub-intervals so that $\boldsymbol{R}_q = (j_{m_h+1,u_{m_h+1}}, \ldots, j_{m,u_m})_q$, where $q = 1, \ldots, \rho$. We can thus write

$$F_{\mathcal{G}}(\chi_k, \boldsymbol{\chi}_{hard}, \boldsymbol{I}, \boldsymbol{I}(\mathcal{S}))$$

$$= P_{\mathcal{G}}\left[x_k \leq \chi_k, \boldsymbol{x}_{hard} = \boldsymbol{\chi}_{hard}, \boldsymbol{x}_{soft} \in \bigcup\nolimits_{q=1}^{\rho} \boldsymbol{R}_q, \boldsymbol{x}_{soft} \in \bigcup\nolimits_{q'=1}^{\rho} \boldsymbol{R}_{q'}(\mathcal{S})\right]$$

$$= \sum\nolimits_{q=1}^{\rho} P_{\mathcal{G}}\left[x_k \leq \chi_k, \boldsymbol{x}_{hard} = \boldsymbol{\chi}_{hard}, \boldsymbol{x}_{soft} \in \boldsymbol{R}_q, \boldsymbol{x}_{soft} \in \boldsymbol{R}_q(\mathcal{S})\right]$$

since $\boldsymbol{R}_q \cap \boldsymbol{R}_{q'}(\mathcal{S}) = \emptyset$ for $q \neq q'$. Furthermore, the events $[x_k \leq \chi_k, \boldsymbol{x}_{hard} = \boldsymbol{\chi}_{hard}, \boldsymbol{x}_{soft} \in \boldsymbol{R}_q]$ and $\boldsymbol{x}_{soft} \in \boldsymbol{R}_q(\mathcal{S})$ are independent (they are associated with independently derived probability laws). Therefore, by letting $\rho \to \infty$

$$F_{\mathcal{G}}(\chi_k, \boldsymbol{\chi}_{hard}, \boldsymbol{I}, \boldsymbol{I}(\mathcal{S}))$$

$$= \lim_{\rho \to \infty} \sum\nolimits_{q=1}^{\rho} P_{\mathcal{G}}\left[x_k \leq \chi_k, \boldsymbol{x}_{hard} = \boldsymbol{\chi}_{hard}, \boldsymbol{x}_{soft} \in \boldsymbol{R}_q\right] P_{\mathcal{G}}\left[\boldsymbol{x}_{soft} \in \boldsymbol{R}_q(\mathcal{S})\right]$$

$$= \lim_{\rho \to \infty} \int_{-\infty}^{\chi_k} d\chi'_k \sum\nolimits_{q=1}^{\rho} \int_{\boldsymbol{R}_q} dF_{\mathcal{S}}(\boldsymbol{\chi}_{soft}) f_{\mathcal{G}}(\chi'_k, \boldsymbol{\chi}_{data})$$

$$= \int_{-\infty}^{\chi_k} d\chi'_k \int_{\boldsymbol{I}} dF_{\mathcal{S}}(\boldsymbol{\chi}_{soft}) f_{\mathcal{G}}(\chi'_k, \boldsymbol{\chi}_{data}) \tag{6.5}$$

and in view of Equations 6.4, 5.6, and 5.35, Equation 6.5 yields Equation 6.3. Similarly, it is found that $A = \int_{\boldsymbol{I}} dF_{\mathcal{S}}(\boldsymbol{\chi}_{soft}) f_{\mathcal{G}}(\boldsymbol{\chi}_{data})$. This equation together with Equations 6.4 and 6.5 gives the corresponding posterior pdf. □

The following example serves a specific instructional purpose: it demonstrates the role of combinations $\boldsymbol{R}_q$ in Proposition 6.2.

EXAMPLE 6.2: Assume that soft data $\boldsymbol{\chi}_{soft} = (\chi_1, \chi_2, \chi_3)$ are available of the probabilistic form (Eq. 3.35, p. 86). The intervals I_i of these soft data are partitioned as follows (see Fig. 6.1),

$$\chi_1 \in I_1 = j_{1,1} \cup j_{1,2}, \ \chi_2 \in I_2 = j_{2,1} \cup j_{2,2} \text{ and } \chi_3 \in I_3 = j_{3,1} \tag{6.6}$$

The possible combinations $\boldsymbol{R}_q$, $q = 1, \ldots, 4$ of subsets for $\boldsymbol{\chi}_{soft}$ are

$$\left.\begin{array}{l} \boldsymbol{R}_1 = (j_{1,1}, j_{2,1}, j_{3,1}), \ \boldsymbol{R}_2 = (j_{1,1}, j_{2,2}, j_{3,1}) \\ \boldsymbol{R}_3 = (j_{1,2}, j_{2,1}, j_{3,1}), \ \boldsymbol{R}_4 = (j_{1,2}, j_{2,2}, j_{3,1}) \end{array}\right\} \tag{6.7}$$

i.e., $\boldsymbol{R}_q$ is a vector of sets which belong to the domain of the soft data. Note that the $\boldsymbol{R}_q$'s are mutually exclusive and exhaustive sets. Each $\boldsymbol{R}_q$ is a vector with elements that are subsets (intervals) of the $\boldsymbol{\chi}_{soft}$ domain; so, even though the $\boldsymbol{R}_1$ and $\boldsymbol{R}_4$, *e.g.*, have one interval in common, they do not

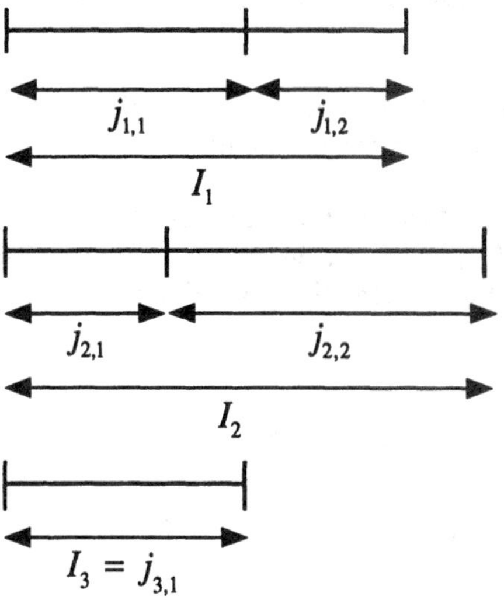

Figure 6.1. Sub-intervals of $\boldsymbol{\chi}_{soft}$.

have a volume in common in the three-dimensional vector space of $\boldsymbol{\chi}_{soft}$. The collection of all combinations of intervals $\boldsymbol{R}_q$ $(q = 1, \ldots, 4)$ constitutes the domain $\boldsymbol{I} = \bigcup_{q=1}^{4} \boldsymbol{R}_q$ of the vector $\boldsymbol{\chi}_{soft}$. □

COMMENT 6.1: *Proposition 6.1 is a special case of Proposition 6.2. Indeed, since the interval (soft) data of Equation 3.32 (p. 85) implies* $d\,F_S\,(\boldsymbol{\chi}_{soft}) = d\boldsymbol{\chi}_{soft}\,U$ *(U denotes uniform pdf), Equation 6.3 reduces to Equation 6.1. In some practical applications, the probability knowledge (Eq. 3.33) may be available for a subset of* $\boldsymbol{\chi}_{soft}$, *while the interval knowledge (Eq. 3.32) is available for the remaining soft data points. In such a case one can obtain the posterior pdf by combining Propositions 6.1 and 6.2. If knowledge is provided about each element of the vector* $\boldsymbol{\chi}_{soft}$, *independently and identically distributed, then we have* $d\,F_S\,(\boldsymbol{\chi}_{soft}) = \prod_{i=m_h+1}^{m} d\,F_S(\chi_i)$.

We already mentioned that the probability models above should always be understood to apply in the appropriate context, which defines the current state of knowledge. In several cases, it may not be possible to articulate an expert's intuition, belief, or evidence $\mathcal{S}$ propositionally. Then, instead of trying to describe the quality of $\mathcal{S}$ itself, one may describe the effects of $\mathcal{S}$ on the expert by saying that after examination of the situation, the expert suggests a specific probability law.

Posterior Operators for Other Forms of Soft Data

Working along the lines of the proofs of Propositions 6.1 and 6.2, the validity of the following two propositions can be demonstrated.

PROPOSITION 6.3: Assume that the specificatory knowledge consists of the hard data (Eq. 3.30, p. 84) and the probabilistic (soft) data of Equation 3.34 (p. 86). Then, the posterior operator is given by

$$\mathcal{Y}_S[\boldsymbol{\chi}_{map};\, \boldsymbol{p}_k] = Z^{-1} \int_{R^1} dF_S(\zeta;\, h) \int_{I(\zeta)} d\boldsymbol{\chi}_{soft} \exp\{\mathcal{Y}_G[\boldsymbol{\chi}_{map};\, \boldsymbol{p}_{map}]\} \tag{6.8}$$

where R^1 is the real line, and $I(\zeta)$ is the domain of $\boldsymbol{\chi}_{soft}$ determined by $h(\boldsymbol{\chi}_{soft}) \leq \zeta$.

PROPOSITION 6.4: Assume that the specificatory knowledge $\mathcal{S}$ consists of the hard data (Eq. 3.30) and the probabilistic (soft) data of Equation 3.35 (p. 86). Then,

$$\mathcal{Y}_S[\boldsymbol{\chi}_{map};\, \boldsymbol{p}_k] = Z^{-1}\Big[\prod_{i=m_h+1}^{m} \sum_{\lambda_i}\Big] p_S(\lambda_{m_h+1}, \dots, \lambda_m) \int_D d\boldsymbol{\chi}_{soft} \exp\{\mathcal{Y}_G[\boldsymbol{\chi}_{map};\, \boldsymbol{p}_{map}]\} \tag{6.9}$$

where $\lambda_i \in \{\chi_i \in I_i,\ \chi_i \in I_i^c\}$, $D = \prod_{i=m_h+1}^{m} e_i$ with $e_i \in \{I_i,\ I_i^c\}$.

Next, Propositions 6.3 and 6.4 above are illustrated with the help of a few examples.

EXAMPLE 6.3: Let $\boldsymbol{\chi}_{soft} = (\chi_1, \chi_2)$ such that

$$\mathcal{S}: f_S(\boldsymbol{\chi}_{soft}) = \begin{cases} \frac{1}{5} & \text{for } \boldsymbol{\chi}_{soft} \in \boldsymbol{R}_1 = \{0 < \chi_1,\ \chi_2 \leq 1\} \\ \frac{2}{5} & \text{for } \boldsymbol{\chi}_{soft} \in \boldsymbol{R}_2 = \{1 < \chi_1,\ \chi_2 \leq 2\} \\ \frac{2}{5} & \text{for } \boldsymbol{\chi}_{soft} \in \boldsymbol{R}_3 = \{2 < \chi_1,\ \chi_2 \leq 3\} \end{cases} \tag{6.10}$$

This is shown diagrammatically in Figure 6.2.
In view of Equation 6.10, the posterior operator is given by

$$\mathcal{Y}_S[\boldsymbol{\chi}_{map};\, \boldsymbol{p}_k] = \tfrac{1}{5} Z^{-1} \int_{R_1} d\boldsymbol{\chi}_{soft} \exp\{\mathcal{Y}_G[\boldsymbol{\chi}_{map};\, \boldsymbol{p}_{map}]\} + \tfrac{2}{5} Z^{-1} \sum_{q=2}^{3} \int_{R_q} d\boldsymbol{\chi}_{soft} \exp\{\mathcal{Y}_G[\boldsymbol{\chi}_{map};\, \boldsymbol{p}_{map}]\} \tag{6.11}$$

As usual, the posterior pdf is as follows: $f_{\mathcal{K}}(\chi_k) = A^{-1}\, \mathcal{Y}_S[\boldsymbol{\chi}_{map}]$. □

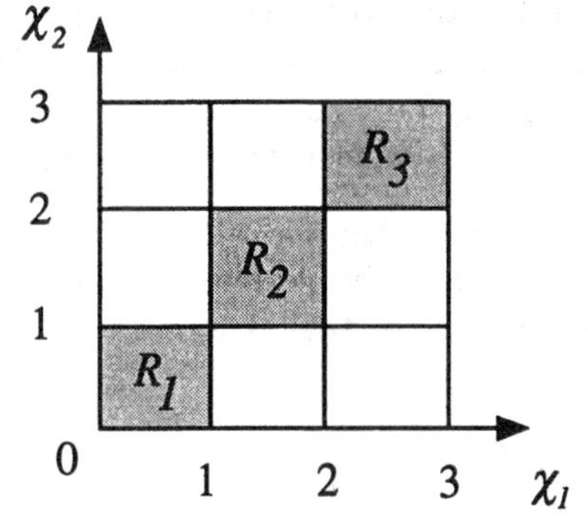

Figure 6.2. The ranges $\boldsymbol{R}_q$ $(q = 1, 2, 3)$ of $\boldsymbol{\chi}_{soft} = (\chi_1, \chi_2)$.

EXAMPLE 6.4: Let $\boldsymbol{\chi}_{soft} = (\chi_1, \chi_2)$ and

$$h(\boldsymbol{\chi}_{soft}) = \chi_1^2 + \chi_2^2 \leq \zeta^2, \quad \zeta \in R^1 \tag{6.12}$$

The specificatory knowledge $\mathcal{S}$ is the following cdf

$$\mathcal{S}: P_S[\chi_1^2 + \chi_2^2 \leq \zeta^2] = F_S(\zeta;\, h) = \begin{cases} \frac{1}{3}\zeta, & 1 \leq \zeta \leq 3 \\ 0, & \text{otherwise} \end{cases} \tag{6.13}$$

The posterior operator is found from Equation 6.8 as follows,

$$\mathcal{Y}_S[\boldsymbol{\chi}_{map};\, \boldsymbol{p}_k] = \tfrac{1}{3}Z^{-1} \int_1^3 d\zeta \int_{I(\zeta)} d\boldsymbol{\chi}_{soft} \exp\left\{\mathcal{Y}_G[\boldsymbol{\chi}_{map};\, \boldsymbol{p}_{map}]\right\} \tag{6.14}$$

where the $\boldsymbol{\chi}_{soft}$ domain is defined by $\boldsymbol{I}(\zeta) = \{(\chi_1, \chi_2) : \chi_1^2 + \chi_2^2 \leq \zeta^2\}$ □

EXAMPLE 6.5: A simple case of Equation 3.35 (p. 86) is when there are only two probabilities $P_S[h(\boldsymbol{x}_{soft}) \leq \zeta] = p_S$ and $P_S[h(\boldsymbol{x}_{soft}) > \zeta] = 1 - p_S$. Then, Equation 6.9 reduces to

$$\begin{aligned}\mathcal{Y}_S[\boldsymbol{\chi}_{map};\, \boldsymbol{p}_k] &= Z^{-1}p_S \int_{I(\zeta)} d\boldsymbol{\chi}_{soft} \exp\left\{\mathcal{Y}_G[\boldsymbol{\chi}_{map};\, \boldsymbol{p}_{map}]\right\} \\ &+ Z^{-1}(1 - p_S) \int_{I^c(\zeta)} d\boldsymbol{\chi}_{soft} \exp\{\mathcal{Y}_G[\boldsymbol{\chi}_{map};\, \boldsymbol{p}_{map}]\}\end{aligned} \tag{6.15}$$

where $\boldsymbol{I}(\zeta) = \{\boldsymbol{\chi}_{soft} : h(\boldsymbol{x}_{soft}) \leq \zeta\}$ and $\boldsymbol{I}^c(\zeta) = \{\boldsymbol{\chi}_{soft} : h(\boldsymbol{x}_{soft}) > \zeta\}$. Consider the simple case $h(\boldsymbol{\chi}_{soft}) = \chi_1 - \chi_2 \leq 1$. Then, Equation 6.15 yields $\mathcal{Y}_S = Z^{-1} \int_{R^1} d\chi_2\{p_S \int_{-\infty}^{1+\chi_2} d\chi_1 \exp[\mathcal{Y}_G] + (1 - p_S) \int_{1+\chi_2}^{\infty} d\chi_1 \exp[\mathcal{Y}_G]\}$, where $\boldsymbol{I}(\zeta) = \{\chi_1 \in (-\infty,\ 1 + \chi_2),\ \chi_2 \in R^1\}$ and $\boldsymbol{I}^c(\zeta) = \{\chi_1 \in (1 + \chi_2,\ \infty),\ \chi_2 \in R^1\}$. □

BME analysis is easily modified to handle various other cases of practical importance arising in physical applications in which soft data are available not only at the data points $\boldsymbol{p}_{soft}$, but also at the estimation point $\boldsymbol{p}_k$ itself (see Example 3.17, p. 86). Such a situation is presented in the following proposition.

PROPOSITION 6.5: Assume that the specificatory knowledge includes the hard data (Eq. 3.30) and the probability law $f_S(\boldsymbol{\chi}_{soft}, \chi_k)$, where $\mathcal{S}_0 : \boldsymbol{\chi}_{soft} \in \boldsymbol{I}$ and $\mathcal{S}_1 : \chi_k \in I_k$ so that $\mathcal{S} = \mathcal{S}_0 \cup \mathcal{S}_1$. Then, the posterior operator has the form

$$\mathcal{Y}_S[\boldsymbol{\chi}_{map}; \boldsymbol{p}_k] = Z^{-1} \int_I d\boldsymbol{\chi}_{soft}\, f_S(\boldsymbol{\chi}_{soft}, \chi_k) \exp\{\mathcal{Y}_G[\boldsymbol{\chi}_{map}]\} \tag{6.16}$$

Similar expressions for the $\mathcal{Y}_S$-operator can be derived in cases where other forms of soft data are available at the estimation points.

EXAMPLE 6.6: From Equation 6.16 one obtains the posterior pdf $f_{\mathcal{K}}(\chi_k) = A^{-1}\mathcal{Y}_S[\boldsymbol{\chi}_{map}; \boldsymbol{p}_k]$, where

$$A = Z^{-1} \int_{I, I_k} d\boldsymbol{\chi}_{soft}\, d\chi'_k\, f_S(\boldsymbol{\chi}_{soft}, \chi'_k) \exp\{\mathcal{Y}_G[\boldsymbol{\chi}_{data}, \chi'_k]\}$$

is the normalization parameter. In the special case that $f_S(\boldsymbol{\chi}_{soft}, \chi_k) = f_{S_0}(\boldsymbol{\chi}_{soft})\, f_{S_1}(\chi_k)$, Equation 6.16 yields the recursive relationship $f_{\mathcal{K}}(\chi_k) = A^{-1} f_{S_1}(\chi_k)\, f_{\mathcal{K}_0}(\chi_k)$, where $f_{\mathcal{K}_0}(\chi_k)$ is the posterior pdf associated with $f_{S_0}(\boldsymbol{\chi}_{soft})$ and $A = \int_{I_k} d\chi'_k\, f_{S_1}(\chi'_k)\, f_{\mathcal{K}_0}(\chi'_k)$. □

Depending on the practical application, various combinations of the preceding results are possible. Furthermore, additional classes of $\mathcal{Y}_S$-operators can be created that incorporate many other kinds of soft data, including the intuition, beliefs, and subjective assessments of laymen. However, these kinds of soft data must be fitted into an organized and coherent system of knowledge before they can attain significance or applicability. Indeed, it is the rich network of physical knowledge bases created by generations of scientists and engineers that gives meaning to such soft data as is occasionally provided by laymen.

Discussion

Certain of the previous results can be summarized in terms of the following useful expressions of the BME posterior operator and pdf

$$\left.\begin{array}{l} f_{\mathcal{K}}(\chi_k) = A^{-1}\mathcal{Y}_S[\boldsymbol{\chi}_{map}; \boldsymbol{p}_k], \\ \text{with } \mathcal{Y}_S[\boldsymbol{\chi}_{map}; \boldsymbol{p}_k] = Z^{-1}B \int_D d\Xi_S(\boldsymbol{\chi}_{soft}) \exp\{\mathcal{Y}_G[\boldsymbol{\chi}_{map}; \boldsymbol{p}_{map}]\} \end{array}\right\} \tag{6.17}$$

where A is the normalization parameter; and the B, $\boldsymbol{D}$ and Ξ_S determine the form of the $\mathcal{Y}_S$-posterior operator (Table 6.1). This form depends, of course, on the specificatory knowledge available.

Table 6.1. Examples of parameters and operators in Equation 6.17.

Soft Data Equation	B	D	Ξ_S	$\mathcal{Y}_S$
3.32	1	I	$\boldsymbol{\chi}_{soft}$	Eq. 6.1
3.33	1	I	$F_S(\boldsymbol{\chi}_{soft})$	Eq. 6.3
3.34	$\int_{R^1} dF_S(\zeta;\, h)$	$I(\zeta)$	$\boldsymbol{\chi}_{soft}$	Eq. 6.8
3.35	$\prod_{i=m_h+1}^{m} \sum_{\lambda_i} p_S(\lambda_{m_h+1}, ..., \lambda_m)$	$\prod_{i=m_h+1}^{m} e_i$	$\boldsymbol{\chi}_{soft}$	Eq. 6.9

COMMENT 6.2: *One should keep in mind that the posterior operator in Equation 6.17 is a mathematical quantity representing our* state of knowledge *of the physical phenomenon under study. Consequently, the* $\mathcal{Y}_S$*-operator has epistemic features that allow it to possess whatever properties the modern geostatistician chooses based on available knowledge, the goals of the study,* etc. *The only requirements are that the corresponding pdf must satisfy the mathematical conditions of a pdf and that the results of any calculations we make with this pdf agree with the physical data, scientific laws,* etc.

The above results certainly do not exhaust all possibilities regarding the $\mathcal{Y}_S$-operators. These results deal with knowledge that can be expressed in terms of the natural variable X. However, knowledge that depends on some other variables Y_1, Y_2, *etc.* can also be studied by BME (*e.g.*, Chapter 9). The reader is encouraged to consider operators that best describe physical knowledge about a scientific or engineering problem of his/her own interest and derive mathematical expressions for the relevant posterior operator. Certain implementations of the BME formulas above could be computationally intensive, especially when multiple integrations over pdf's are involved. However, continuing progress in numerical approximation techniques (Monte Carlo, *etc.*), as well as the fast-developing technology of workstations and parallel computation is expected to handle such computational issues efficiently.

The derivation of the BME posterior pdf (Eq. 6.17) does not involve any of the restrictive assumptions and approximations used by other geostatistical methods, such as the multi-Gaussian and indicator approaches. Limitations of the multi-Gaussian characterization of posterior probability distributions include (Goovaerts, 1997): (a) strong assumptions are made about the multivariate probability distribution, which usually cannot be checked in practice; (b) extremely large and low values are considered spatially uncorrelated, which is often an invalid assumption; (c) the corresponding variance is data-independent. The indicator approach suffers from theoretical and practical limitations such as (Olea, 1999): (*i*) unfeasible values for cdf may be obtained; (*ii*) the derived cdf sometimes fail order-relationship requirements; (*iii*) the large number of indicator thresholds involved require heavy computational and inference efforts (to reduce these efforts approximations are introduced, which may make things worse); (*iv*) the approach can account for only a few cases of imprecise data;

(v) in order to account for hard and soft data, complicated combinations of the various types of kriging techniques are considered, many of which are cumbersome and involve arbitrary approximations while lacking a sound theoretical background; and (vi) variograms for very high and very low thresholds are customarily difficult to model. BME does not need to make any of the above assumptions and approximations and, thus, it does not suffer from any of the limitations of the multi-Gaussian and indicator approaches. In addition, while both of these approaches calculate local (*i.e.*, single-point) probability distributions, an important feature of BME is that it can calculate local as well as global (*i.e.*, multipoint) pdf. Instead of complicated combinations of various types of estimation techniques, BME provides a unified general framework for integrating and processing various kinds of hard and soft data.

COMMENT 6.3: *Some geostatisticians may prefer to concentrate on the purely mathematical formulation of the BME space/time approach. In that case, the five main steps involved in the formal BME approach are:*

(i.) *In light of the general physical knowledge $\mathcal{G}$ available, formulate the corresponding equations of the stochastic moments (see Eq. 3.1, p. 74).*

(ii.) *Assume a pdf of the general form of Equation 5.6 (p. 106) and (depending on the kind of moments involved in Eq. 3.1) select the g_α-functions.*

(iii.) *Substitute Equation 5.6 into Equation 3.1 and solve for the multipliers μ_α. Insert these multipliers back into Equation 5.6 to find the exact form of the $\mathcal{G}$-based pdf of the map.*

(iv.) *In light of the specificatory knowledge $\mathcal{S}$ available, develop the corresponding hard and soft data parameters and operators (p. 82*ff *and Table 6.1, p. 133).*

(v.) *Insert the $\mathcal{S}$-operators of* (iv.) *together with the $\mathcal{G}$-based pdf of* (iii.) *into Equation 5.35 (p. 120) [or, Eq. 6.17, p. 132] to find the $\mathcal{K}$-based pdf of the map. Select appropriate space/time estimates, depending on the goals of the study (see also Chapter 7).*

One may notice that the principle of maximum expected information (Postulate 5.2, p. 106) is not mentioned in the mathematical derivation of the space/time equations according to steps (i.)–(v.) *above. In fact, Equation 5.6 of* (ii.) *may be considered as a reasonable mathematical assumption consistent with the $\mathcal{G}$-knowledge available. Thus, Postulate 5.2 may be viewed by some geostatisticians as optional,* i.e., *to be used only if they wish to provide an epistemic justification for the choice of the pdf form (Eq. 5.6).*

In the following pages, the analytical results we obtained in the preceding sections will be tested by means of synthetic examples in a controlled environment and by real-world case studies. The distinction between the two somehow resembles that between experimental and observational tests: While observational data (or real-world studies) may be fraught with many uncontrolled variables, experiments (or synthetic examples, in this case) have the crucial advantage that the scientist can control most of the variables except the ones that are of particular interest.

7

THE CHOICE OF A SPATIOTEMPORAL ESTIMATE

"Natural science does not simply describe and explain nature, it is part of the interplay between nature and scientists." W. Heisenberg

Versatility of the BME Approach

At the integration (or posterior) stage of the BME analysis, estimates must be determined at the space/time mapping points. BME is a *versatile* approach that allows for a variety of possibilities regarding the choice of the appropriate spatiotemporal estimate at the integration stage. As a matter of fact, an interesting interpretation of the integration stage is obtained in the context of *scientific demonstration*, as the latter is understood in natural sciences. Scientific demonstration—in the wide sense—stands for any experiential evidence that has a large measure of cogency or suasive power relative to a predictive map. In other words, scientific demonstration may be associated with specificatory data that lead to a BME estimate with high posterior probability, or some other desirable feature. Indeed, since the posterior pdf is rigorously determined through the BME analysis, a large number of options become available, depending on the physical, economical, and other characteristics of the application considered. In other words, the BME maps obtained are *case-specific*.

Many people will agree that a *cogent* choice of an estimate is the map that maximizes the posterior pdf. This choice leads to a *BMEmode* map which is described in the following section; a case study involving an extensive porosity data set is discussed in the section entitled "The West Lyons Porosity Field" (p. 143). Other choices include maps that optimize the stochastic expectation (with respect to the posterior pdf) of a function of the natural variable of interest. Typical examples are the *conditional mean* estimate (which minimizes the mean squared estimation error), and the *median* estimate (which minimizes the absolute estimation error). These, as well as other choices, are discussed in the section, "Other BME Estimates," on p. 147. Just as in Chapter 6, the present chapter is also concerned with *single-point* estimation; multipoint mapping is considered in a later chapter.

The BMEmode Estimate

Consider first the estimate $\hat{\chi}_k$ that maximizes the posterior pdf; this is the mode of the BME posterior pdf. For such a choice, the spatiotemporal single-point estimation procedure is summarized as follows.

> **BMEmode mapping concept:** Derive estimates $\hat{\chi}_k$ of a natural variable $X(\boldsymbol{p})$ at space/time points $\boldsymbol{p}_k$ given data (hard and/or soft) at points $\boldsymbol{p}_i$ $(i = 1, \ldots, m;\ i \neq k)$ such that: (a) the expected information (Eq. 5.2, p. 105) is maximized with respect to the pdf $f_{\mathcal{G}}$ subject to the general knowledge base $\mathcal{G}$; and (b) the posterior pdf (Eq. 5.35, p. 120) is maximized with respect to $\chi_k = \hat{\chi}_k$.

The outcome of requirement (a) was already given in Equation 5.6 (p. 106). Formally, maximization of the posterior pdf in requirement (b) involves solving the equation

$$\frac{\partial}{\partial \chi_k} f_{\mathcal{K}}(\chi_k)|_{\chi_k=\hat{\chi}_k} = 0 \tag{7.1}$$

or, in light of Equation 5.35, solving the equation

$$\frac{\partial}{\partial \chi_k} \mathcal{Y}_S[\boldsymbol{\chi}_{map};\ \boldsymbol{p}_k]_{\chi_k=\hat{\chi}_k} = 0 \tag{7.2}$$

The estimate provided by Equation 7.2 is the *mode* of the posterior pdf, for short

$$\hat{\chi}_{k,\,mode} : \max_{\chi_k} f_{\mathcal{K}}(\chi_k) \tag{7.3}$$

Equation 7.2 is, therefore, a basic BMEmode equation. Strictly speaking, in order for the solution of Equation 7.2 to represent a pdf maximum (rather than a minimum or an inflexion point), the latter must be concave in the neighborhood of $\chi_k = \hat{\chi}_{k,\,mode}$, *i.e.*, it must also hold that

$$\frac{\partial^2}{\partial \chi_k^2} f_{\mathcal{K}}(\chi_k)|_{\chi_k=\hat{\chi}_k} < 0 \tag{7.4}$$

Since concavity is a property of the logarithmic function, $\ell og\, f_{\mathcal{K}}$ is sometimes used in place of simply $f_{\mathcal{K}}$.

COMMENT 7.1: *Another way of saying the general BMEmode equation in words is that Equation 7.2 leads to a choice of an estimate* $\hat{\chi}_{k,mode}$ *that adds the minimum possible amount of information to that already extracted from the data* $\boldsymbol{\chi}_{data}$; i.e., *the choice* $\hat{\chi}_{k,mode}$ *should minimize the following* information gain

$$\text{Info}_{\mathcal{K}}[\chi_k] = \text{Info}_{\mathcal{K}}[\boldsymbol{\chi}_{map}] - \text{Info}_{\mathcal{K}}[\boldsymbol{\chi}_{data}] \tag{7.5}$$

Equation 7.2 leads to a BME mapping equation, the form of which depends on the form of the $\mathcal{Y}_S$-operator assumed (and, thus, the hard and soft data used). The following examples analyze a few common cases of BMEmode mapping.

EXAMPLE 7.1: By substituting Equations 5.35 (p. 120) and 6.1 (p. 126) into Equation 7.1 we find

$$\int_I d\boldsymbol{\chi}_{soft} \{\exp[\mathcal{Y}_G] \frac{\partial}{\partial \chi_k} \mathcal{Y}_G\}_{\chi_k = \hat{\chi}_k} = 0 \tag{7.6}$$

By substituting Equations 5.35 and 6.3 (p. 127) into Equation 7.1 we get

$$\int_I dF_S(\boldsymbol{\chi}_{soft}) \{\exp[\mathcal{Y}_G] \frac{\partial}{\partial \chi_k} \mathcal{Y}_G\}_{\chi_k = \hat{\chi}_k} = 0 \tag{7.7}$$

By substituting Equations 5.35 and 6.8 (p. 130) into Equation 7.1 we find

$$\int_{R^1} dF_S(\zeta;\, h) \int_{I(\zeta)} d\boldsymbol{\chi}_{soft} \{\exp[\mathcal{Y}_G] \frac{\partial}{\partial \chi_k} \mathcal{Y}_G\}_{\chi_k = \hat{\chi}_k} = 0 \tag{7.8}$$

and by substituting Equations 5.35 and 6.9 (p. 130) into Equation 7.1 we obtain

$$\left[\prod_{i=m_h+1}^{m} \sum_{\lambda_i}\right] p_S(\lambda_{m_h+1}, \ldots, \lambda_m) \int_D d\boldsymbol{\chi}_{soft} \{\exp[\mathcal{Y}_G] \frac{\partial}{\partial \chi_k} \mathcal{Y}_G\}_{\chi_k = \hat{\chi}_k} = 0 \tag{7.9}$$

All the above equations must be solved for $\hat{\chi}_k = \hat{\chi}_{k,mode}$. □

The results of Equations 7.6–7.9 above are summarized by means of the following proposition.

PROPOSITION 7.1: For the operators and parameters considered in Table 6.1 (p. 133), the BMEmode equation is as follows

$$B \int_D d\Xi_S(\boldsymbol{\chi}_{soft}) \{\exp \mathcal{Y}_G[\boldsymbol{\chi}_{map};\, \boldsymbol{p}_k] \frac{\partial}{\partial \chi_k} \mathcal{Y}_G[\boldsymbol{\chi}_{map};\, \boldsymbol{p}_{map}]\}_{\chi_k = \hat{\chi}_k} = 0 \tag{7.10}$$

where the B, D, and Ξ_S determine the form of the $\mathcal{Y}_S$ posterior operator considered, as in Equation 6.17 (p. 132).

The BME equation (Eq. 7.10) is a concise, general, and in some sense beautiful representation of the spatiotemporal mapping problem. In addition, the solution of the BME equation can be accomplished efficiently with the current computing technology (in many cases the computational time required is only a fraction of that needed for space/time regression methods; *e.g.*, Serre *et al.*, 1998). As we shall see later, the BME equation (Eq. 7.10) can be extended to include several estimation points p_{k_j} $(j = 1, \ldots, \rho)$ simultaneously (*multipoint* mapping). BME Equations 7.6–7.10 are, of course, all mathematical consequences of the general form of Equation 5.35 (p. 120). But there are a few aspects here that are not obvious to the unaided intuition. The BME equations, *e.g.*, include a mechanism that allows them to distinguish between hard (more accurate) data and soft (less accurate) data, and then assign the appropriate weight to them. Furthermore, it should be remembered that, at this stage, the Lagrange multipliers μ_α have known values which are determined from the solution of the system of Equations 5.10 and 5.11 (p. 107) and incorporate general knowledge $\mathcal{G}$. Equation 7.10 is, in general, a nonlinear equation of the estimate $\hat{\chi}_k$, and may have more than one solution that includes more than one local maximum. In this case, the estimate is equal to the largest local maximum of the pdf. The verification of the condition in Equation 7.4 and the search for the largest local maximum can be accomplished by analytical means or numerical procedures. In order to study such aspects in more detail, as well as to obtain explicit analytical expressions and numerical results for the BME estimators, we will focus mainly on Equations 7.6 and 7.7 in the following examples. Of course, the analysis can be extended to any other $\mathcal{Y}_S$-operator, as well.

Statistics—Hard and soft data

The next example serves to illustrate the step-by-step implementation of the BMEmode approach in light of statistical knowledge and hard/soft data.

EXAMPLE 7.2: Consider the simple but instructive case of three points p_1, p_2, and p_k. It is assumed that the prior knowledge includes the mean and the centered ordinary covariance. Also, assume that there is a hard datum (measurement) at point p_1 and a soft datum (interval) at point p_2. Based on this knowledge, an estimate is sought at the point p_k. The constraint functions g_α $(\alpha = 0, 1, \ldots, 9)$ are shown in Table 7.1. The Lagrange multipliers μ_α should typically be found from the solution of the system of Equations 5.10 and 5.11, which in this case can be written as

$$\left.\begin{array}{l} Z = \int d\boldsymbol{\chi}_{map} \exp[\mathcal{Y}_G] \\ \overline{x_i} = Z^{-1} \int d\boldsymbol{\chi}_{map}\, \chi_i \exp[\mathcal{Y}_G] \\ \sigma_i^2 = Z^{-1} \int d\boldsymbol{\chi}_{map} (\chi_i - \overline{x_i})^2 \exp[\mathcal{Y}_G], \quad i = 1, 2, k \\ c_{ij} = Z^{-1} \int d\boldsymbol{\chi}_{map} (\chi_i - \overline{x_i})(\chi_j - \overline{x_j}) \exp[\mathcal{Y}_G], \quad i \neq j = 1, 2, k \end{array}\right\} \quad (7.11)$$

Table 7.1. The g_α functions.

α	g_α	$\overline{g_\alpha}$
	Normalization constraint	
0	$g_0 = 1$	$\overline{g_0} = 1$
	Mean constraints	
1	$g_1(\chi_1) = \chi_1$	$\overline{g_1} = \overline{x_1}$
2	$g_2(\chi_2) = \chi_2$	$\overline{g_2} = \overline{x_2}$
3	$g_3(\chi_k) = \chi_k$	$\overline{g_3} = \overline{x_k}$
	Covariance constraints	
4	$g_4(\chi_1, \chi_1) = (\chi_1 - \overline{x_1})^2$	$\overline{g_4} = \sigma_1^2$
5	$g_5(\chi_2, \chi_2) = (\chi_2 - \overline{x_2})^2$	$\overline{g_5} = \sigma_2^2$
6	$g_6(\chi_k, \chi_k) = (\chi_k - \overline{x_k})^2$	$\overline{g_6} = \sigma_k^2$
7	$g_7(\chi_1, \chi_2) = (\chi_1 - \overline{x_1})(\chi_2 - \overline{x_2})$	$\overline{g_7} = c_{12}$
8	$g_8(\chi_1, \chi_k) = (\chi_1 - \overline{x_1})(\chi_k - \overline{x_k})$	$\overline{g_8} = c_{1k}$
9	$g_9(\chi_2, \chi_k) = (\chi_2 - \overline{x_2})(\chi_k - \overline{x_k})$	$\overline{g_9} = c_{2k}$

where $d\boldsymbol{\chi}_{map} = \prod_{i=1,2}^{k} d\chi_i$ and the exponent $\mathcal{Y}_G = \mathcal{Y}_G[\chi_1, \chi_2, \chi_k] = \sum_{\alpha=1}^{9} \mu_\alpha\, g_\alpha(\chi_1, \chi_2, \chi_k)$ is given by

$$\mathcal{Y}_G[\chi_1, \chi_2, \chi_k] = \sum_{i=1}^{2,k} \mu_i\, \chi_i + \sum_{i,j=1}^{2,k} \mu_{ij}(\chi_i - \overline{x_i})\,(\chi_j - \overline{x_j}) \tag{7.12}$$

in which the Lagrange multipliers μ_{ij} $(= \mu_{ji})$ are directly mapped onto the constraints of Table 7.1. This becomes more obvious using the following notation: $\mu_{11} = \mu_4$, $\mu_{22} = \mu_5$, $\mu_{kk} = \mu_6$, $\mu_{12} = \mu_7$, $\mu_{1k} = \mu_8$, and $\mu_{2k} = \mu_9$. Also, $\mu_1 = \mu_2 = \mu_3 = 0$ and Equation 7.11 lead to the following expressions for the remaining multipliers

$$\begin{bmatrix} \mu_{11} & \mu_{12} & \mu_{1k} \\ \mu_{21} & \mu_{22} & \mu_{2k} \\ \mu_{k1} & \mu_{k2} & \mu_{kk} \end{bmatrix} = -\tfrac{1}{2} \begin{bmatrix} \sigma_1^2 & c_{12} & c_{1k} \\ c_{21} & \sigma_2^2 & c_{2k} \\ c_{k1} & c_{k2} & \sigma_k^2 \end{bmatrix}^{-1}$$

$$= -\frac{1}{2\,|\boldsymbol{c}|} \begin{bmatrix} \sigma_2^2\sigma_k^2 - c_{2k}^2 & c_{k1}c_{k2} - \sigma_k^2 c_{21} & c_{21}c_{k2} - \sigma_2^2 c_{k1} \\ c_{1k}c_{2k} - \sigma_k^2 c_{12} & \sigma_1^2\sigma_k^2 - c_{k1}^2 & c_{k1}c_{12} - \sigma_1^2 c_{k2} \\ c_{12}c_{2k} - \sigma_2^2 c_{1k} & c_{1k}c_{21} - \sigma_1^2 c_{2k} & \sigma_1^2\sigma_2^2 - c_{12}^2 \end{bmatrix} \tag{7.13}$$

where $|\boldsymbol{c}| = c_{1k}(c_{21}c_{k2} - \sigma_2^2 c_{k1}) - c_{2k}(\sigma_1^2 c_{k2} - c_{12}c_{k1}) + \sigma_k^2(\sigma_1^2\sigma_2^2 - c_{12}^2)$.

Estimating the value of the variable at p_k is equivalent to solving the BMEmode of Equation 7.6, which in this case reduces to

$$\int d\chi_2 \{\exp[\mathcal{Y}_G] \frac{\partial}{\partial \chi_k} [\sum_{i=1}^{2,k} c_{ii}^{-1}(\chi_i - \overline{x_i})^2 + \sum_{\substack{i,j=1\\ i\neq j}}^{2,k} c_{ij}^{-1}(\chi_i - \overline{x_i})(\chi_j - \overline{x_j})]\}_{\chi_k = \hat{\chi}_k} = 0, \quad \text{or}$$

$$c_{kk}^{-1}(\hat{\chi}_k - \overline{x_k}) + c_{1k}^{-1}(\chi_1 - \overline{x_1}) + c_{2k}^{-1} \frac{\int d\chi_2 (\chi_2 - \overline{x_2}) \exp \mathcal{Y}_G[\chi_1, \chi_2, \hat{\chi}_k]}{\int d\chi_2 \exp \mathcal{Y}_G[\chi_1, \chi_2, \hat{\chi}_k]} = 0 \tag{7.14}$$

where the integration is over the range $[\chi_2^l, \chi_2^u]$ of the soft datum. After the evaluation of the partial derivatives and the substitution of the c_{ij}^{-1}'s from Equation 7.13, Equation 7.14 gives

$$(\sigma_1^2\sigma_2^2 - c_{12}^2)(\hat{\chi}_k - \overline{x_k}) + (c_{21}c_{k2} - \sigma_2^2 c_{k1})(\chi_1 - \overline{x_1}) + (c_{k1}c_{k2} - \sigma_1^2 c_{k2}) \frac{\int_{\chi_2^l}^{\chi_2^u} d\chi_2(\chi_2 - \overline{x_2}) \exp[\mathcal{Y}_G(\chi_1, \chi_2, \hat{\chi}_k)]}{\int_{\chi_2^l}^{\chi_2^u} d\chi_2 \exp[\mathcal{Y}_G(\chi_1, \chi_2, \hat{\chi}_k)]} = 0 \tag{7.15}$$

The desired BMEmode estimate $\hat{\chi}_k$ is obtained as the solution to Equation 7.15. □

As was mentioned in previous chapters, BME formalism is very general, and one has considerable freedom in choosing covariance, variogram, and generalized covariance models. Homogeneous/stationary or nonhomogeneous/ nonstationary, separable or nonseparable, *etc.* models can be used depending on one's understanding of the basic features of the problem (see also "Spatiotemporal Covariance and Variogram Models" in Chapter 11, p. 224). The next numerical example is concerned with the effect of incorporating *skewness* into the mapping calculations.

EXAMPLE 7.3: Based upon the spatiotemporal hard data configuration of Figure 7.1, $\chi^{(\ell)}(p_k)$-realizations ($\ell = 1, \ldots, 1000$) were generated assuming the same statistics as in Example 12.7 (p. 238). For each one of these realizations, the BMEmode estimate $\hat{\chi}_{BME}^{(\ell)}(p_k)$ was obtained at point p_k, assuming various skewness values (*i.e.*, $0, 2, 3$, and 3.5) as prior knowledge. Then, the corresponding estimation errors $e_{BME}^{(\ell)}(p_k) = \chi^{(\ell)}(p_k) - \chi_{BME}^{(\ell)}(p_k)$ ($\ell = 1, ..., 1000$) were calculated. The pdf's of these estimation errors are plotted in Figure 7.2. The error pdf's change considerably as the skewness values vary, thus showing the importance in estimation accuracy of the skewness values incorporated by BME. □

Furthermore, Bogaert *et al.* (1999) have suggested a computational technique that incorporates into BME analysis information available about the non-Gaussian *shape* of the univariate pdf of the random field. This technique involves a suitable transformation of the posterior pdf to the Gaussian space. Then, the pdf is back-transformed to the original space using a smoothed estimate of the transformation.

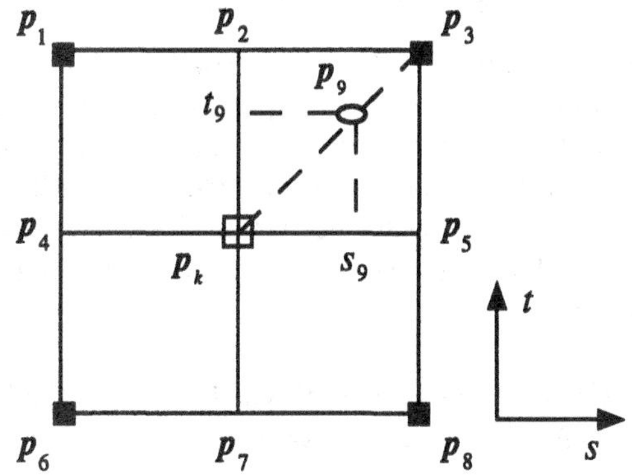

Figure 7.1. Data configuration in $R^1 \times T$; measurements (hard data) are available at points p_1 through p_8; soft datum available at point p_9. Estimates are sought at point p_k.

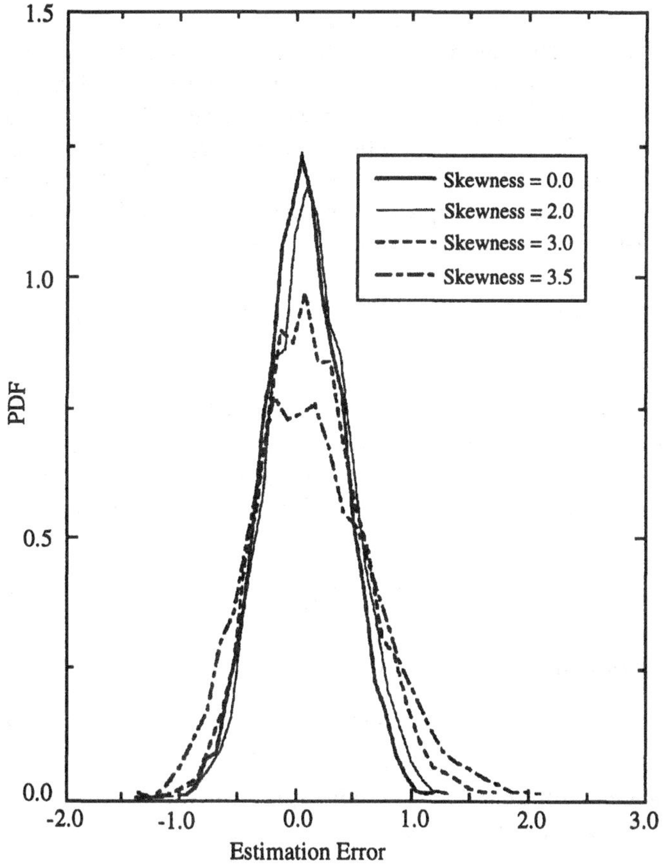

Figure 7.2. Plots of pdf of BME estimation error showing the effect of incorporating knowledge about skewness into estimation.

Physical laws—Hard and soft data

As we discussed in Chapter 3 ("General knowledge in terms of physical laws," p. 76), an important general knowledge base is expressed in terms of laws of nature (physical, biological, *etc.*). In this section, therefore, we present some analytical and numerical results in which the BME estimates obtained incorporate this sort of physical knowledge. The first example is concerned with a situation in which the physical law has the form of a stochastic ordinary differential equation.

EXAMPLE 7.4: Consider the general knowledge base in the form of the physical law (Eq. 3.16, p. 79). As we saw in Example 5.4 (p. 112–114), the corresponding $\mathcal{Y}_{\mathcal{G}}$-operator is given by

$$\mathcal{Y}_{\mathcal{G}}[\boldsymbol{\chi}_{map};\, \boldsymbol{t}_{map}] = \sum_{\alpha=1}^{m,k} \sum_{\lambda=1}^{2} \mu_{\alpha,\lambda}(t_\alpha)\, \chi_\alpha^\lambda \tag{7.16}$$

where $\boldsymbol{t}_{map} = (t_1, \ldots, t_m, t_k)$ and $\mu_{\alpha,\lambda}(t_\alpha)$ ($\alpha = 1, \ldots, m, k$ and $\lambda = 1, 2$) are the solutions of Equations 5.20 and 5.21 (p. 113). Also, assume that there are hard data at times $\boldsymbol{t}_{hard} = (t_1, \ldots, t_{m_h})$ and interval (soft) data at times $\boldsymbol{t}_{soft} = (t_{m_h+1}, \ldots, t_m)$. Based on this knowledge, an estimate is sought at the future time t_k. The BMEmode estimate at time t_k is simply the solution of Equation 7.6 above, where $\mathcal{Y}_{\mathcal{G}}$ is given by Equation 7.16. For illustration purposes, let us consider the simple case in which $\alpha = k$, so that $\chi_\alpha = \chi_k$. In this case, as we saw in Example 5.4, the $\mathcal{Y}_{\mathcal{G}}$ is given by

$$\mathcal{Y}_{\mathcal{G}}[\chi_k;\, t_k] = \overline{X(0)}\{\sigma_x^2(0)\exp[bt_k]\}^{-1}\chi_k - \{2\sigma_x^2(0)\exp[2bt_k]\}^{-1}\chi_k^2 \tag{7.17}$$

The moments involved in the prior stage were implied by the physical law and did not need to be calculated experimentally from the data. The BMEmode estimate (which, due to the symmetry of the pdf, is the same as the BME conditional mean estimate) is the solution of the following equation:

$$\frac{\partial}{\partial \chi_k} \mathcal{Y}_{\mathcal{G}}[\chi_k;\, t_k\,]\,|_{\chi_k=\hat{\chi}_k} = 0 \tag{7.18}$$

which, in view of Equation 7.17, gives

$$\hat{\chi}_k(t_k) = \overline{X(0)}\, \exp[b\, t_k] \tag{7.19}$$

As should be expected, Equation 7.19 is in agreement with the mean solution of the physical law in Equation 3.16 (p. 79). □

Furthermore, some numerical results involving Darcy's law of groundwater flow are discussed in the following example.

EXAMPLE 7.5: Serre and Christakos (1999b) considered Darcy's law governing one-dimensional groundwater flow,

$$K(s)\,\frac{dH(s)}{ds} = -q(s) \tag{7.20}$$

The specific discharge $q(s)$ was assumed to be deterministic; the hydraulic head $H(s)$ and the hydraulic conductivity $K(s)$ were considered as random fields; and the value of the hydraulic head was assumed known at the spatial origin $s = 0$. Then, a $K(s)$ profile was generated that had an asymmetric distribution. On the basis of this profile and Darcy's law, the actual head fluctuation profile $H(s) - \overline{H(s)}$ was calculated and compared to the estimated profiles as follows. In Figure 7.3a the estimated head fluctuation profile was derived from simple kriging (SK) using only hard head data; the actual head fluctuation profile is also shown for comparison. Note the poor SK estimates at unobserved locations. The head fluctuation profile in Figure 7.3b was obtained from BME using hard and soft (interval) head data as well as Darcy's law. Note that the BME estimated profile is a substantial improvement over the classical SK method. Indeed, by being able to incorporate Darcy's law, the BME method allowed us to account for hard and soft (interval) hydraulic conductivity data, as well. Serre and Christakos (1999b) also considered the problem of estimating the hydraulic resistivity profile using sparse head and resistivity measurements (this is sometimes called the *inverse* problem; Kitanidis and Vomvoris, 1983). Again, the BME approach led to a considerably more accurate estimate of the hydraulic resistivity profile than the kriging techniques. □

The West Lyons Porosity Field

Porosity data were collected in the West Lyons field in west-central Kansas (Olea, 1999). This comprises a spatial data set on a reservoir occurring in Mississippian (Lower Carboniferous) sediments deposited in the shallow epicontinental seas that covered much of North America in the Late Paleozoic. The study area is approximately 2.5 × 4.5 miles^2. A total of 76 data values were available, as shown in Figure 7.4. The general knowledge considered consists of the porosity mean and covariance functions, also plotted in Figure 7.4.

COMMENT 7.2: *In some practical applications, hard data may be involved in both the prior stage (indirectly) and the meta-prior stage (directly) of BME. Particularly in cases in which a physical model is not available, the prior constraints (*e.g.*, mean and covariance) may be calculated from hard data sets alone. Thus, at the prior stage, hard data provide a partial characterization of the underlying random field—since there are several realizations sharing the same statistics. Hard data considered at the meta-prior stage are used in the subsequent integration stage, this time with the purpose of providing the desired posterior estimate. This double role of hard*

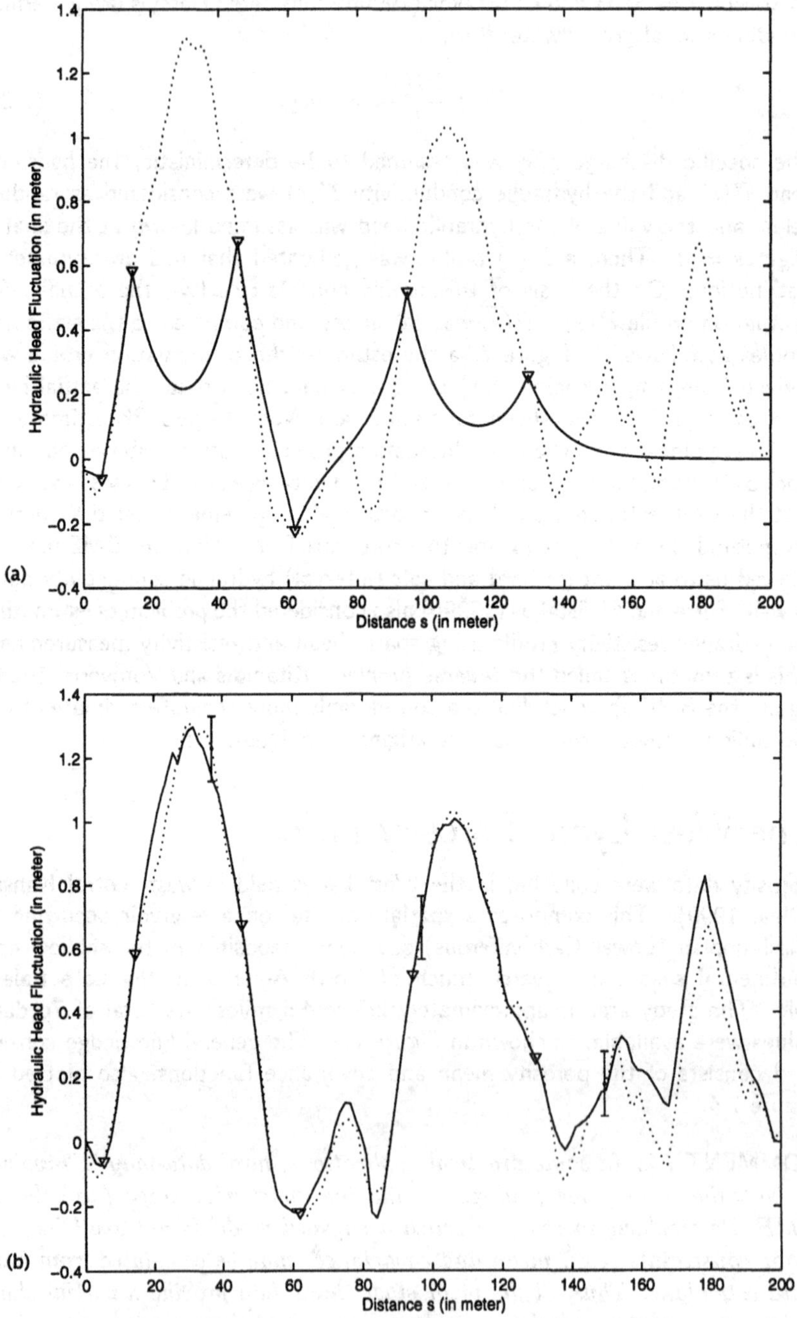

Figure 7.3. Estimated head fluctuation profiles obtained from: (a) SK using hard head data (triangles); (b) BME using hard and soft (interval) head data and Darcy's law. Dotted lines represent the actual head fluctuation profile.

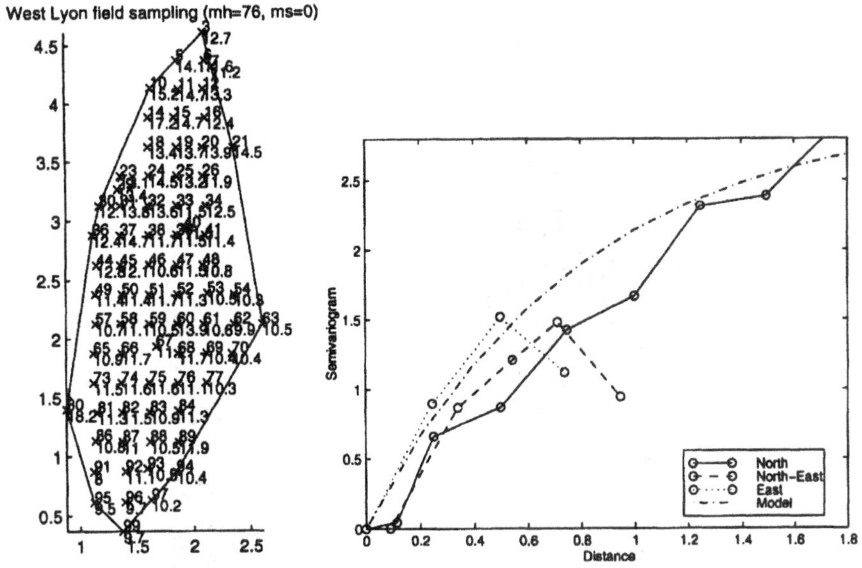

Figure 7.4. The complete porosity data set (porosity values in percent; data available at locations denoted by "x") is shown at left; on the right, the covariances of the porosity data (distances in miles).

data is common in kriging methods too (Davis, 1986; Cressie, 1991). The hard data are first used to calculate the statistics, which are then inserted into the kriging system to obtain the kriging weights; finally, a weighted summation of the same data is used to obtain the kriging estimate (see also "circular problem of geostatistics" in Comment 5.4, p. 110).

Using all 76 hard data, BMEmode produces the porosity map in Figure 7.5. [As we shall see in Chapter 12, Example 12.8 (p. 239), this is, in fact, the map that is obtained by simple kriging (SK) using the same hard data, mean and covariance model.] Now, suppose that at the beginning of this study we wanted to save some money (drill fewer wells, *etc.*) and, as a consequence, we decided to collect hard data at only 56 locations. The hydrogeologic knowledge of the region provided interval (soft) data of varying widths at the remaining 20 locations. The BME approach can take into consideration both the hard data and the soft data in an efficient manner, thus producing the map in Figure 7.6. It is worth noticing that, despite the uncertainty introduced by the soft data, the map of Figure 7.6 closely resembles the spatial structure of the map of Figure 7.5 at a considerably lower cost (fewer wells need to be installed). The analysis above deserves some additional comments. The problem-solving power of BME comes from the knowledge it processes, and not only from the mathematical formalisms and inference schemes it uses. At the prior stage, the pdf is assigned in a way that is consistent with the general knowledge $\mathcal{G}$. This involves the maximization of the expected prior

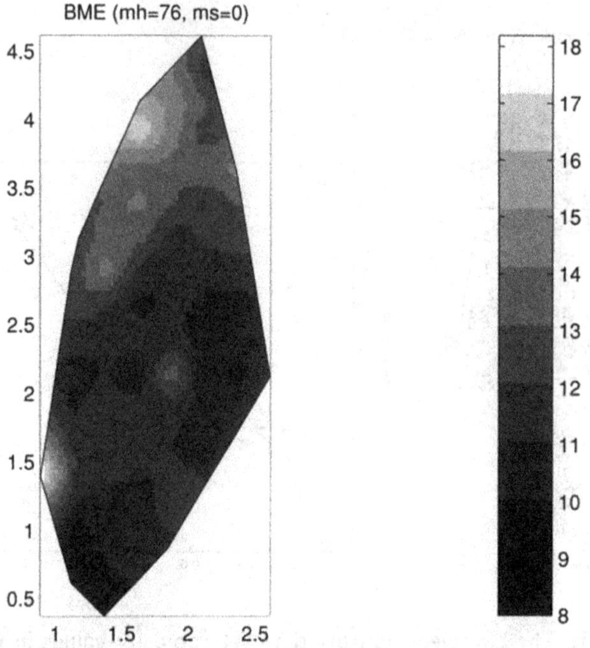

Figure 7.5. BMEmode porosity map using 76 porosity data (%).

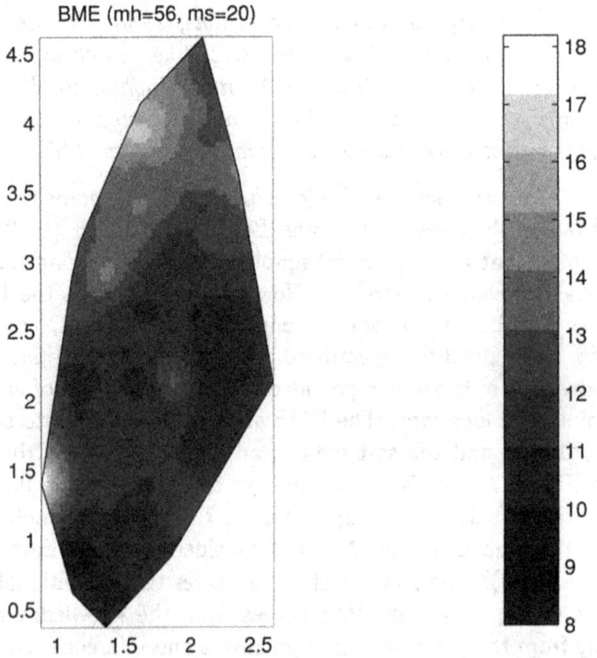

Figure 7.6. BMEmode porosity map using 56 hard and 20 soft data (%).

information subject to constraints representing general knowledge. The integration stage leads to the maximization of the posterior pdf which incorporates the specificatory knowledge $\mathcal{S}$. As a consequence, the BME map offers a body of information as well as a point of view. Surely, BME's efficiency in practical situations depends on the amount and quality of the physical knowledge objectively available which, however, does not need to be limited to the forms processed by the methods of classical geostatistics.

Other BME Estimates

Since the posterior pdf is available through the BME approach, in addition to the mode estimate, several other choices of an estimate $\hat{\chi}_k$ are possible as well (see Fig. 7.7). In fact, as was emphasized in a previous section, the choice of the appropriate estimate is *case specific* rather than universal (*i.e.*, it depends on the physical variables involved in the specific case study, the economic and topographic constraints, personal preferences, multiple objectives, *etc.*). Given the posterior pdf $f_{\mathcal{K}}(\chi_k)$, one may choose as an estimate the median (*BMEmedian*)

$$\hat{\chi}_{k,\, median} = F_{\mathcal{K}}^{-1}(0.5;\ \boldsymbol{p}_k) \tag{7.21}$$

at each point $\boldsymbol{p}_k$; or, the conditional mean (*BMEmean*)

$$\hat{\chi}_{k,\, mean} = \overline{X(\boldsymbol{p}_k) \mid \boldsymbol{\chi}_{data}(\mathcal{S})} = \int d\chi_k\, \chi_k\, f_{\mathcal{K}}(\chi_k) \tag{7.22}$$

at each point $\boldsymbol{p}_k$. The BMEmedian estimate may be used when one is interested in assigning equal weights to the underestimation and overestimation cases. The BMEmean estimate is suitable, *e.g.*, for mapping situations where one seeks to penalize large errors more than smaller ones. *Percentiles*, *quantiles*, *etc.* may also be defined from the posterior pdf.

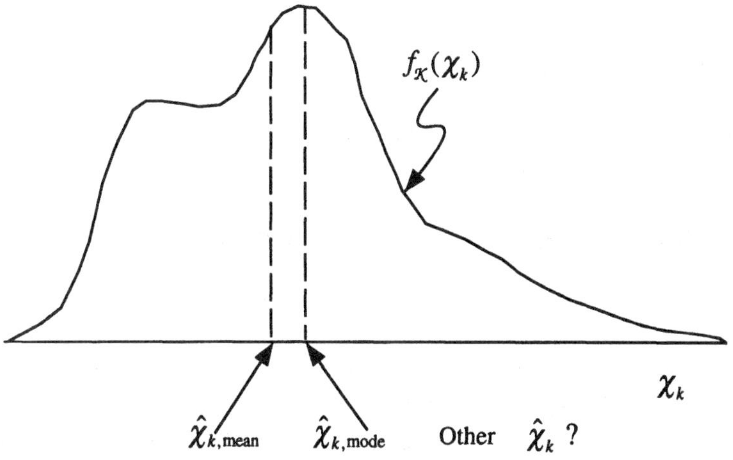

Figure 7.7. Choosing a BME estimate from the posterior pdf.

A general way to search for an appropriate estimate is by *optimizing* (with respect to the estimate sought) the expected value of some function of the natural variable and its estimate at each point $\boldsymbol{p}_k$, *i.e.*, a function of the form

$$\overline{L}(\boldsymbol{p}_k) = \int d\chi_k \, L(\chi_k, \hat{\chi}_k; \boldsymbol{p}_k) \, f_{\mathcal{K}}(\chi_k) \tag{7.23}$$

where the shape of the L-function depends on the physical, economical, and other features of the situation considered. Assume, *e.g.*, that L is some sort of a *loss function* of the estimation error $x_k - \hat{x}_k$, specific to the problem at hand. Then, minimization of Equation 7.23 with respect to $\hat{\chi}_k$ leads to the following integral equation

$$\int d\chi_k \frac{\partial L(\chi_k - \hat{\chi}_k; \boldsymbol{p}_k)}{\partial \hat{\chi}_k} f_{\mathcal{K}}(\chi_k) = 0 \tag{7.24}$$

where the required conditions for a minimum are assumed valid. A variety of estimators can be derived from Equation 7.24, as is demonstrated in the following example.

EXAMPLE 7.6: If the loss function L is chosen to be equal to the absolute estimation error, the estimate obtained from Equation 7.24 is the BMEmedian (Eq. 7.21). Assuming that L is of the quadratic error form, the corresponding estimate is the BMEmean (Eq. 7.22). Also, if L has the form of a (0, 1)-loss function, Equation 7.24 gives the BMEmode estimate (Christakos, 1992). □

Note that in the case of a symmetric unimodal posterior pdf, the estimates (Eqs. 7.3, 7.21, and 7.22) coincide. This fact can be used to significantly improve the computational efficiency of the BME approach. In Chapter 8 (p. 155), BMEmode and BMEmean estimates are calculated for a real data set representing water-level elevations in the Equus Beds aquifer in the State of Kansas.

A Matter of Coordination

By way of a summary, one can argue that the choice of a spatiotemporal estimate (mode, median, mean, *etc.*) in BME analysis depends on the successful *coordination* of four essential factors:

1. the satisfactory *track record* of the specific kind of estimate with similar situations;
2. the *theoretical background* that promotes its use against other possible estimates;
3. the *explanatory rationale* provided by the estimate; and
4. the case-specific *objectives* of the map.

The above approach is consistent with the fact that BME follows the hypothetico-deductive conception of scientific mapping discussed in Chapter 5 (p. 122), rather than the linear (or pure inductive) paradigm.

8

UNCERTAINTY ASSESSMENT

"Chance is beloved of Art, and Art of Chance." Aristotle

Mapping Accuracy

One of the hallmarks of geostatistical analysis is the determination and reporting of the *uncertainty* attached to our scientific conclusions. This statement reflects the fact that due to incomplete knowledge, what we actually determine is a posterior pdf for different realizations of the natural variable. In the stochastic description of a natural variable in which uncertainty plays a major role, it is often important to supplement our pdf-based estimates with a summary assessment of the *mapping accuracy*. The uncertainty of the BMEmode estimate, *e.g.*, is a measure of the dispersion of values around the maximum of the posterior pdf. Assessing the accuracy of maps representing the spatiotemporal distribution of natural variables is also of considerable importance to subsequent stages of the scientific or engineering process related to, *e.g.*, environmental risk assessment, multi-objective optimization, and decision making.

The general formula for the posterior pdf is Equation 6.17 (p. 132), which gives the pdf at each spatiotemporal point $\boldsymbol{p}_k$. The definition of the mapping accuracy depends on the shape of the pdf. Usually, it can be defined in a straightforward manner when the posterior pdf has a single maximum. If this is not the case, the estimation uncertainty should be considered separately for each maximum. In the case of a pdf with a single maximum, one can further distinguish between symmetric and asymmetric cases. If the posterior pdf is symmetric around the maximum, then the maximum coincides with the mean, *etc.* In more complex cases, we may need to consider the complete picture provided by the BME pdf.

Symmetric Posteriors

For many single maximum pdf's, a measure of the accuracy of the BME estimate $\hat{\chi}_k$ may be obtained by means of the *standard deviation* at each point $\boldsymbol{p}_k$, *i.e.*,

$$\sigma_x(\boldsymbol{p}_k) = \mathrm{StDev}\left[f_{\mathcal{K}}(\chi_k)\right] \tag{8.1}$$

Equation 8.1 offers a particularly good measure of mapping accuracy in cases of symmetric or approximately symmetric pdf. In these cases, we can also write

$$\sigma_x(\boldsymbol{p}_k) = \left\{ \overline{\left[X(\boldsymbol{p}_k) - \hat{X}(\boldsymbol{p}_k)\right]^2} \right\}^{1/2} = \left[\int d\chi_k \, (\chi_k - \hat{\chi}_k)^2 \, f_{\mathcal{K}}(\chi_k)\right]^{1/2} \tag{8.2}$$

where $\hat{\chi}_k = \hat{\chi}_{k,\,mode} = \hat{\chi}_{k,\,mean}$ (due to symmetry), and the expectation is now with respect to the posterior pdf rather than a realization average (as is the case with Eq. 8.5 below). Equation 8.2 is an accuracy measure typically reported by traditional mapping methods. As numerical simulations show, the $\sigma_x(\boldsymbol{p}_k)$ of BME analysis does an excellent job in approximating the actual estimation error. On the basis of the estimation uncertainty (Eq. 8.2), confidence intervals can be defined. In the case, *e.g.*, of a Gaussian pdf there is a 95% confidence that $X(\boldsymbol{p}_k)$ lies in the interval $\hat{\chi}_k \pm 1.96\,\sigma_x(\boldsymbol{p}_k)$.

COMMENT 8.1: *Traditionally, confidence intervals are taken such that the probability of the estimated value falling on the left part of the interval is equal to the probability of it falling on the right part. The BME confidence intervals are as small as possible. BME confidence intervals are calculated in the study of the Equus Beds aquifer later in this chapter (p. 155).*

EXAMPLE 8.1: Consider the case of a symmetric $f_{\mathcal{K}}(\chi_k)$, such that a second-order Taylor series expansion of $\log f_{\mathcal{K}}(\chi_k)$ is a good approximation within a neighborhood $|\chi_k - \hat{\chi}_k| < \varepsilon$ around $\chi_k = \hat{\chi}_k$, *i.e.*

$$\log f_{\mathcal{K}}(\chi_k) \approx \log f_{\mathcal{K}}(\hat{\chi}_k) + \tfrac{1}{2}\,(\chi_k - \hat{\chi}_k)^2 \frac{d^2}{d\chi_k^2} \log f_{\mathcal{K}}(\chi_k)\big|_{\chi_k=\hat{\chi}_k} \tag{8.3}$$

The first derivative of the logarithm is zero (Eq. 7.1, p. 136) and, hence, it does not appear in the Taylor expansion. By exponentiating, we find the following approximation for the posterior pdf

$$f_{\mathcal{K}}(\chi_k) \approx f_{\mathcal{K}}(\hat{\chi}_k)\, \exp\left[-\,(2\upsilon^2)^{-1}\,(\chi_k - \hat{\chi}_k)^2\right] \tag{8.4}$$

where υ is a constant given by $\upsilon^2 = \left[-\, d^2 \log f_{\mathcal{K}}(\chi_k)\,/d\chi_k^2\,\big|_{\chi_k=\hat{\chi}_k}\right]^{-1} = -f_{\mathcal{K}}(\chi_k)\left[d^2 f_{\mathcal{K}}(\chi_k)\,/d\chi_k^2\,\right]^{-1}\big|_{\chi_k=\hat{\chi}_k}$. Then, Equation 8.4 is a Gaussian pdf with standard deviation υ. Provided that the approximation range is sufficiently large $\varepsilon > \sigma_x(\boldsymbol{p}_k)$, a measure of the estimation uncertainty is $\sigma_x(\boldsymbol{p}_k) \approx \upsilon$. □

Minimum mean squared error (MMSE) techniques (see, *e.g.*, Davis, 1986) offer a measure of mapping accuracy in terms of the estimation root-squared-error average. In theory, a large number of realizations is assumed and the right-hand side of Equation 8.2 is determined as the average

$$\left\{N_r^{-1}\sum_{i=1}^{N_r}\left[\chi_i(\boldsymbol{p}_k)-\hat{\chi}_i(\boldsymbol{p}_k)\right]^2\right\}^{1/2} \tag{8.5}$$

over all realizations N_r. In practice, however, Equation 8.5 is usually calculated in terms of the means and correlation functions (covariances, variograms, *etc.*), which have been approximated on the basis of the single realization (measured values) available, using some ergodicity assumption. In many cases, this approach can lead to questionable approximations (*e.g.*, Stein, 1999). Also, while most traditional MMSE techniques do not have the mechanism that would allow them to incorporate most forms of physical knowledge, the BME method accounts for both general and specificatory knowledge bases.

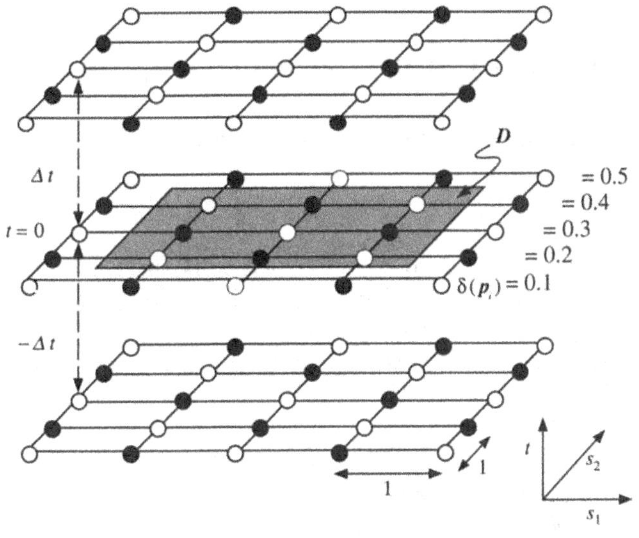

Figure 8.1. $R^2 \times T$ data configuration; hard data are available at points indicated by •; soft data at points indicated by ∘. Estimates are sought at points within region D (for numerical calculations, $\Delta t = 0.5$ units).

EXAMPLE 8.2: In Figure 8.1, measurements (hard data) of a natural variable $X(\boldsymbol{p})$ are obtained at the space/time points $\boldsymbol{p} = (s_1, s_2, t)$ indicated by solid circles (•). While no actual measurements are available at the points $\boldsymbol{p}_i$ indicated by the open circles (∘), soft data are available in the form of observation intervals $\varpi_i = [\chi_i - u\,\delta(\boldsymbol{p}_i),\ \chi_i + (1-u)\,\delta(\boldsymbol{p}_i)]$; $u \in (0,\ 1)$ is a random number and the $\delta(\boldsymbol{p}_i)$-values increase in the direction of the increasing s_2 coordinate.

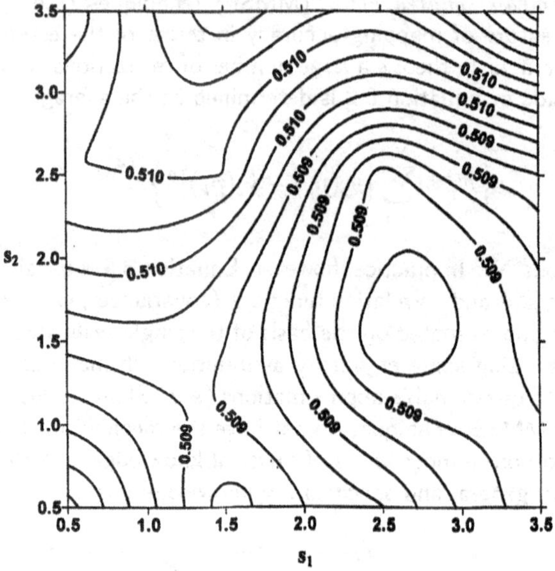

Figure 8.2. Map of the BME standard deviation $\sigma_{BME}(\boldsymbol{p}_k)$; $\Delta t = 0.5$.

Suppose that $X(\boldsymbol{p})$ has a zero mean and a covariance of the form

$$c_x(\boldsymbol{h}, \tau) = \exp\left[-\tfrac{1}{2}\pi(h_1^2/4 - h_2^2 - \tau^2)\right] \tag{8.6}$$

Using a space/time LU decomposition method (Christakos, 1992), 200 $X(\boldsymbol{p})$-realizations were simulated, which provided the "actual" values of the variable at the space/time points within the region D of Figure 8.1. The simulations obtained at the points indicated by the solid circles are considered to be the hard data for the numerical analysis. While the actual values of $X(\boldsymbol{p})$ are assumed unknown at the points indicated by open circles, soft data of varying uncertainty levels (determined by the widths of the ϖ_i intervals) are provided. For each realization ℓ $(= 1, 2, \ldots, 200)$, BME estimates $\hat{\chi}^{(\ell)}_{BME}(\boldsymbol{p}_k)$ were obtained at all $\boldsymbol{p}_k \in D$. With each estimation map, BME associates an accuracy map in terms of the estimated standard deviation $\sigma^{(\ell)}_{BME}(\boldsymbol{p}_k)$ of Equation 8.1. In Figure 8.2, the average estimation standard deviation over all 200 realizations, *i.e.*, $\sigma_{BME}(\boldsymbol{p}_k) = < \sigma^{(\ell)}_{BME}(\boldsymbol{p}_k) >$, is plotted for all $\boldsymbol{p}_k \in D$ $(\langle\cdot\rangle = 200^{-1}\sum_{\ell=1}^{200}[\cdot])$. The $\sigma_{BME}(\boldsymbol{p}_k)$ increases with increasing s_2, reflecting an increase in the width of the data intervals ϖ_i (and, thus, an increase in the uncertainty level). Also, for each realization $\chi^{(\ell)}(\boldsymbol{p}_k)$ the estimation errors

$$e^{(\ell)}_{BME}(\boldsymbol{p}_k) = \chi^{(\ell)}_{BME}(\boldsymbol{p}_k) - \hat{\chi}^{(\ell)}(\boldsymbol{p}_k) \tag{8.7}$$

were computed at all $\boldsymbol{p}_k \in D$. Since the actual values $\chi^{(\ell)}(\boldsymbol{p}_k)$ are known for each realization, the average estimation error standard deviation for BME

(Fig. 8.3) was calculated over the 200 realizations, *i.e.*, $\overline{e_{BME}}(\boldsymbol{p}_k) = \sqrt{< e_{BME}^{(\ell)}(\boldsymbol{p}_k)^2 >}$. Note that the (actual) $\overline{e_{BME}}(\boldsymbol{p}_k)$-values are of the same magnitude as the $\sigma_{BME}(\boldsymbol{p}_k)$-values (Fig. 8.2), indicating that the latter offer a very good measure of mapping accuracy. The practical advantage of the accuracy measure $\sigma_{BME}(\boldsymbol{p}_k)$ over $\overline{e_{BME}}(\boldsymbol{p}_k)$ is that while the former can be calculated in real-world applications, the latter cannot (because the actual values are unknown). In Chapter 12 (Example 12.6, p. 236), the BME results above are compared to the corresponding maps obtained using simple kriging (SK). □

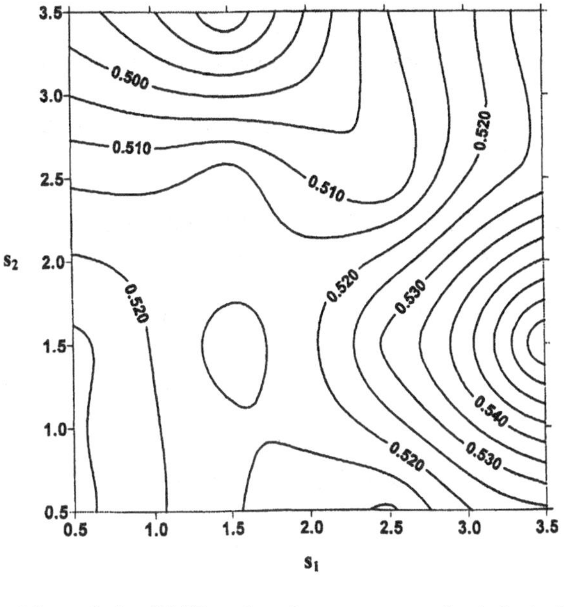

Figure 8.3. Map of the BME estimation error standard deviation $\overline{e_{BME}}(\boldsymbol{p}_k)$; $\Delta t = 0.5$.

Asymmetric Posteriors

Not surprisingly, in the case of an asymmetric posterior pdf, the best estimate and the standard deviation may not necessarily offer a satisfactory description of the situation; we then need to look at the *complete picture* as provided by the BME pdf at each point $\boldsymbol{p}_k$. There are, nevertheless, several less complicated situations in which alternative descriptions are still possible in practice. One such description is discussed next in terms of χ_k-value intervals that correspond to acceptable probabilities.

In the case of an asymmetric pdf with a single maximum, a reasonable measure of accuracy of the BME estimate is provided by the *confidence width* $w_{x,\eta}(\boldsymbol{p}_k)$ determined as follows: Choose an appropriate value η for the probability $P\left[\hat{\chi}_k - a \leq X(\boldsymbol{p}_k) \leq \hat{\chi}_k + b\right]$. The choice of η clearly depends on the situation under consideration. Given η, the values a and b must be determined so that

$$\eta = P\left[\hat{\chi}_k - a \leq X(\boldsymbol{p}_k) \leq \hat{\chi}_k + b\right] = \int_{\hat{\chi}_k - a}^{\hat{\chi}_k + b} d\chi_k \, f_{\mathcal{K}}(\chi_k) \tag{8.8}$$

Then, the confidence width is given by

$$w_{x,\eta}(\boldsymbol{p}_k) = a + b \tag{8.9}$$

and, with probability η, $X(\boldsymbol{p}_k) \in [\hat{\chi}_k - a,\, \hat{\chi}_k + b]$. In modern spatiotemporal geostatistics, confidence widths are a popular way to communicate quickly and efficiently the amount of uncertainty associated with any estimate $\hat{\chi}_k$. Certainly, there are several choices of a and b values. A specific choice is that which has the smallest width (Eq. 8.9) for a given η.

PROPOSITION 8.1: Given η, the a and b that have the smallest width $w_{x,\eta}(\boldsymbol{p}_k) = a + b$ correspond to the posterior pdf value

$$f_{\mathcal{K}}(\hat{\chi}_k + b) = f_{\mathcal{K}}(\hat{\chi}_k - a) \tag{8.10}$$

Proof: We are seeking the minimization of the width $a + b$ subject to $\eta = F_{\mathcal{K}}(\hat{\chi}_k + b) - F_{\mathcal{K}}(\hat{\chi}_k - a)$. This is equivalent to minimizing

$$L = a + b + \mu\left[F_{\mathcal{K}}(\hat{\chi}_k + b) - F_{\mathcal{K}}(\hat{\chi}_k - a) - \eta\right] \tag{8.11}$$

(where μ is a multiplier) with respect to a and b, which implies

$$\partial L / \partial(\hat{\chi}_k + b) = 1 + \mu\, f_{\mathcal{K}}(\hat{\chi}_k + b) = 0 \tag{8.12}$$

or $f_{\mathcal{K}}(\hat{\chi}_k + b) = -\mu^{-1}$. Similarly, $\partial L / \partial(\hat{\chi}_k - a) = -1 - \mu\, f_{\mathcal{K}}(\hat{\chi}_k - a) = 0$ or $f_{\mathcal{K}}(\hat{\chi}_k - a) = -\mu^{-1}$, and the proposition is proven. □

While in the case of asymmetric pdf, we generally have $a \neq b$, in the case of a symmetric pdf, Equation 8.10 gives $a = b$. Typical BME confidence intervals obtained using Equation 8.8 are shown in Figure 8.4. Also, temporal profiles of various BME confidence intervals of the Equus Beds water-level estimates are plotted in Figures 8.8 and 8.9 in the following section (p. 159–160). Generalizations of the above uncertainty measures in the context of multipoint analysis will be considered in Chapter 9.

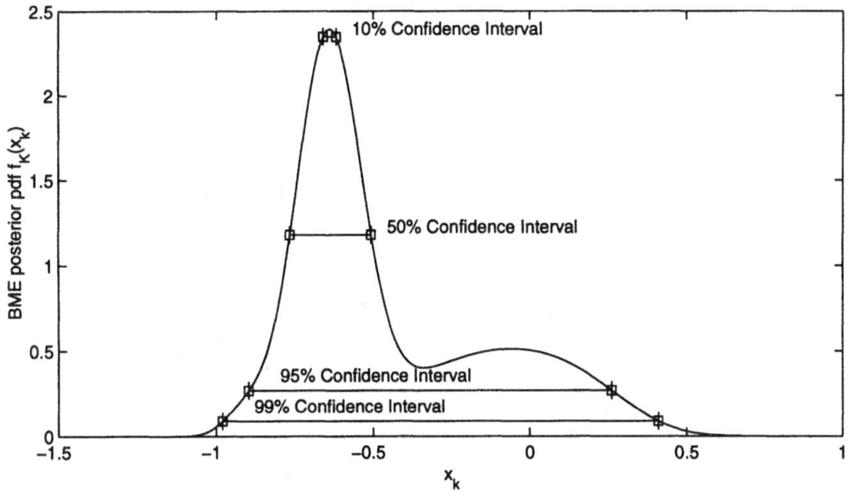

Figure 8.4. BME confidence intervals plotted on a typical single-point posterior pdf.

The Equus Beds Aquifer

The Equus Beds aquifer is an alluvial deposit near the city of Wichita in south-central Kansas. The Wichita well field was developed in the Equus Beds to supply water for the city. Pumping started in 1940. Groundwater pumping from the well field and droughts during the 1950's and late 1980's resulted in a substantial decline of water levels over a large area. The decline of this vital resource has motivated regulatory agencies to monitor the water levels using a network of groundwater observation wells (Olea, 1982). Measurements of the water level at the observation wells were made throughout the years. However, because of recording errors and due to the difficulties of accurately measuring fluctuating water level in a pumping well field, the available knowledge base $\mathcal{S}$ consists of a combination of hard and soft (uncertain) data. The purpose of this case study is to incorporate both hard and soft data into BME analysis in order to produce accurate water-level elevation spatiotemporal maps and, also, to make a reliable uncertainty assessment.

The study area

The study area covers approximately 4,000 km^2 (Fig. 8.5). A total of 70 observation wells were used, numbered from 1 to 70 by increasing Northing. The Little Arkansas River flows from northwest to southeast through the study area, and the land surface slopes towards the river, as depicted by the contour lines of equal land-surface elevation (in feet) above sea level (Fig. 8.5). The water-level decline in the Equus Beds aquifer has been documented in previous works.

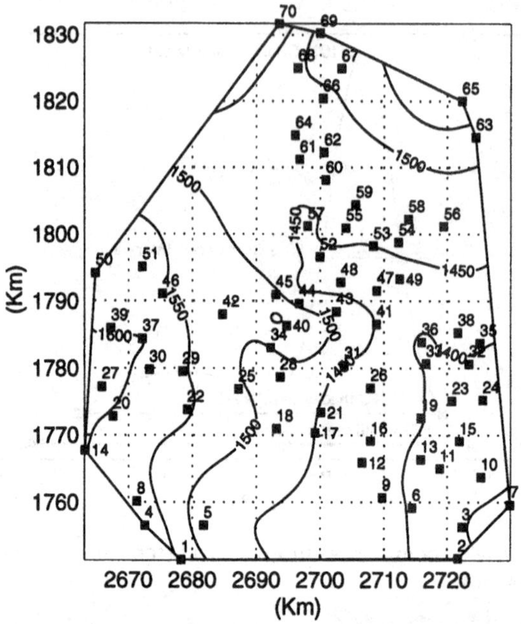

Figure 8.5. Locations of monitoring wells in the Equus Beds aquifer, shown by squares with well numbers. Contour lines for ground elevation are in feet. Northing (vertical axis) and Easting (horizontal axis) are in kilometers.

Much of this decline occurred from 1940 to early 1957 (Stramel, 1966). Water levels stabilized in the 1960's and 1970's and continued to decline between the late 1970's and the drought of 1988–92, reaching their maximum decline to date of as much as 40 ft or more during the period 1991–1993. The water level recovered moderately during the time period 1993–1998, primarily as a result of decreased city withdrawal (Aucott and Myers, 1998). Since 1995, the city of Wichita has investigated the possibility of artificial groundwater recharge in the well field in order to meet future needs and to protect the aquifer from saltwater intrusion from natural and anthropogenic sources to the west.

Data collection

Water-level data were collected by Wichita city personnel at the 70 wells mentioned above. Data collection started in 1940 and, as well-field development proceeded, water levels were measured in additional wells. Data were stored by the city in paper and electronic form, and by the U.S. Geological Survey in electronic form. In order to minimize the drawdown cone effects due to pumping in the well itself or at surrounding wells, measurements were taken during the winter when irrigation is discontinued for a few months. Measuring frequency varied from well to well, resulting in duplicate measurements for some winters and no measurements during other winters. In our study, a data set of 1,573

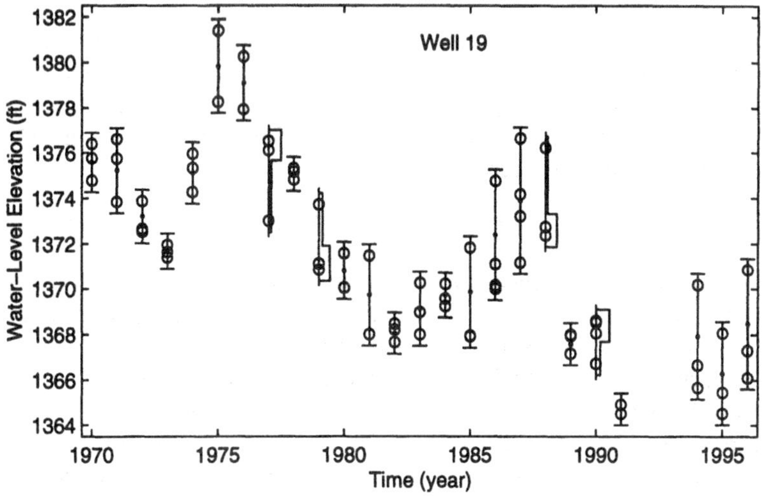

Figure 8.6. Water-level elevations measured in the Equus Beds aquifer at well no. 19. Circles depict available readings, error bars show interval (soft) data, pulse-shaped curves represent probabilistic (soft) data.

water-level elevation measurements provided by the Kansas Geological Survey (KGS) has been used. For the observation wells shown in Figure 8.5, the data set covers the period 1970–1998.

An aspect of particular importance to the KGS is measurement quality. Measurements of water-level elevations at each observation well are obtained by measuring the depth to water relative to a fixed measuring point, which is usually located a few feet above the land surface. Experience has shown that some measurements contain random errors due to several factors, *e.g.*, inaccurate readings, recording errors, uncertainty about the measuring point used for early measurements, and fluctuation of the water-level elevation during the monitoring season. Observations free of attached measurement error were considered *hard* data. Observations that had associated uncertainties were treated as *soft* data of either interval or probabilistic type by the BME approach (soft data of the *interval* and *probabilistic* types were generally assigned on the basis of a review of the available readings and the previous experience of the KGS in collecting similar data at other sites). For illustration, consider the time profile of well no. 19 (Fig. 8.6). In this case, the duplicate measurements taken during the same monitoring season showed variations that were attributed to inaccurate readings and pumping-induced fluctuations. Limits on the measurements were assigned, leading to the soft data of interval and probabilistic types shown in Figure 8.6. A similar approach was implemented for the remaining wells.

The water-level elevation model

The water-level elevation in the Equus Beds study area was modeled as an S/TRF $W(\boldsymbol{p})$, with coordinates $\boldsymbol{p} = (\boldsymbol{s},\, t)$; the spatial coordinates $\boldsymbol{s} = (s_1,\, s_2)$ specify Northing and Easting (in kilometers), and t is the temporal coordinate (in years). Physical considerations suggested (Serre and Christakos, 1999a) that $W(\boldsymbol{p})$ is adequately represented as

$$W(\boldsymbol{p}) = \mu(\boldsymbol{s}) + X(\boldsymbol{p}) \tag{8.13}$$

where $X(\boldsymbol{p})$ is a homogeneous S/TRF with zero mean and separable covariance, and the mean value of the water-level elevation $\mu(\boldsymbol{s}) = \overline{W(\boldsymbol{p})}$ is a function of the spatial location $\boldsymbol{s}$ only. The $\mu(\boldsymbol{s})$ strongly depends on the land-surface elevation, and it is estimated at each well by taking the mean value of the water-level elevation for that well (on average, more than 35 observations were used to calculate the mean water-level elevation at each well). The $X(\boldsymbol{p})$ field contains all the randomness associated with the water-level elevation. The covariance of $X(\boldsymbol{p})$ is estimated using the values obtained by subtracting the mean $\mu(\boldsymbol{s})$ from the observations of the water-level elevation $W(\boldsymbol{p})$. An isotropic covariance model $c_x(r,\, \tau)$ was chosen, where $r = |\boldsymbol{s} - \boldsymbol{s}'|$ is the spatial lag and $\tau = |t - t'|$ is the time lag. The estimated values of the covariance are shown in Figure 8.7 as functions of the spatial and temporal lags. The covariance decreases smoothly as r and τ increase, and tends to zero for large r and τ, thus suggesting that the water-level elevation model (Eq. 8.13) was indeed a reasonable choice. In Figure 8.7 we also plot the space/time separable covariance model

$$c_x(r,\, \tau) = c_0 \exp\left[-r/a_r\right] \exp\left[-\tau/a_\tau\right] \tag{8.14}$$

with sill $c_0 = 8$ ft^2, spatial range $a_r = 20$ km, and temporal range $a_\tau = 2$ yr. The exponential model (Eq. 8.14) offers a reasonably good fit to the estimated covariance values.

BME water-level elevation mapping

The BME approach produces spatiotemporal $X(\boldsymbol{p})$ estimates. Then, the corresponding water-level elevations $W(\boldsymbol{p})$ are obtained by adding the (known) spatial mean $\mu(\boldsymbol{s})$ to the $X(\boldsymbol{p})$ estimates. In addition to hard data, both interval and probabilistic soft data were used by the BME mapping method. For illustration, in Figure 8.8, the BMEmode estimates for well no. 64 are plotted as a function of time; also shown are the BME 10%, 50%, and 90% confidence intervals. For this representative hydrograph, the confidence intervals are consistent with the probabilistic soft data (denoted by pulse-shaped curves at the observation times). As should be expected, these confidence intervals are wider at times between observations.

Figure 8.9 shows the BMEmode estimates and the 90% confidence intervals of the water-level elevation for a representative set of wells. The cross-like

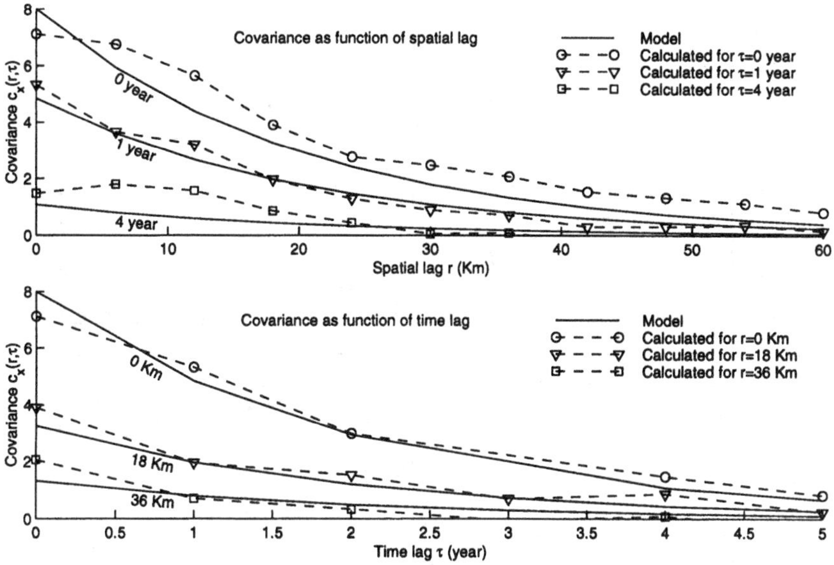

Figure 8.7. Space/time covariances of water-level elevation shown as functions of spatial lag r (top) and time lag τ (bottom). Markers show covariance values calculated from actual measurements. Solid lines show the fitted covariance model.

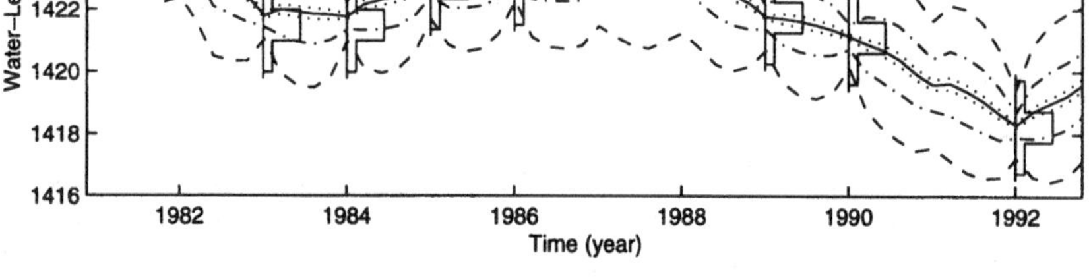

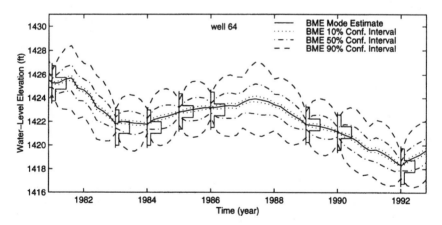

Figure 8.8. Temporal profile of BMEmode estimates of water-level elevation at well no. 64. The corresponding BME 10%, 50%, and 90% confidence intervals are also shown. Probabilistic soft data are depicted by pulse-shaped curves.

symbols ($\times$) denote hard data; the bar-like and pulse-like curves denote interval and probabilistic data, respectively. While the BMEmode estimates express the most likely changes of the water-level elevation over the time period 1970–1998,

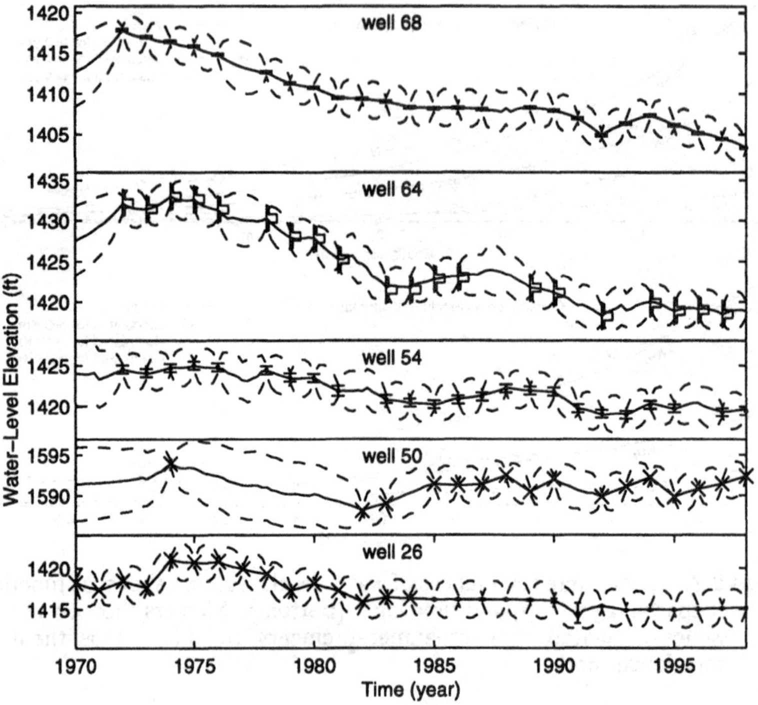

Figure 8.9. Temporal profiles of BMEmode estimates (solid line) and 90% confidence interval (dashed line) of the water-level elevation at selected wells in the Equus Beds aquifer. Hard data are shown by ×'s; soft data of interval and probabilistic types are depicted by error bars and pulse-shaped curves, respectively.

the BME 90% confidence intervals provide an assessment of the associated estimation uncertainty. One can see from these hydrographs that most wells show a decline in the water-level elevation from the late 1970's to the drought of 1988–1992. The water-level elevation reached its lowest level during the period 1991–1993, followed by a moderate recovery in the following years.

BME also produces *space/time* conditional mean estimates of water-level elevation which possess high estimation accuracy and are physically meaningful. Figure 8.10 shows the BMEmean estimates of the water-level elevations during the years 1975 and 1998 (in feet above sea level). The locations where hard data points, interval soft data points, and probabilistic soft data points were available during each of the years 1975 and 1998 are shown with triangles, circles, and stars, respectively. In order to construct the maps, water-level elevations were estimated on a 40 × 40 regular grid covering the study area. At each space/time estimation point, a local neighborhood of hard and soft data points was used. Each neighborhood includes data points located at spatial distances $< 4\,a_r$ and temporal lags $< 4\,a_\tau$ from the estimation point (if a local neighborhood included too many points, only the ten hard data

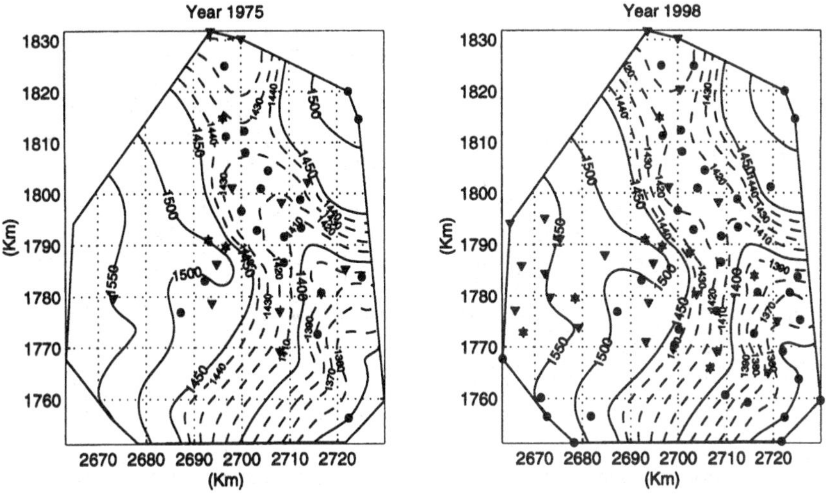

Figure 8.10. Maps of BMEmean estimates of water-level elevations (in feet) in the Equus Beds aquifer for the years 1975 and 1998. Observation wells for which hard, interval soft, and probabilistic soft data were available in space/time are depicted by triangles, circles, and stars, respectively.

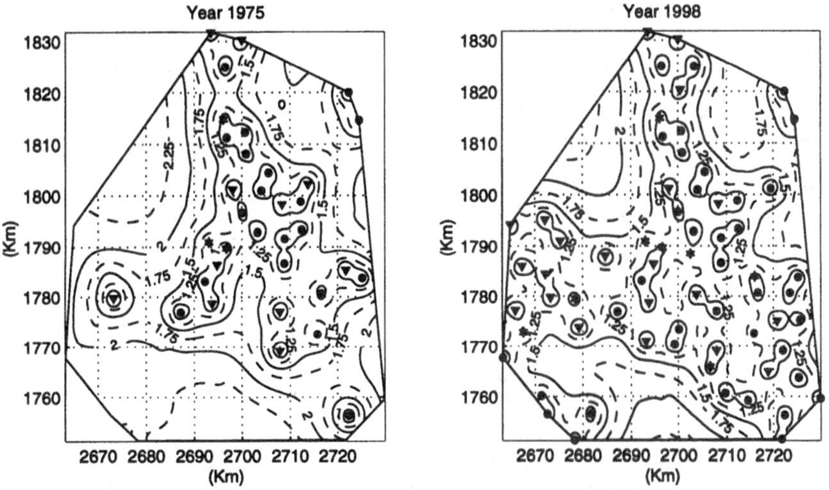

Figure 8.11. Maps of standard deviation error of the BMEmean estimates shown in Figure 8.10.

points and five soft data points most highly correlated to the estimation point were selected). The *BMEmean* maps of water-level elevation (Fig. 8.10) show that, while the water level stayed unchanged at higher ground elevations, it dropped substantially in the valley along the Little Arkansas River (see also Fig. 8.12). Estimation uncertainty for the BME map is assessed from the associated estimation error standard deviation map (Fig. 8.11). The estimation

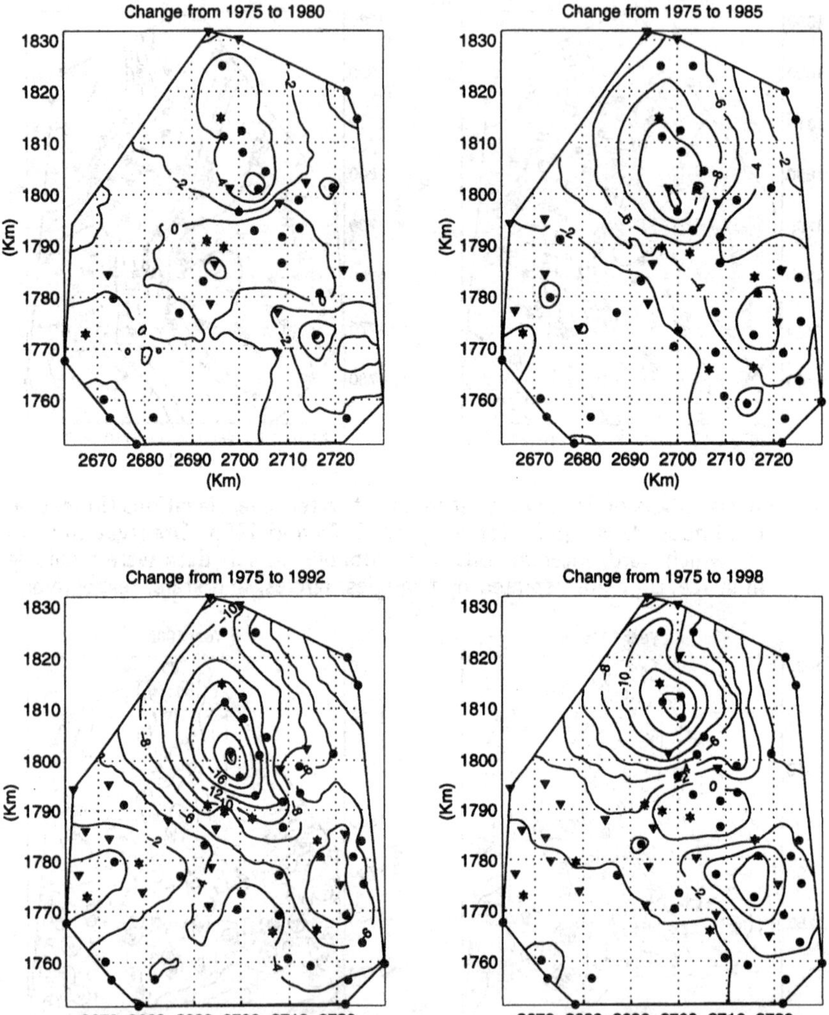

Figure 8.12. Maps of the change in water-level elevation (in feet) in the Equus Beds aquifer from 1975 to selected years obtained using BMEmean estimates. Observation wells where hard, soft interval, and soft probabilistic data were available in space/time are depicted by triangles, circles, and stars, respectively.

error standard deviation for the 1975 map is larger than for the 1998 map, due to the smaller number of observations available in 1975. Also, as a result of the significant number of observation wells located in the valley, the 1975 and 1998 water-level maps are more accurate in the valley than at higher ground elevations.

The changes in water-level elevations between 1975 and a few selected years are plotted in Figure 8.12. These maps show a continuing depletion during the periods 1975–1980, 1975–1985, and 1975–1992; a modest recovery occurred in the period 1992–1998. A zone of water-level decline developed during the period 1975–1980 in a region generally encompassing the location of the city pumping wells in the north of the study area (Aucott and Myers, 1998). The zone of water-level depletion extended further in the periods ending in 1985 and 1992, encompassing the entire well field. The maximum decline was more than 12 ft during the period 1975–1985 and more than 18 ft during the period 1975–1992. This water-level decline was due to increased water pumping for municipal use and greatly increased agricultural withdrawal, combined with the effect of the 1988–1992 drought.

The result of groundwater depletion in the Equus Beds aquifer includes loss of water saturated thickness, increased pumping costs to lift water from greater depths, and greater exposure to saltwater intrusion from natural and anthropogenic sources to the west. This situation generally represents a decrease in water resources available for use. The period 1992–1998 is characterized by some recovery in water levels, due primarily to a decrease in pumping for municipal use. However, as shown in the maps of water-level change, the extent of the water-level decline since 1975 is still large, with a maximum decline exceeding 12 ft during the period 1975–1998.

COMMENT 8.2: *In the context of the Equus Beds study, the BMEmean and BMEmode estimates above have also been compared to kriging estimates of classical geostatistics (see Chapter 12, Example 12.10 on p. 242).*

Optimal decision making

BME maps can improve the hydrogeologic understanding of the entire Equus Beds region and also optimize local *decision making* regarding the operation or extension of the Wichita well field. A well may be considered worth drilling if the anticipated net benefit function is given by, *e.g.*, $\mathfrak{B}(\hat{\chi}_k;\ \boldsymbol{p}_k) = b_k - c_k(\zeta_k - \hat{\chi}_k) > 0$, where c_k is the drilling cost (in \$/m of well depth) at point $\boldsymbol{p}_k$; b_k (in \$) is the corresponding benefit if water is found at $\boldsymbol{p}_k$; ζ_k is the elevation of the ground surface; and $\hat{\chi}_k$ is the water-level elevation estimate obtained from the BME maps above.

It is clear that in the above setup the value of the anticipated benefit $\mathfrak{B}$ could be affected by the kind of $\hat{\chi}_k$ estimate considered (mean, mode, *etc.*). Depending on the situation, more complicated (*e.g.*, nonlinear) functions $\mathfrak{B}$ may need to be considered (Christakos and Killam, 1993). This should not be a problem, given that a complete stochastic characterization of the situation is available by means of the BME posterior pdf.

Doing Progressive Guesswork

Unlike Arthur Golden's novel, *Memoirs of a Geisha*, which describes a world in which appearances are paramount (Golden, 1997), in the geostatistical world one seeks to go beyond appearances, adding scientific substance to the maps obtained. This can happen if not only the physical knowledge bases are integrated into space/time analysis, but the uncertainties associated with these bases are adequately assessed and visualized, as well.

The preceding measures of mapping accuracy (or, if you prefer, measures of mapping uncertainty) are most valuable in the vast majority of applications in geostatistical practice in which the phenomenon being studied cannot be isolated from a host of confounding influences that introduce a chance component of considerable size into the quantitative analysis. In many situations, while the amount of uncertainty estimated solely on the basis of the hard data may initially be substantial, there is usually a significant amount of physical knowledge currently being ignored that could improve considerably the mapping accuracy of the phenomenon.

BME analysis establishes a *progressive* process, during which the maps produced using the new knowledge show a substantial improvement in accuracy over the maps previously obtained. As the process tends towards the "ultimate" map, the application-specific goals of map making allow a specified level of uncertainty relative to the actual (but unknown) map. Assessing this uncertainty is the important task of the quantitative measures discussed in this chapter.

9

MODIFICATIONS OF FORMAL BME ANALYSIS

"Common sense has the very curious property of being more correct retrospectively than prospectively. It provides a kind of ultimate validation after science has completed its work; common sense seldom anticipates what science is going to discover." R.L. Ackoff

Versatility and Practicality

In the preceding chapters, strong emphasis has been laid on the view that the construction of a spatiotemporal map representing a natural phenomenon—which is characterized by considerable variability and uncertainty—requires both: (a) broad-based *physical knowledge*; and (b) *scientific reasoning* strategies to operate over such knowledge. This double requirement is fully appreciated by the BME theory, which combines a variety of physical knowledge sources with a sound scientific reasoning framework that revolves around meaningful epistemic ideals and goals, multiple estimation options, and cogent predictions.

An important feature of the theoretical BME formulation is its *versatility*. Indeed, several modifications of the formal BME approach can be made in a rather straightforward manner. Some of these modifications are examined in this chapter. In particular, in the following section, we are concerned with the study of functionals such as block average concentration, temporally averaged exposure, and population damage indicator. Multivariable or vector mapping (*i.e.*, estimation of several natural variables jointly) is the topic of the third section. Then, the fourth section introduces the main framework of multipoint BME analysis (*i.e.*, estimation at several space/time points simultaneously). This topic is of considerable practical importance, which is the reason that it is revisited in Chapter 11. Finally, in the last section, we discuss the important role that BME can play in the context of *systems analysis*. In many applications the outcomes of BME investigations (predictive maps, uncertainty measures, *etc.*) are the input parameters to subsequent steps of scientific decision making, engineering design, *etc.*

Functional BME Analysis

There is a plethora of applications involving some kind of functional analysis. This is, *e.g.*, the case of ore mining that involves the spatial estimation of large mining blocks from smaller core samples. Waste-site characterization also depends on the analysis of contaminant processes at various scales and the establishment of suitable quantitative connections between the results obtained at each one of these scales. In this section we study functional analysis from the perspective of modern spatiotemporal geostatistics.

General formulation

In natural sciences we are often seeking a spatiotemporal map of the following general functional

$$X_{\Lambda}(\boldsymbol{p}_k) = \mathfrak{f}\,[X(\boldsymbol{p}), \Lambda] \tag{9.1}$$

where Λ is a space/time domain, and the form of the functional $\mathfrak{f}$ may depend on the physics or the economics of the problem. As usual, general knowledge as well as specificatory data $\boldsymbol{\chi}_{data}$ are available at points $\boldsymbol{p}_i$ $(i = 1, \ldots, m)$, and the data available are considered as point samples. Physically, the meaning of the term "point sample" is that its size is much smaller than the space/time distances considered in geostatistical analysis and, certainly, much smaller than the size of Λ. Accordingly, the $X(\boldsymbol{p})$ in Equation 9.1 is usually called a "point" natural variable.

Important special cases of Equation 9.1 in Earth sciences and environmental health engineering include:

(a) The *V-block average* of the natural variable $X(\boldsymbol{p})$ is given by

$$X_V(\boldsymbol{p}_k) = |V(\boldsymbol{p}_k)|^{-1} \int_{V(\boldsymbol{p}_k)} d\boldsymbol{u}\, X(\boldsymbol{s}_k - \boldsymbol{u}, t_k) \tag{9.2}$$

where $\Lambda(\boldsymbol{p}_k) = V(\boldsymbol{p}_k)$, and $\boldsymbol{p}_k = (\boldsymbol{s}_k, t_k)$.

(b) The *temporally averaged exposure*

$$X_T(\boldsymbol{p}_k) = |\tau_e(\boldsymbol{p}_k)|^{-1} \int_{\tau_e(\boldsymbol{p}_k)} dt'\, f_e(\boldsymbol{s}_k, t_k - t')\, X(\boldsymbol{s}_k, t_k - t') \tag{9.3}$$

where $X(\boldsymbol{p}_k)$ is the exposure rate, $\Lambda = \tau_e$ denotes the exposure duration, and f_e is the exposure frequency (*i.e.*, the fraction of total exposure time during which the receptor is actually exposed; in %).

(c) The *health damage indicator* of a population

$$X_V(\boldsymbol{p}_k) = |V(\boldsymbol{p}_k)|^{-1} \int_{V(\boldsymbol{p}_k)} d\boldsymbol{u}\, \theta(\boldsymbol{s}_k - \boldsymbol{u}, t_k)\, X(\boldsymbol{s}_k - \boldsymbol{u}, t_k) \qquad (9.4)$$

where $\theta(\boldsymbol{p}_k)$ is the density of receptors in the neighborhood of $V(\boldsymbol{p}_k)$, and $X(\boldsymbol{p}_k)$ is a specified health effect (see "Human-exposure systems" on p. 182).

In the functional case of Equation 9.1, the basic equations of BME analysis are easily modified as follows. The Lagrange multipliers μ_α are the solutions of the system of moment equations

$$\overline{h_\alpha} = \int d\boldsymbol{\chi}_{data} \int d\chi_\Lambda\, g_\alpha(\boldsymbol{\chi}_{data}, \chi_\Lambda)\, \exp\left[\mu_0 + \mathcal{Y}_G\right], \quad \alpha = 0, 1, ..., N_c \quad (9.5)$$

where χ_Λ is a realization of the functional random field (Eq. 9.1); the BMEmode equation is

$$\frac{\partial}{\partial \chi_\Lambda} \mathcal{Y}_S[\boldsymbol{\chi}_{map}; \Lambda]\Big|_{\chi_\Lambda = \hat{\chi}_\Lambda} = 0 \qquad (9.6)$$

and the posterior pdf is

$$f_{\mathcal{K}}(\chi_\Lambda) = A^{-1}\, \mathcal{Y}_S[\boldsymbol{\chi}_{map}; \Lambda] \qquad (9.7)$$

where $\mathcal{Y}_G = \sum_{\alpha=1}^{N_c} \mu_\alpha\, g_\alpha$. The form of the $\mathcal{Y}_S$-operator in Equations 9.6 and 9.7 depends on the knowledge available. In the case of the knowledge described in Proposition 6.1 (p. 126), the $\mathcal{Y}_S$-operator has the form used in Equation 6.1 (p. 126); then, Equations 9.6 and 9.7 reduce to

$$\sum_{\alpha=1}^{N_c} \mu_\alpha \int_I d\boldsymbol{\chi}_{soft} \left\{\frac{\partial g_\alpha(\boldsymbol{\chi}_{data}, \chi_\Lambda)}{\partial \chi_\Lambda} \exp\left[\mathcal{Y}_G\right]\right\}_{\chi_\Lambda = \hat{\chi}_\Lambda} = 0 \qquad (9.8)$$

and

$$f_{\mathcal{K}}(\chi_\Lambda) = A^{-1} \int_I d\boldsymbol{\chi}_{soft}\, \exp\left[\mu_0 + \mathcal{Y}_G\right] \qquad (9.9)$$

The g_α's are now functions of the point and the functional random fields (*e.g.*, they may include point and block covariances and point-block covariances as well). The analysis can be generalized in terms of the operators shown in Table 6.1 (p. 133) or any other posterior operator available.

EXAMPLE 9.1: The average ozone exposure $X_T = E$ during each day was calculated from Equation 9.3 for a geographical region in the eastern U.S. that includes New York City and Philadelphia. It was assumed that $\tau_e = 24$ hr, $f_e = 100\%$, and $X = E_{1-h}$ is the 1-hr ozone exposure (in ppm). In Figure 9.1, the temporally averaged 1-hr ozone exposure maps are plotted for a few selected days in July of 1995. These maps can help us detect variations in

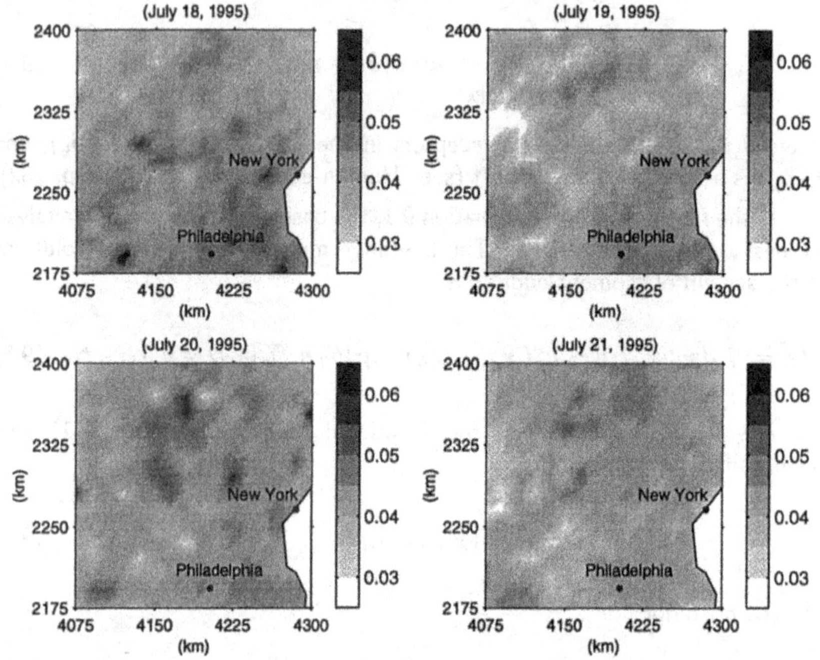

Figure 9.1. Daily-averaged ozone exposure maps (ppm) at a region in the eastern U.S.

the daily-averaged ozone exposure across areas and identify spatiotemporal exposure patterns of considerable importance in environmental health studies. Certain differences between spatial and temporal exposure variations may be due to extra-continua properties of space and time, topographic characteristics of the region, urban activities, meteorological conditions, *etc.* The exposure maps honor the data values at points in which monitoring stations exist. □

The support effect

In many geostatistical applications, the *support effect* (see also Chapter 3, "The Specificatory Knowledge Base," p. 82) is due to the difference in practice of the sizes of the samples and the domains (blocks, *etc.*) to be estimated. It has been well documented that the support effect can be the cause of incorrect mapping estimates (*e.g.*, Rivoirard, 1994). The functional BME analysis can account for the support effect in a rigorous and systematic manner, as is illustrated by means of the following simple example.

EXAMPLE 9.2: Consider the case of one hard datum χ_1 at point $\boldsymbol{p}_1$ and one soft (interval) datum χ_2 at $\boldsymbol{p}_2$ in the domain of interest. An estimate is sought of the V-block average $X_V(\boldsymbol{p}_k)$ defined by Equation 9.2. The general knowledge consists of the variograms between all points of the domain, including the two

Table 9.1. The g_α functions corresponding to case presented in Example 9.2.

α	g_α
0	$g_0 = 1$
1	$g_1(\chi_1, \chi_V) = \frac{1}{2}(\chi_1 - \chi_V)^2$
2	$g_2(\chi_2, \chi_V) = \frac{1}{2}(\chi_2 - \chi_V)^2$

samples (assumed to have point support) and the block. In view of the support effect between samples (χ_i) and block (χ_V), the BME system of equations consists of Equations 9.5, 9.8, and 9.9 with $\chi_\Lambda = \chi_V = V^{-1} \int_V d\boldsymbol{u}\, \chi(\boldsymbol{u})$. The corresponding g_α functions ($\alpha = 0,\ 1,\ 2$) are shown in Table 9.1.

The Lagrange multipliers μ_α are found from the solution of the following system of equations

$$\left.\begin{array}{l} \overline{h_0} = \int d\boldsymbol{\chi}_{map} \exp\left[\mu_0 + \mathcal{Y}_G\right] = 1,\ \overline{h_1} = \frac{1}{2}\int d\boldsymbol{\chi}_{map}(\chi_1 - \chi_V)^2 \exp\left[\mu_0 + \mathcal{Y}_G\right] \\ \overline{h_2} = \frac{1}{2}\int d\boldsymbol{\chi}_{map}(\chi_2 - \chi_V)^2 \exp[\mu_0 + \mathcal{Y}_G] \end{array}\right\} \tag{9.10}$$

where $d\boldsymbol{\chi}_{map} = \prod_{i=1}^{2,V} d\chi_i$ and $\mathcal{Y}_G = \mu_1 g_1(\chi_1, \chi_V) + \mu_2 g_2(\chi_2, \chi_V)$. The left-hand sides of Equation 9.10 are computed experimentally in terms of the variograms as follows ($\overline{h_0} = 1$)

$$\left.\begin{array}{l} \overline{h_1} = \frac{1}{2}\overline{(x_1 - x_V)^2} = \frac{1}{2}(2\overline{\gamma}_{1,V} - \overline{\gamma}_{V,V}) \\ \overline{h_2} = \frac{1}{2}\overline{(x_2 - x_V)^2} = \frac{1}{2}(2\overline{\gamma}_{2,V} - \overline{\gamma}_{V,V}) \end{array}\right\} \tag{9.11}$$

where $\overline{\gamma}_{i,V}$ and $\overline{\gamma}_{V,V}$ are the point (sample)-block and the block-block averaged variograms, respectively. Equation 9.8 now becomes

$$\int_I d\chi_2 \{[\mu_1 \frac{\partial g_1(\chi_1, \chi_V)}{\partial \chi_V} + \mu_2 \frac{\partial g_2(\chi_2, \chi_V)}{\partial \chi_V}] \exp\left[\mathcal{Y}_G\right]\}_{\chi_V = \hat{\chi}_V} = 0$$

and using the relationships in Table 9.1, the last equation reduces to the following integral representation

$$\int_I d\chi_2 \left\{[\mu_1(\chi_1 - \chi_V) + \mu_2(\chi_2 - \chi_V)] \exp\left[\mathcal{Y}_G\right]\right\}_{\chi_V = \hat{\chi}_V} = 0 \tag{9.12}$$

which is solved with respect to the $\hat{\chi}_V$ that is the desired BME block estimate. □

BME's treatment of the support effect does not suffer from the limitations of classical geostatistics approaches (*e.g.*, the problematic inference of the point-block indicator covariance involved in indicator kriging; Journel, 1989). BME analysis also provides a method for studying *change-of-scale problems*, as well as applications related to effective parameters of random media (Christakos and Hristopulos, 1998).

Multivariable or Vector BME Analysis

The BME analysis can be easily extended to include several natural variables represented by a vector S/TRF. The primary natural variables are related to certain secondary variables by means of a physical law, a theory, or an empirical relationship (Chapter 3). In geochemical exploration, *e.g.*, the grade of an element (Fe_2O_3 or Ag, *etc.*; a primary variable) could be related to geologic characteristics of the region (lithological factors, *etc.*, secondary variables). Hard and/or soft data are usually available for some of these variables. This sort of BME analysis is called *multivariable BME*, or *vector BME* or *co-BME analysis*.

General formulation

The general formulation of the multivariable BME is as follows: Assume that the primary natural variable $X(\boldsymbol{p})$ is related to the $N-1$ secondary variables $\boldsymbol{Y} = (Y_2(\boldsymbol{p}), \ldots, Y_N(\boldsymbol{p}))$. We seek to estimate $X(\boldsymbol{p})$ at point $\boldsymbol{p} = \boldsymbol{p}_k$. The basic BME equations for the situation are formulated as follows. The Lagrange multipliers μ_α are solutions of the system of moment equations

$$\overline{h_\alpha} = \int d\boldsymbol{\chi}_{map} \int d\boldsymbol{\Psi}_{soft}\, g_\alpha(\boldsymbol{\chi}_{map}, \boldsymbol{\Psi}_{data}) \exp[\mu_0 + \mathcal{Y}_G], \quad \alpha = 0, 1, \ldots, N_c \tag{9.13}$$

where $\boldsymbol{\Psi}_{data} = (\boldsymbol{\psi}_{2,\,data}, \ldots, \boldsymbol{\psi}_{N,\,data})$, $\boldsymbol{\Psi}_{soft} = (\boldsymbol{\psi}_{2,\,soft}, \ldots, \boldsymbol{\psi}_{N,\,soft})$, and $\mathcal{Y}_G = \sum_{\alpha=1}^{N_c} \mu_\alpha\, g_\alpha$. The BMEmode estimation equation is

$$\frac{\partial}{\partial \chi_k} \mathcal{Y}_S[\boldsymbol{\chi}_{map}, \boldsymbol{\Psi}_{data}; \boldsymbol{p}_k]_{\chi_k = \hat{\chi}_k} = 0 \tag{9.14}$$

and the posterior pdf is given by

$$f_{\mathcal{K}}(\chi_k) = A^{-1}\, \mathcal{Y}_S[\boldsymbol{\chi}_{map}, \boldsymbol{\Psi}_{data}; \boldsymbol{p}_k] \tag{9.15}$$

The g_α are the important functions that incorporate general knowledge in terms of $\boldsymbol{\chi}_{data}$ and $\boldsymbol{\Psi}_{data}$ (like in other kinds of BME analysis, the general knowledge of vector BME may involve multiple-point statistics). To understand these equations it is better to work out some special cases.

For illustration consider only two fields $X(\boldsymbol{p})$ and $Y(\boldsymbol{p})$. Assume that hard data (Eq. 3.30, p. 84) and soft (interval) data (Eq. 3.32, p. 85) are available for $X(\boldsymbol{p})$ and $Y(\boldsymbol{p})$ at points $\boldsymbol{p}_i$ ($i = 1, \ldots, m$; data for the two fields may be available at the same or at different sets of points). An estimate of $X(\boldsymbol{p})$ is sought at the point p_k $(k \neq i)$. Equations 9.13–9.15 reduce to the following system of equations

$$\overline{h_\alpha} = \int d\boldsymbol{\chi}_{map} \int d\boldsymbol{\psi}_{data}\, g_\alpha(\boldsymbol{\chi}_{map}, \boldsymbol{\psi}_{data}) \exp[\mu_0 + \mathcal{Y}_G] \quad \alpha = 0, 1, \ldots, N_c \tag{9.16}$$

and

$$\sum_{\alpha=1}^{N_c} \mu_\alpha \int_I d\boldsymbol{\chi}_{soft} \int_I d\boldsymbol{\psi}_{soft} \left\{ \frac{\partial g_\alpha(\boldsymbol{\chi}_{map}, \boldsymbol{\psi}_{data})}{\partial \chi_k} \exp[\mathcal{Y}_G] \right\}_{\chi_k = \hat{\chi}_k} = 0 \tag{9.17}$$

and the posterior pdf (given $\boldsymbol{\chi}_{data}$ and $\boldsymbol{\psi}_{data}$)

$$f_{\mathcal{K}}(\chi_k) = A^{-1} \int_I d\boldsymbol{\chi}_{soft} \int_I d\boldsymbol{\psi}_{soft} \exp[\mu_0 + \mathcal{Y}_G] \tag{9.18}$$

where, as usual, $\mathcal{Y}_G = \sum_{\alpha=1}^{N_c} \mu_\alpha\, g_\alpha$. Equation 9.17 is the vectorial form of the BME equation (Eq. 7.6, p. 137). Other posterior operators $\mathcal{Y}_S$ (as described in Chapters 6 and 7) could also be implemented. In fact, any possible combination of hard and soft data for $X(\boldsymbol{p})$ and $Y(\boldsymbol{p})$ could be considered in a similar fashion. This is a good point to pause and discuss a simple example.

EXAMPLE 9.3: Consider the points $\boldsymbol{p}_i$ $(i = 1, 2, 3, 4)$. The hard and soft data available include, $\boldsymbol{\chi}_{hard} = (\chi_1, \chi_2)$, $\boldsymbol{\chi}_{soft} = \chi_3$, and $\boldsymbol{\chi}_{data} = (\boldsymbol{\chi}_{hard}, \boldsymbol{\chi}_{soft})$ for the primary field $X(\boldsymbol{p})$; and $\boldsymbol{\psi}_{hard} = (\psi_1, \psi_2)$, $\boldsymbol{\psi}_{soft} = (\psi_3, \psi_4)$ and $\boldsymbol{\psi}_{data} = (\boldsymbol{\psi}_{hard}, \boldsymbol{\psi}_{soft})$ for the secondary field $Y(\boldsymbol{p})$. An estimate of $X(\boldsymbol{p})$ is sought at point $\boldsymbol{p}_k$, so that $\boldsymbol{\chi}_{map} = (\boldsymbol{\chi}_{data}, \chi_k)$. Known statistics are the means, variances, (centered) covariances, and cross-covariances between all points considered. The g_α functions $(\alpha = 0, 1, \ldots, 44)$ are shown in Table 9.2. The $\mathcal{Y}_G$ is given by

$$\begin{aligned} \mathcal{Y}_G[\boldsymbol{\chi}_{map}, \boldsymbol{\psi}_{data}] &= \sum_{i=1,2}^{3,k} \mu_i^x\, g_i(\chi_i) + \sum_{i=1,2}^{3,k} \mu_{ii}^x\, g_{ii}(\chi_i, \chi_i) \\ &+ \sum_{i,j=1,2}^{3,k} \mu_{ij}^x\, g_{ij}(\chi_i, \chi_j) + \sum_{i=1,2}^{3,4} \mu_i^y\, g_i(\psi_i) \\ &+ \sum_{i=1,2}^{3,4} \mu_{ii}^y\, g_{ii}(\psi_i, \psi_i) + \sum_{i,j=1,2}^{3,4} \mu_{ij}^y\, g_{ij}(\psi_i, \psi_j) \\ &+ 2 \sum_{i=1,2}^{3,k} \sum_{j=1,2}^{3,4} \mu_{ij}^{xy}\, g_{ij}(\chi_i, \psi_j) \end{aligned} \tag{9.19}$$

Finally, the BMEmode equation is written as

$$\begin{aligned} \int_I d\chi_3\, d\psi_3\, d\psi_4 \Big[\mu_{kk}^x(\hat{\chi}_k - \overline{x_k}) &+ \sum_{i=1,2}^{3} \mu_{ik}^x\, (\chi_i - \overline{x_i}) \\ + \sum_{i=1,2}^{3,4} \mu_{ik}^{xy}\, (\psi_i - \overline{y_i}) \Big] &\exp\left\{ \mathcal{Y}_G[\boldsymbol{\chi}_{map}, \boldsymbol{\psi}_{data}] \right\}_{\chi_k = \hat{\chi}_k} = 0 \end{aligned} \tag{9.20}$$

which is solved with respect to $\hat{\chi}_k$. □

Table 9.2. The g_α functions corresponding to case presented in Example 9.3.

α	g_α	$\overline{g_\alpha}$
	Normalization constraint	
0	$g_0 = 1$	$\overline{g_0} = 1$
	Mean constraints for $X(\boldsymbol{p})$	
1–4	$g_i(\chi_i) = \chi_i$ $i = 1, 2, 3$ and k	$\overline{g_i} = \overline{x_i}$
	Variance constraints for $X(\boldsymbol{p})$	
5–8	$g_{ii}(\chi_i, \chi_i) = (\chi_i - \overline{x_i})^2$ $i = 1, 2, 3$ and k	$\overline{g_{ii}} = \sigma^2_{x,i}$
	Covariance constraints for $X(\boldsymbol{p})$	
9–14	$g_{ij}(\chi_i, \chi_j) = (\chi_i - \overline{x_i})(\chi_j - \overline{x_j})$ $i, j = 1, 2, 3$ and k; $i < j$	$\overline{g_{ij}} = c_{x,ij}$
	Mean constraints for $Y(\boldsymbol{p})$	
15–18	$g_i(\psi_i) = \psi_i$ $i = 1, 2, 3$ and 4	$\overline{g_i} = \overline{y_i}$
	Variance constraints for $Y(\boldsymbol{p})$	
19–22	$g_{ii}(\psi_i, \psi_i) = (\psi_i - \overline{y_i})^2$ $i = 1, 2, 3$ and 4	$\overline{g_{ii}} = \sigma^2_{y,i}$
	Covariance constraints for $Y(\boldsymbol{p})$	
23–27	$g_{ij}(\psi_i, \psi_j) = (\psi_i - \overline{y_i})(\psi_j - \overline{y_j})$ $i, j = 1, 2, 3$ and 4; $i < j$	$\overline{g_{ij}} = c_{y,ij}$
	Cross-covariance constraints for $X(\boldsymbol{p})$ and $Y(\boldsymbol{p})$	
28–44	$g_{ij}(\chi_i, \psi_j) = (\chi_i - \overline{x_i})(\psi_j - \overline{y_j})$ $i = 1, 2, 3, k$ and $j = 1, 2, 3, 4$	$\overline{g_{ij}} = c_{xy,ij}$

Physical laws

Consider the situation in which a physical law relating $X(\boldsymbol{p})$ and $Y(\boldsymbol{p})$ is available (for a discussion of such situations see Chapter 3). Then, the $\mathcal{Y}_\mathcal{G}$-function should include a term incorporating the knowledge of the physical law into the BME analysis. As we saw in Chapter 5, depending on the form of the physical law available, there are two ways to proceed: We either start from the continuous-domain formulation of the physical law, define the corresponding g_α-statistics equations, and then solve for the Lagrange multipliers, or else we can formulate the physical law in the discrete domain and then incorporate it into the expressions of the g_α-statistics (see Chapter 5, "General knowledge in the form of physical laws," p. 109). Below we examine the second option by means of an example.

EXAMPLE 9.4: As general knowledge we consider the groundwater flow law (Eq. 3.25, p. 81). The flow equation offers a physical basis for relating the stochastic moments of hydraulic head and hydraulic log conductivity. A possible discretization of this law is given in Equation 3.26. Specificatory data include the hard data $\{\chi_{i,j-1,k}, \chi_{i-1,j,k}, \chi_{i,j+1,k}, \psi_{i-1/2,j,k}, \psi_{i,j+1/2,k}\}$ and the soft (interval) data $\{\chi_{i+1,j,k}, \psi_{i,j-1/2,k}, \psi_{i+1/2,j,k}\}$. An estimate of $X(\boldsymbol{p})$ is sought at the point $\boldsymbol{p} = (i, j, k)$. Then, a possible term of the corresponding BME equation may be as follows

$$\begin{aligned}\mu_k \int_I d\chi_{i+1,j,k}\, d\psi_{i,j-1/2,k}\, d\psi_{i+1/2,j,k}\, [S\,\Delta t^{-1} \\ - \Delta\, s_1^{-2}(\psi_{i+1/2,j,k} + \psi_{i-1/2,j,k}) \\ - \Delta\, s_2^{-2}(\psi_{i,j-1/2,k} + \psi_{i,j+1/2,k})]\, \exp\,[\mathcal{Y}_{\mathcal{G}}]_{\chi_{i,\,j,\,k}=\hat{\chi}_{i,\,j,\,k}}\end{aligned} \tag{9.21}$$

where the vector $\boldsymbol{I}$ denotes the interval data domains. As was mentioned in Example 3.10 (p. 81), in a realistic flow analysis several terms of the form of Equation 9.21 will be included in the BME equation. Depending on the situation (objectives of the analysis, statistics available, *etc.*), other forms of the $\mathcal{G}$-operator are also possible. The BME equation should be solved for $\hat{\chi}_{i,j,k}$, usually numerically. □

The analysis above suggests an interesting approach of studying stochastic algebraic and differential equations representing physical laws. Generally, given that a natural variable X obeys an equation of the form $D(X, \boldsymbol{Y}) = 0$, where $\boldsymbol{Y} = (Y_1, \ldots, Y_k)$ are observed variables, we wish to find solutions of the corresponding stochastic expectation equation, say $\overline{D(X, \boldsymbol{Y})} = 0$. The traditional approach is either to solve the original physical equation for X and then take the expected value of the solution, or to solve the corresponding expectation equation directly for the moment of interest. Alternatively, BME analysis suggests another way, as follows: Include the expectation equation in the BME mapping process together with any other form of general and specificatory knowledge available. BME will search for solutions such that the expectation equation is an identity. A similar approach can be used in the case of a series of physical equations $D_j(X, \boldsymbol{Y}) = 0$ $(j = 1, \ldots, \zeta)$.

Transformation laws

Now we consider another interesting application of multivariable BME analysis. Assume that a natural variable $X(\boldsymbol{p})$ can be expressed in terms of a secondary variable $Y(\boldsymbol{p})$ by means of a transformation law of the form

$$X(\boldsymbol{p}) = \mathcal{T}[Y(\boldsymbol{p})] \tag{9.22}$$

where $\mathcal{T}[\cdot]$ is one-to-one. The BME posterior pdf of $X(\boldsymbol{p})$ can be written in terms of that of $Y(\boldsymbol{p})$ as follows

$$f_{\mathcal{K}}(\chi_k) = |d\ \mathcal{T}/d\ \psi_k|^{-1}\ f_{\mathcal{K}}(\psi_k\ )|_{\psi_k=\mathcal{T}^{-1}(\chi_k)} \quad (9.23)$$

An immediate consequence of Equation 9.23 is that the BME posterior pdf of a natural variable can be calculated by means of the corresponding pdf of a suitable transformation of the variable. A straightforward application of the analysis in spatiotemporal mapping is demonstrated by Example 9.5.

EXAMPLE 9.5: Consider a map $\boldsymbol{x}_{map}$—which is generally characterized by a non-Gaussian multivariate pdf—and assume that a transformation $\mathcal{T}[\cdot]$ can be established such that the $\boldsymbol{y}_{map} = \mathcal{T}^{-1}[\boldsymbol{x}_{map}]$ has a multivariate Gaussian pdf. Then, the BME posterior pdf $f_{\mathcal{K}}(\chi_k)$ can be calculated from the Gaussian $f_{\mathcal{K}}(\psi_k)$ using Equation 9.23. As we shall see later in Chapter 12 ("Other sorts of kriging," p. 249), in the case that only hard data are available as specificatory knowledge, the multi-Gaussian kriging method is a special case of the above procedure. □

Decision making

There exist various applications where the variable Y used in the decision problem is influenced by another variable X and, therefore, the estimation of X directly affects the decision one makes. Let $\mathfrak{B}(\psi_k, \hat{\chi}_k, d_k;\ \boldsymbol{p}_k)$ be the benefit if ψ_k does occur at point $\boldsymbol{p}_k$ and decision d_k was made. Also, assume that $\hat{\chi}_k$ is the estimate provided by BME analysis. One seeks an optimal decision d_k^* such that the expected benefit is maximized. In mathematical terms, the decision problem is written as

$$\begin{aligned} d_k = d_k^*(\hat{\chi}_k) :\ & \max_{d_k} \overline{\mathfrak{B}(y_k, \hat{\chi}_k, d_k;\ \boldsymbol{p}_k)} \\ & = \max_{d_k} \int d\psi_k\, \mathfrak{B}(\psi_k, \hat{\chi}_k, d_k;\ \boldsymbol{p}_k)\, f_{\mathcal{K}}(\psi_k) \end{aligned} \quad (9.24)$$

where $d_k^*(\hat{\chi}_k)$ denotes that the optimal decision d_k^* is a function of the BME estimate $\hat{\chi}_k$, and the total knowledge base $\mathcal{K}$ takes into account the relationship between X and Y. Decision analysis incorporates: (a) information about the spatiotemporal distribution of the natural variables (represented by BME maps, *etc.*), as well as (b) information about the application-specific goals and the decision maker's preferences. The contribution of modern spatiotemporal geostatistics can be considered at several levels: integrating geographical information systems (GIS) in the decision making process, striking a reasonable compromise among constraints (*e.g.*, physical, economical, political, and ecological constraints), simulating alternative decision scenarios, *etc.* (see also section on "Modern Spatiotemporal Geostatistics and GIS Integration Technologies" on p. 261).

Multipoint BME Analysis

In Chapters 6–8, we mainly considered single-point mapping, *i.e.*, we were seeking an estimate $\hat{\chi}_k$ of χ_k at a single point $\boldsymbol{p}_k$, given the physical knowledge bases $\mathcal{G}$ (general) and $\mathcal{S}$ (specificatory). The *multipoint* mapping approach, on the other hand, is seeking the estimates $\hat{\chi}_{k_j}$ of χ_{k_j} at several points $\boldsymbol{p}_{k_j}$ $(j = 1, \ldots, \rho)$ simultaneously, given the physical knowledge available (Fig. 5.4, p. 121). When practically possible, multipoint mapping could be a considerable improvement over single-point mapping. Indeed, while multipoint mapping offers several interrelated estimates at a time, single-point mapping gives one estimate at a time, independently of its neighboring estimates. The former involves a multivariate pdf that is more informative than the univariate pdf used in the latter.

Multipoint BME estimation

In principle, the BME approach can provide multipoint estimates by replacing the single χ_k with the vector

$$\boldsymbol{\chi}_k = (\chi_{k_1}, \ldots, \chi_{k_\rho}) \tag{9.25}$$

and the vector $\boldsymbol{\chi}_{map} = (\chi_1, \ldots, \chi_m, \chi_k)$ with the vector

$$\boldsymbol{\chi}_{map} = (\chi_1, \ldots, \chi_m, \chi_{k_1}, \ldots, \chi_{k_\rho}) \tag{9.26}$$

Then, the basic BME equations of multipoint space/time mapping are derived as follows. The Lagrange multipliers μ_α are calculated from the solution of the following system of (ρ-point) equations

$$\overline{h_\alpha}(\boldsymbol{p}_{map}) = \int d\boldsymbol{\chi}_{map}\, g_\alpha(\boldsymbol{\chi}_{map}) \exp\,[\mu_0 + \mathcal{Y}_{\mathcal{G}}], \quad \alpha = 0, 1, \ldots, N_c \tag{9.27}$$

The multipoint (ρ-point) posterior pdf is given by

$$f_{\mathcal{K}}(\boldsymbol{\chi}_k) = A^{-1}\, \mathcal{Y}_{\mathcal{S}}[\boldsymbol{\chi}_{map};\, \boldsymbol{p}_k] \tag{9.28}$$

where $\mathcal{Y}_{\mathcal{G}} = \sum_{\alpha=1}^{N_c} \mu_\alpha\, g_\alpha$ and, as usual, the forms of A and $\mathcal{Y}_{\mathcal{S}}$ depend on the specificatory knowledge considered (see also Example 9.7). We now have a system of ρ BMEmode equations

$$\left.\begin{array}{l} \frac{\partial}{\partial \chi_{k_\ell}}\, \mathcal{Y}_{\mathcal{S}}[\boldsymbol{\chi}_{map};\, \boldsymbol{p}_k]_{\boldsymbol{\chi}_k = \hat{\boldsymbol{\chi}}_k} = 0 \\ \ell = 1, \ldots, \rho \end{array}\right\} \tag{9.29}$$

Solution of the system (Eq. 9.29) provides the estimates $\hat{\boldsymbol{\chi}}_k = (\hat{\chi}_{k_1}, \ldots, \hat{\chi}_{k_\rho})$. The number ρ of points to be estimated simultaneously may depend on theoretical, computational, and physical, as well as on decision-making considerations.

Equations 9.27 through 9.29 permit the body of available knowledge to reach out and seek relations beyond its present state.

EXAMPLE 9.6: In the case of the knowledge described in Proposition 6.1, Equations 9.28 and 9.29 reduce to

$$f_{\mathcal{K}}(\boldsymbol{\chi}_k) = A^{-1} \int_I d\boldsymbol{\chi}_{soft} \exp\left[\mu_0 + \mathcal{Y}_{\mathcal{G}}\right] \tag{9.30}$$

and

$$\int_I d\boldsymbol{\chi}_{soft} \left\{\exp\left[\mathcal{Y}_{\mathcal{G}}\right] \frac{\partial}{\partial \chi_{k_\ell}} \mathcal{Y}_{\mathcal{G}}\right\}_{\boldsymbol{\chi}_k = \hat{\boldsymbol{\chi}}_k} = 0, \quad \ell = 1, ..., \rho \tag{9.31}$$

An obvious difference between Equations 7.6 (p. 137) and 9.31 is that while the former determines the χ_k-value at a single point $\boldsymbol{p}_k$ that maximizes the corresponding univariate posterior pdf, the latter provides the combination of $\chi_{k_1}, \ldots, \chi_{k_\rho}$ values at a set of points $\boldsymbol{p}_{k_1}, \ldots, \boldsymbol{p}_{k_\rho}$ that maximize the multivariate posterior pdf. □

EXAMPLE 9.7: For the specificatory knowledge described in Table 6.1 (p. 133), Equations 9.28 and 9.29 yield

$$f_{\mathcal{K}}(\boldsymbol{\chi}_k) = A^{-1} \mathcal{Y}_{\mathcal{S}}[\boldsymbol{\chi}_{map}; \boldsymbol{p}_k] = A^{-1} B \int_{\mathcal{D}} d\Xi_{\mathcal{S}}(\boldsymbol{\chi}_{soft}) \exp\left[\mu_0 + \mathcal{Y}_{\mathcal{G}}\right] \tag{9.32}$$

and

$$\frac{\partial}{\partial \chi_{k_\ell}} \mathcal{Y}_{\mathcal{S}}|_{\chi_{k_\ell} = \hat{\chi}_{k_\ell}} = B \int_D d\Xi_{\mathcal{S}}(\boldsymbol{\chi}_{soft}) \left\{ \exp[\mathcal{Y}_{\mathcal{G}}] \frac{\partial}{\partial \chi_{k_\ell}} \mathcal{Y}_{\mathcal{G}}\right\}_{\chi_{k_\ell} = \hat{\chi}_{k_\ell}}$$

$$= 0, \quad \ell = 1, \ldots, \rho \tag{9.33}$$

which will be used in the derivation of analytical results in Chapter 11. □

Mathematically, in order to assure that the solution of the system of equations (Eq. 9.29) provides the maximum multipoint posterior pdf estimates, the Hessian of $f_{\mathcal{K}}(\boldsymbol{\chi}_k)$ at $\boldsymbol{\chi}_k = \hat{\boldsymbol{\chi}}_k$,

$$H f_{\mathcal{K}}(\hat{\boldsymbol{\chi}}_k)(\boldsymbol{q}) = \tfrac{1}{2} \sum\nolimits_{i,j=k_1}^{k_\rho} \frac{\partial^2 f_{\mathcal{K}}(\hat{\boldsymbol{\chi}}_k)}{\partial \chi_i \, \partial \chi_j} \, q_i \, q_j \tag{9.34}$$

must be negative-definite, *i.e.*, $H f_{\mathcal{K}}(\hat{\boldsymbol{\chi}}_k)(\boldsymbol{q}) \leq 0$ for all $\boldsymbol{q} = (q_{k_1}, \ldots, q_{k_\rho})$ $[H f_{\mathcal{K}}(\hat{\boldsymbol{\chi}}_k)(\boldsymbol{q}) = 0$ for $\boldsymbol{q} = 0$ only$]$. Some numerical examples of multipoint BME analysis are included in Chapter 11.

As discussed in Chapter 1, scientific explanation and prediction are to some extent parallel processes. In this context, the posterior pdf (Eq. 9.28) is useful not only because we wish to make predictions. Another reason for using Equation 9.28 pertains to the goal of capturing significant *generalizations* which are important in the *explanatory* description of the natural phenomenon the map represents.

EXAMPLE 9.8: If we say that a porosity map of the West Lyons field (Kansas) is given by the vector

$$\hat{\boldsymbol{\chi}}_k = \underbrace{(9.5,\ 13.7,\ \ldots,\ 10.6)}_{\rho\,=\,5,000 \text{ points}} \tag{9.35}$$

with a probability of 0.7, we are making a generalization that goes beyond the small data set available (76 data points, see p. 143–147). In assigning this probability, we understand that the little we actually observe is part of a much larger natural process. □

Multipoint BME uncertainty assessment

A complete characterization of mapping uncertainty is provided by the multipoint BME pdf. In many applications, a realistic assessment of the multipoint mapping error is achieved using the concept of the *multipoint BME confidence set*, which is an extension of the confidence width of single-point mapping (Chapter 8, "Asymmetric Posteriors," p. 153). As we did in previous situations, we choose an appropriate confidence level η (with $0 \leq \eta \leq 1$). The choice of η depends on the mapping situation at hand. Then, a multipoint confidence set Φ_η is determined such that

$$P[\boldsymbol{\chi}_k \in \Phi_\eta] = \eta \tag{9.36}$$

Clearly, for a given level of probability η there are several sets Φ_η satisfying Equation 9.36 (intuitively, as the η increases, so should the corresponding Φ_η). The multipoint BME confidence set is defined as the set with the smallest size

$$\| \Phi_\eta \| = \int_{\boldsymbol{\chi}_k \in \Phi_\eta} d\boldsymbol{\chi}_k \tag{9.37}$$

for a given η. These concepts are best illustrated by means of an example.

EXAMPLE 9.9: If, *e.g.*, $\rho = 1$ (single-point analysis), the BME confidence set is simply the confidence interval $\Phi_\eta = [\chi_{k,l}, \chi_{k,u}]$ with the smallest length $\| \Phi_\eta \| = \chi_{k,u} - \chi_{k,l}$ such that Equation 9.36 is satisfied. □

A multipoint confidence set requires a multidimensional (ρ-dimensional) plot; *e.g.*, a contour map in the case of two points ($\rho = 2$). A confidence set takes into consideration dependencies between all variables, as expressed by the posterior pdf. The following result is useful in determining BME confidence sets (Serre and Christakos, 1999a).

PROPOSITION 9.1: The Λ_β-probability density set defined as

$$\Lambda_\beta = \{\boldsymbol{\chi}_k : f_{\mathcal{K}}(\boldsymbol{\chi}_k)/f_{\mathcal{K}}(\hat{\boldsymbol{\chi}}_k) \geq 1 - \beta\} \tag{9.38}$$

is a BME confidence set Φ_η with a confidence level given by

$$\eta(\beta) = P[\boldsymbol{\chi}_k \in \Lambda_\beta] \tag{9.39}$$

The Λ_β probability density sets are easily constructed from the posterior pdf, in which case Equation 9.38 above provides the corresponding BME confidence set. Note that if the inverse function for $\eta(\beta)$ exists, then one can directly obtain a BME confidence set Φ_η for a selected level η. In this case, the Φ_η is given by

$$\Phi_\eta = \Lambda_\beta : P[\boldsymbol{\chi}_k \in \Lambda_\beta] = \eta \tag{9.40}$$

A useful corollary of Proposition 9.1 for $\rho = 1$ is when the BME confidence set is a single interval. In this case, the

$$\Phi_\eta = [\chi_{k,l}, \chi_{k,u}] \tag{9.41}$$

is a BME confidence interval such that $f_{\mathcal{K}}(\chi_{k,l}) = f_{\mathcal{K}}(\chi_{k,u})$ and $P[\chi_{k,l} \leq \chi_k \leq \chi_{k,u}] = \eta$ (see also Chapter 8). A result that is useful in computational applications is suggested by the following proposition (Serre *et al.*, 1998).

PROPOSITION 9.2: If the multipoint posterior pdf $f_{\mathcal{K}}(\boldsymbol{\chi}_k)$ is approximated by a Gaussian law, the probability function $P[\Lambda_\beta]$ is efficiently calculated by means of the following expression

$$P[\Lambda_\beta] \approx (2\pi)^{-\rho/2} \, |\boldsymbol{c}_{map}|^{-1/2} \int_{Q(\boldsymbol{\chi}_k) \leq -2\ln(1-\beta)} d\boldsymbol{\chi}_k \, exp\left[-\tfrac{1}{2}Q(\boldsymbol{\chi}_k)\right] \tag{9.42}$$

where $Q(\boldsymbol{\chi}_k) = [\boldsymbol{\chi}_k - \hat{\boldsymbol{\chi}}_k]^T \boldsymbol{c}_{map}^{-1} [\boldsymbol{\chi}_k - \hat{\boldsymbol{\chi}}_k]$.

COMMENT 9.1: *As mentioned in Comment 2.8 (p. 58), when dealing with matrix multiplications, a vector $\boldsymbol{x} = (x_1, \ldots, x_n)$ will be considered as a column vector* $\boldsymbol{x} = \begin{bmatrix} x_1 \\ \vdots \\ x_n \end{bmatrix}$; i.e., *the two forms of writing vectors are identical. Then, we can also write* $\boldsymbol{x}^T = [x_1, \ldots, x_n]$, etc.

EXAMPLE 9.10: In the case of the two-point posterior pdf ($\rho = 2$), the simplification $P[\Lambda_\beta] = \eta \approx \beta$ is obtained. Hence, the confidence set $\Phi_{\eta \approx \beta}$ coincides with the index set Λ_β. □

COMMENT 9.2: *As we already mentioned, the BME confidence sets are a multipoint generalization of the single-point confidence intervals (in which several estimated values are considered simultaneously). The same concepts can easily be extended jointly to the vector estimation of several random fields.*

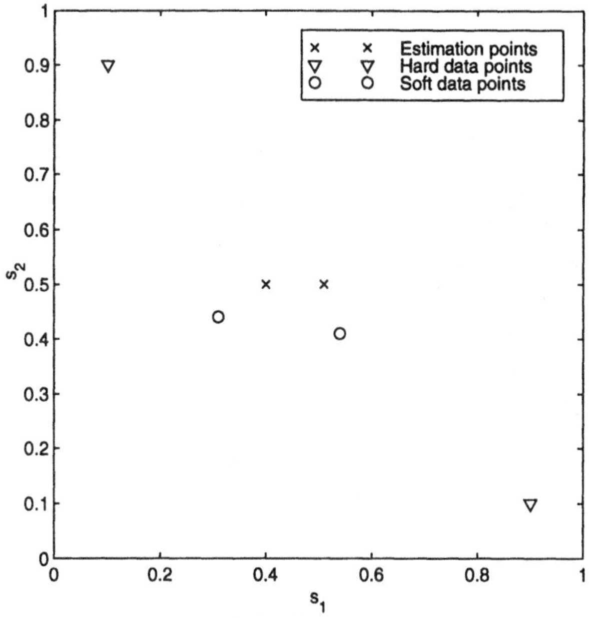

Figure 9.2. Location of estimation points and hard and soft (interval) data points for simulation study in Example 9.11.

A simulation study is discussed next, which provides an illustration of uncertainty assessment in terms of confidence sets. The study includes a comparison of single-point *vs*. multipoint BME analysis.

EXAMPLE 9.11: It is impossible to do serious modeling in natural sciences without assessing uncertainty. In order to obtain a numerical illustration of uncertainty assessment by means of BME confidence sets, consider the data and estimation point configuration of Figure 9.2. Estimates were sought for the estimation point on the left of the figure (x_{k_1}) and for the estimation point on the right of the figure (x_{k_2}). A zero mean and the covariance model

$$c_x(r) = c_o \, exp\,[-r^2/a_r^2] \tag{9.43}$$

($a_r = c_o = 1$), were assumed valid for the underlying random field. Typical BME confidence sets for x_{k_1} and x_{k_2} are obtained using Equation 9.40. These sets are shown in Figure 9.3a.

The lines delineate the contours of the confidence sets and the labels indicate the confidence probability η (ranging from 0.1 to 0.9, with 0.1 increments). Each contour represents the smallest set of x_{k_1} and x_{k_2} values for the η-value shown on the label. These sets are small, because they take into consideration the correlations between x_{k_1} and x_{k_2}. For comparison purposes, in Figure 9.3b we plot the confidence sets which would be obtained if we were to replace the

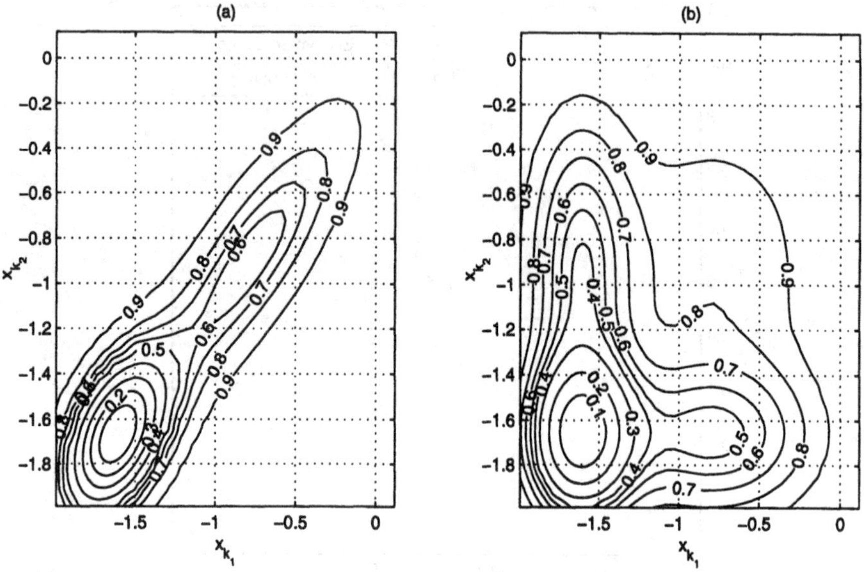

Figure 9.3. BME confidence sets Φ_η for x_{k_1} and x_{k_2} constructed from (a) multipoint analysis, and (b) single-point analysis using the BME confidence intervals.

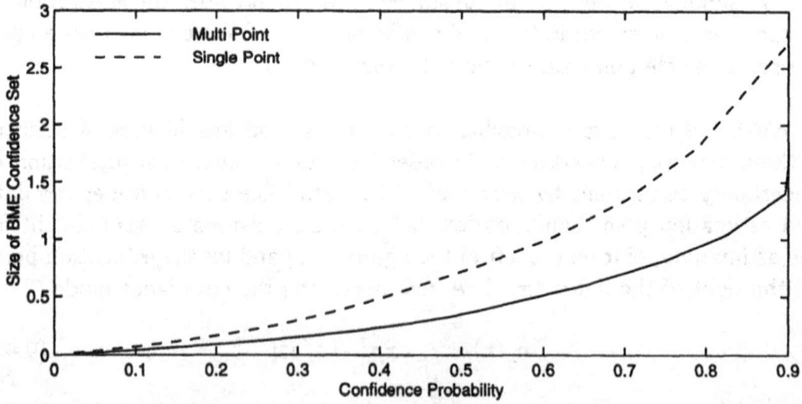

Figure 9.4. Variation of the size $\| \Phi_\eta \|$ of the BME confidence sets as a function of the confidence probability η for the two-point and single-point analyses.

bivariate pdf $f_{\mathcal{K}}(\chi_{k_1}, \chi_{k_2})$ of the two-point analysis with the product of the univariate pdf's $f_{\mathcal{K}}(\chi_{k_1})$ and $f_{\mathcal{K}}(\chi_{k_2})$ of the single-point analysis—which corresponds to assuming (for the sake of analysis) that the variables x_{k_1} and x_{k_2} at the estimation points are independent. It is evident from Figure 9.3 that the shapes of the confidence sets are significantly different; indeed Figure 9.3b provides a poor representation of the real situation. Furthermore, the BME confidence set size $\|\Phi_\eta\|$ defined in Equation 9.37 is equal to the area surrounded by the corresponding confidence probability η-contour in Figure 9.3. For comparison, the variation of $\|\Phi_\eta\|$ *vs.* η is plotted in Figure 9.4 for both the multipoint case of Figure 9.3a and the single-point case of Figure 9.3b. From Figure 9.4 it is interesting to note that for any given confidence probability η, the size of the corresponding multipoint confidence set of Figure 9.3a is consistently smaller than the size of the single-point confidence set of Figure 9.3b. In other words, in this simulation study the multipoint confidence sets seem to offer a considerable improvement over the single-point confidence sets. □

Other applications of the BME uncertainty analysis include the construction of maps displaying the space/time fluctuations in the probability that the values of a natural process: (*i.*) do not exceed a certain threshold (*e.g.*, the concentration of a soil contaminant does not exceed a threshold derived on the basis of environmental health standards); (*ii.*) are between two given boundaries (*e.g.*, the temperature in a region stays within the boundaries that allow the growth of a specified agricultural product); *etc.*

BME in the Context of Systems Analysis

As we have already noted, in many applications the outcomes of BME analysis (predictive maps, uncertainty measures, *etc.*) serve as the precious inputs to subsequent steps of scientific investigations, engineering designs, decision making, *etc.* Many examples can be given, which essentially depend on the scientific background and area of expertise of the person using modern spatiotemporal geostatistics. Below, we will focus our attention on three application areas of general systems analysis: (*i.*) risk analysis of natural systems; (*ii.*) human-exposure assessment; and (*iii.*) environmental exposure–health effect associations.

Risk analysis of natural systems

Several applications in the risk analysis of natural systems can be described in terms of a general *performance function* $\mathcal{H}(\boldsymbol{X})$ such that

$$\mathcal{H}(\boldsymbol{X}) = \begin{cases} > 0, & \text{system is safe} \\ < 0, & \text{system fails} \\ \;\;\; 0, & \text{limiting state} \end{cases} \tag{9.44}$$

where the variables $\boldsymbol{X} = (X_1, \ldots, X_\lambda)$ describe the environmental conditions of interest. The corresponding failure probability of the system can be expressed in terms of the BME posterior pdf and the function $\mathcal{H}$ as follows,

$$P_F = \int_{\mathcal{H}<0} d\boldsymbol{\chi}\, f_{\mathcal{K}}(\boldsymbol{\chi}) \tag{9.45}$$

The $\mathcal{H} = 0$ defines a critical hypersurface in the λ-dimensional space. The *reliability index* β of the system may be generally defined as the minimum distance from the origin to the critical hypersurface. The meaning of β is better illustrated by means of an example.

EXAMPLE 9.12: The case of thermal pollution in a river, *e.g.*, may be modeled in terms of Equation 9.44 with $\mathcal{H}(\boldsymbol{X}) = X_1 - X_2$, where X_1 represents a fraction of the natural flow in the river, and X_2 denotes the discharge from the cooling system of a thermal power plant flowing into the river (Kottegoda and Rosso, 1997). In this simple case we find $\beta = \overline{\mathcal{H}}/\sigma_{\mathcal{H}}$, which may be interpreted as the number of $\sigma_{\mathcal{H}}$'s between $\overline{\mathcal{H}}$ and the critical value $\mathcal{H} = 0$. In the more general case in which

$$\mathcal{H}(\boldsymbol{X}) = c_0 + \sum\nolimits_{i=1}^{\lambda} c_i\, X_i \tag{9.46}$$

the corresponding reliabilty index is given by

$$\beta = \frac{c_0 + \sum_{i=1}^{\lambda} c_i \overline{X_i}}{\sqrt{\sum_{i=1}^{\lambda} \sum_{j=1}^{\lambda} c_i\, c_j\, \rho_{ij}\, \sigma_i\, \sigma_j}} \tag{9.47}$$

where σ_i and σ_j are the standard deviations of X_i and X_j, respectively, and ρ_{ij} are correlation functions between X_i and X_j $(i, j = 1, \ldots, \lambda)$. All the statistics involved in Equation 9.47 are calculated in terms of BME analysis. □

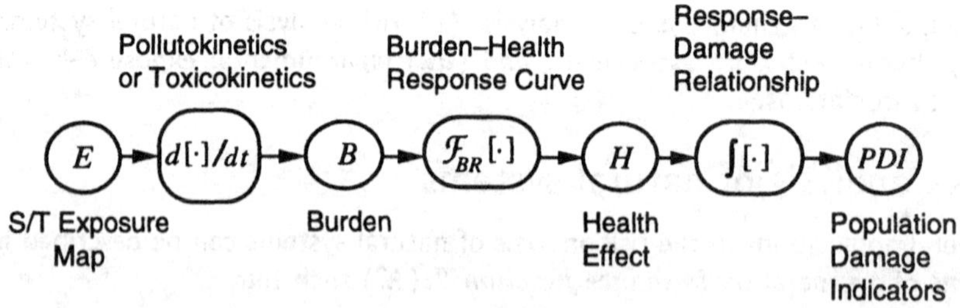

Figure 9.5. The holistic human-exposure system.

Human-exposure systems

BME maps are very valuable in environmental health studies. Figure 9.5 presents an outline of the holistic human-exposure system (proposed by Christakos and Kolovos, 1999). The main parts of this system are as follows.

• The pollutant exposure map E obtained from BME analysis serves as the input to the physiologically based pollutokinetic (or toxicokinetic) model

$d[\cdot]/dt$, which takes into consideration exposure variations in space/time as well as variabilities linked to the biological and physiological characteristics of the individual. The output of the pollutokinetic model is a burden map B for representative receptors (*i.e.*, receptors sharing the same biological and physiological characteristics). Burden provides a measure of the fraction of the pollutant that reaches the target organs and tissues of a receptor and is capable of affecting them.

• The burden map is then used in combination with an empirical burden-response curve $\mathcal{F}_{BR}[\cdot]$ in order to obtain predictive maps of the exposure's impact H on human health.

• The health effect H is, finally, substituted into a response–damage model $\mathcal{f}[\cdot]$ to derive health damage indicators HDI (see, *e.g.*, Eq. 9.4 above). Naturally, each step of the approach above incorporates knowledge from a variety of sources, including exposure monitoring, biologic monitoring, and health damage surveillance. The example below is concerned with the human-exposure assessment of ozone concentrations over eastern U.S.

EXAMPLE 9.13: The ozone burden maps of Figure 9.6 are produced from the solution of a first-order pollutokinetic model (Christakos and Hristopulos, 1998; see also Example 1.14, p. 18) and are associated with a class of representative receptors. The pollutokinetic model provides the ozone burden on a receptor at each space/time point $\boldsymbol{p}$. This model is a function of an absorption rate, a removal rate constant, and the ozone exposure maps of Figure 9.1. Studies have shown that ozone burden and receptor response are correlated, and a knowledge of ozone burden is prerequisite to an unambiguous evaluation of health effects. Using the appropriate burden-response curve $\mathcal{F}_{BR}[\cdot]$ and response–damage model $\mathcal{f}[\cdot]$, the burden map of Figure 9.6 leads to HDI maps for any receptor that belongs to a specific cohort (*i.e.*, a group of individuals with similar time or activity profiles). Such is the health damage indicator map plotted in Figure 1.6 on p. 8 (in which case the HDI expressed no. of representative receptors affected/km^2). While the burden maps represent the actual exposure an individual representative receptor may receive in space/time, the HDI maps offer an assessment of the absolute or relative impact of exposure on the population as a whole. In other words, the HDI maps possess a social policy dimension that burden maps do not. □

Associations between environmental exposure and health effect

Human-exposure analysis generally involves both physical and epidemiologic variables, which means that techniques capable of integrating knowledge from both the physical and epidemiologic sciences are needed. The application of the random field model in human-exposure analysis involves some modeling decisions. A common modeling decision is the spatiotemporal continuity of the exposure and health-effect variables. While most environmental exposures

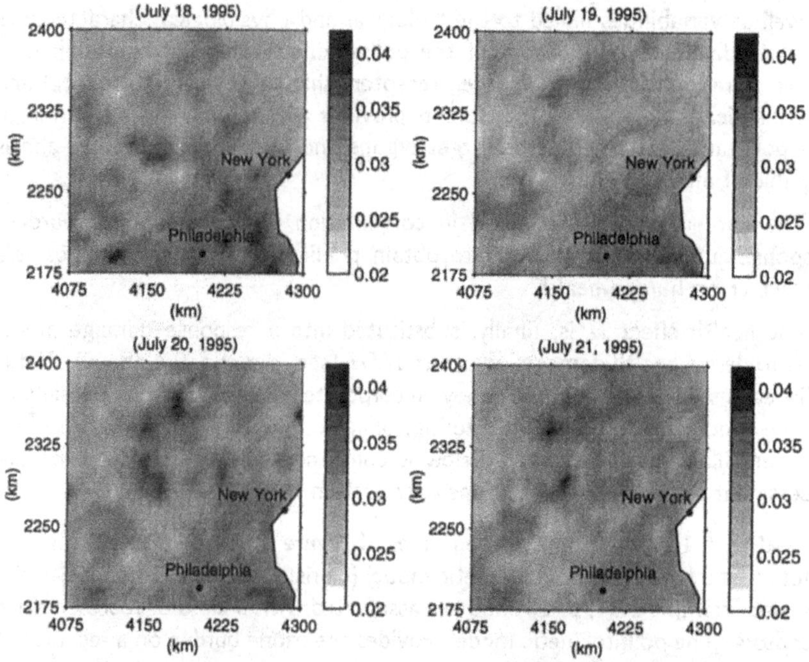

Figure 9.6. Maps of daily accumulated burden (ppm) on representative receptors at a region in the eastern U.S.

can be measured continuously in space and time, many health-effect variables (*e.g.*, death rates) are not measurable at all spatial locations. In such cases, modeling may proceed as follows (Fig. 9.7): Suppose that death-rate D data are available at regions R_i ($i = 1, 2, \ldots, m$), but no data are available at regions R_i^* ($i = 3, 5, \ldots, m-2$). The death rate D_i observed within each region R_i is assigned at a geographical location $\boldsymbol{s}_i$ of the region R_i that is selected on the basis of statistical and health administrative criteria (*e.g.*, the centroid of R_i; Fig. 9.7a). Using the random field techniques, continuously distributed death rates $D(\boldsymbol{s}, t)$ can be generated in space and time (Fig. 9.7b). Furthermore, death-rate values D_i^* can be assigned at the centroids of the unobserved regions in terms of the average value of $D(\boldsymbol{s}, t)$ within each region R_i^* (Fig. 9.7c).

The investigation of human-exposure–health-effect associations is a very complicated yet extremely important issue in environmental health studies, leading to several criteria for testing (*i.e.*, supporting or rejecting) such an association (Hill, 1965; Hoel and Landrigan, 1987; Blot and McLaughlin, 1995). In fact, there exist various sorts of association, including *deterministic causation* in which the causes are necessary and sufficient for their effects, as well as *stochastic causation* which includes causes that raise the chances of their effects. Deterministic exposure–effect relationships refer to the biology of

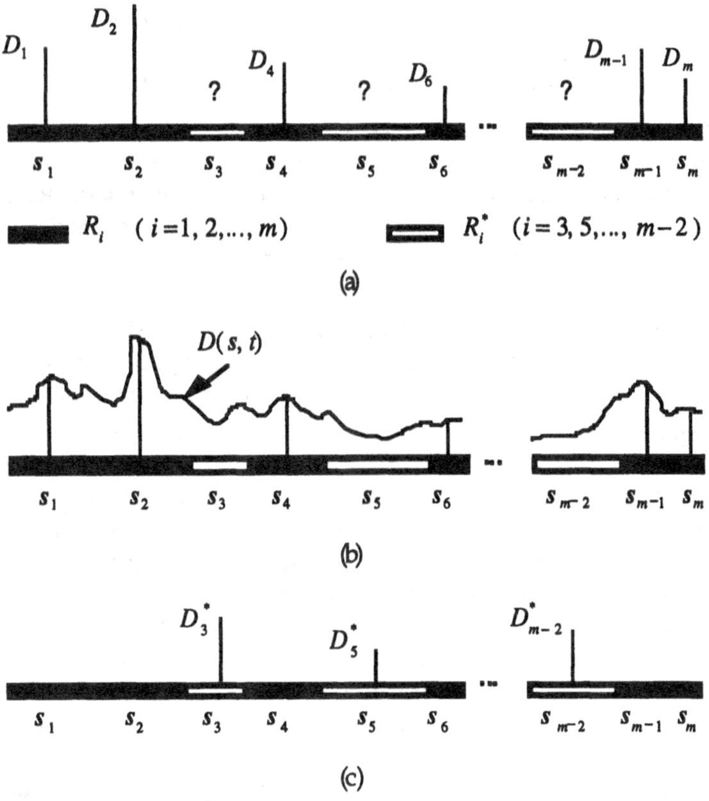

Figure 9.7. (a) Death-rate D data available at regions R_i. (b) Continuously distributed death rates. (c) Death-rate values D_i^* assigned at the centroids of unobserved regions.

causation at the individual level and are generally very difficult to establish. Most studies of environmental epidemiology are concerned with stochastic exposure–effect associations at the population level. While stochastic associations do not usually imply necessary and sufficient causation criteria, they offer useful insight into a very complicated situation of tremendous public-health significance (see, *e.g.*, Rothman and Greenland, 1998). Epidemiologic studies of the determinants of disease in a population involve a variety of techniques, including visual comparisons of the patterns exhibited by the disease and environmental factors and statistical analysis of population exposure–disease occurrence across a geographic area (Glattre, 1989; Krewski *et al.*, 1989; Blot and McLaughlin, 1995). A useful means for evaluating exposure–effect associations is the stochastic *physico-epidemiologic predictability* (PEP) criterion—also called the "scalar *vs.* vector prediction (SVP) criterion"—that is expressed by the following postulate.

POSTULATE 9.1: An exposure–effect association at the population level is supported if the health-effect predictions obtained from vector

(multivariable) BME analysis using both the exposure and the health-effect data are superior to the effect predictions obtained from scalar (single-variable) BME analysis using only health-effect data.

The PEP criterion is well suited for exposure and health distributions with pronounced spatial and temporal characteristics and provides a meaningful representation of the exposure–effect association in spatiotemporal domains by way of graphs and maps. Before the PEP criterion is applied in practice, certain conditions must be satisfied (see Christakos and Hristopulos, 1998; Christakos and Serre, 2000a):

(*i.*) Exposure *precedes* health effect (*e.g.*, there may exist a history of regularity in such a precedence, or there is a biological possibility of the precedence in light of existing knowledge about disease etiology).

(*ii.*) Exposure and health effect are *contiguous* in the spatiotemporal domain (*i.e.*, there is a clear link in time and place of the exposure and health effect that we are connecting causally). Condition (*ii.*) requires the existence of some spatiotemporal connection between exposure and effect (*e.g.*, when we say that a pollutant caused a group of receptors to become ill, we imply that the pollutant and the receptors both are located in the same geographical area). In many cases this contiguity is not a trivial aspect, for biological or organic systems are in a constant state of exchange with their surrounding environmental conditions.

(*iii.*) The necessary adjustments for *confounding* variables (*i.e.*, variables that may be closely associated with both exposure and effect) have been made, so that their effects can be clearly distinguished from those of the exposure under investigation. Nevertheless, several studies have shown that strong associations are highly unlikely to be due entirely to a hidden confounding variable, unless this variable is closely associated with the health effect and the risk factor (*e.g.*, Flanders and Khoury, 1990; Khoury and Yang, 1998). Also, Rothman and Greenland (1998) have suggested that, given one's ignorance regarding the hidden causal components, the best possible approach to health-risk assessment is to classify people according to measured causal risk indicators and then assign the average risk observed within a class to persons within the class.

While these three conditions are commonsense in epidemiologic investigations (*e.g.*, Hill, 1965), no one of them is an all-sufficient basis for judgment. A novel condition introduced by Postulate 9.1 is that the existence and strength of an exposure–effect association is judged on the basis of the successful space/time predictions to which the combined physico-epidemiologic analysis leads. Thus, a central feature of the scientific status of the PEP criterion is its *testability*, *i.e.*, the predictions made by the PEP criterion are testable. The better the vector health-effect predictions (*i.e.*, BME predictions made on the basis of physical exposure and epidemiologic data) compare to the scalar health-effect predictions (*i.e.*, BME predictions made on the basis of epidemiologic data only), the stronger is the exposure–effect association, thus offering a measure of the strength and consistency of the association. An exposure may not be considered as a possible causal factor if the vector health-effect predictions provide no improvement in the rate prediction compared to the scalar health-effect predictions.

The quantitative assessment of the prediction accuracy of Postulate 9.1 can be made in terms of the *prediction errors* at a set of "control points," *i.e.*, points where the actual health effects (*e.g.*, death rates) are known and can be compared with the health-effect predictions obtained from scalar *vs.* vector BME techniques. Analysis of an exposure–mortality association, *e.g.*, is based on the successful prediction of mortality at a set of control points using the combined exposure (X) and death-rate (D) data. Providing that there are no strong effects due to hidden confounding factors, an improved mortality prediction obtained from the combination of death rate and exposure data as compared to that obtained merely from death-rate data should support the existence of an association between X and D. Hence, a suitable PEP measure of the strength of the exposure–mortality association is defined as

$$\beta_{DX} = (E_{DX} - E_D)/E_D \tag{9.48}$$

(in %), where E_{DX} is the mortality prediction error (in the stochastic sense) based on death-rate D and exposure data X, and E_D, the mortality prediction error based on death-rate data only [the extension of Equation 9.48 in the case of confounding factors is discussed later]. A consistently negative β_{DX}-value supports the existence of an exposure–mortality association. In terms of mathematical logic, the analysis above may be expressed by the material conditional of the form "X implies $\beta_{DX} < 0$," for short, $X \to (\beta_{DX} < 0)$. The material conditional is a logical structure that is based on truth-functional concepts (see Chapter 4, "Material and strict map conditionals," p. 98) and is equivalent to the statement "it is not the case that X and not $\beta_{DX} < 0$," in short, $\neg[X \wedge \neg(\beta_{DX} < 0)]$. It is worth mentioning that the scientific reasoning of the PEP criterion violates neither Hume's nor Mill's rules of cause and effect (as discussed, *e.g.*, in Harris, 1996). Also, since the essential structure underlying PEP is *predictability*, it is in agreement with Popper's concept of scientific reasoning (Popper, 1962).

COMMENT 9.3: *In addition to suggesting possible exposure–effect associations, the PEP criterion can be helpful in testing or confirming association hypotheses developed from other investigations (medical, biological, toxicological,* etc.*). In some cases sufficient information may be obtained on the relation between exposure and effect to set realistic exposure standards.*

The following example presents an application of the PEP criterion by Christakos and Serre (2000a), in the case of cold temperature data (which is considered as the exposure variable in this case) *vs.* mortality data (which is the health-effect variable) in the State of North Carolina. This example illustrates the use of the PEP criterion to obtain a quantitative assessment of the association between cold temperature and mortality distributions in space/time, if we were to assume that there are no confounding factors. In Example 9.15, below, the incorporation of a confounder into PEP analysis is demonstrated by using particulate matter (PM_{10}) data.

EXAMPLE 9.14: Temperature–mortality relationships are known to be V-shaped, with both cold and hot temperature extremes resulting in higher death rates (Saez *et al.*, 1995; Christophersen, 1997). However, the North Carolina region did not experience any substantial hot temperature extremes for the time period considered (season 5, which corresponds—roughly—to the winter of 1996); the daily mean temperature during season 5 did not exceed 65° F, which is less than the 91° F threshold suggested by Honda *et al.* (1995) for mortality increase induced by high temperature. Therefore, only low temperatures were considered as the environmental exposure that could lead to higher death rates.

The temperature field $T(\boldsymbol{s}, t)$ is a physical variable that is measured in degrees Fahrenheit (° F) at the weather stations. These temperature measurements are considered as hard data (*i.e.*, there is a high degree of confidence that the data obtained are not contaminated by errors). For the purposes of human-exposure analysis, we defined the exposure to temperature field $X(\boldsymbol{s}, t) = -K \times T(\boldsymbol{s}, t)$, where $K = 4.4729 \times 10^{-2}$ is a constant chosen such that the measured temperature exposure X-values have the same stochastic mean as the measured death rates D; the negative sign ensures a positive correlation between D and X (which, in this case, is considered as the temperature exposure). Clearly, as the temperature T drops, the exposure to cold temperature X increases and the death rate D is expected to increase. The information available for mortality $D\,(\boldsymbol{s}, t)$ consists of daily death counts for 14 representative counties and their aggregated neighbors. Death counts provide an uncertain information about the death rate that can be expressed in terms of soft intervals of the form $\left(\frac{d_i-1}{n_i}, \frac{d_i+1}{n_i}\right)$, where d_i is the daily death count in County i and n_i is the population (in 100,000-people units). Consider, *e.g.*, the soft data for County 2 and County 12 as shown in Figure 9.8. Since County 12 has a much smaller population than County 2, the death-rate measurements at the centroid of County 12 have a larger uncertainty than the death rates obtained in County 2.

The space/time correlation structure of winter season 5 is modeled using homogeneous/stationary covariance functions. The study of these covariances indicated that, while death rate includes a considerable component of randomness, the temperature distribution is a much smoother random field. Also, the spatial correlation range of temperature exposure is much larger than that of death rate. The cross-covariance between exposure and death rate reaches its maximum value for an exposure–death-rate time lag $\tau_0 = 2$ days during the winter time period. This means that there is a time delay between a cold temperature episode and the resulting increase in death rate, indicative of a possible causal association. The precedence and contiguity conditions (*i.*) and (*ii.*), respectively, are clearly satisfied. The no-confounder condition (*iii.*) might

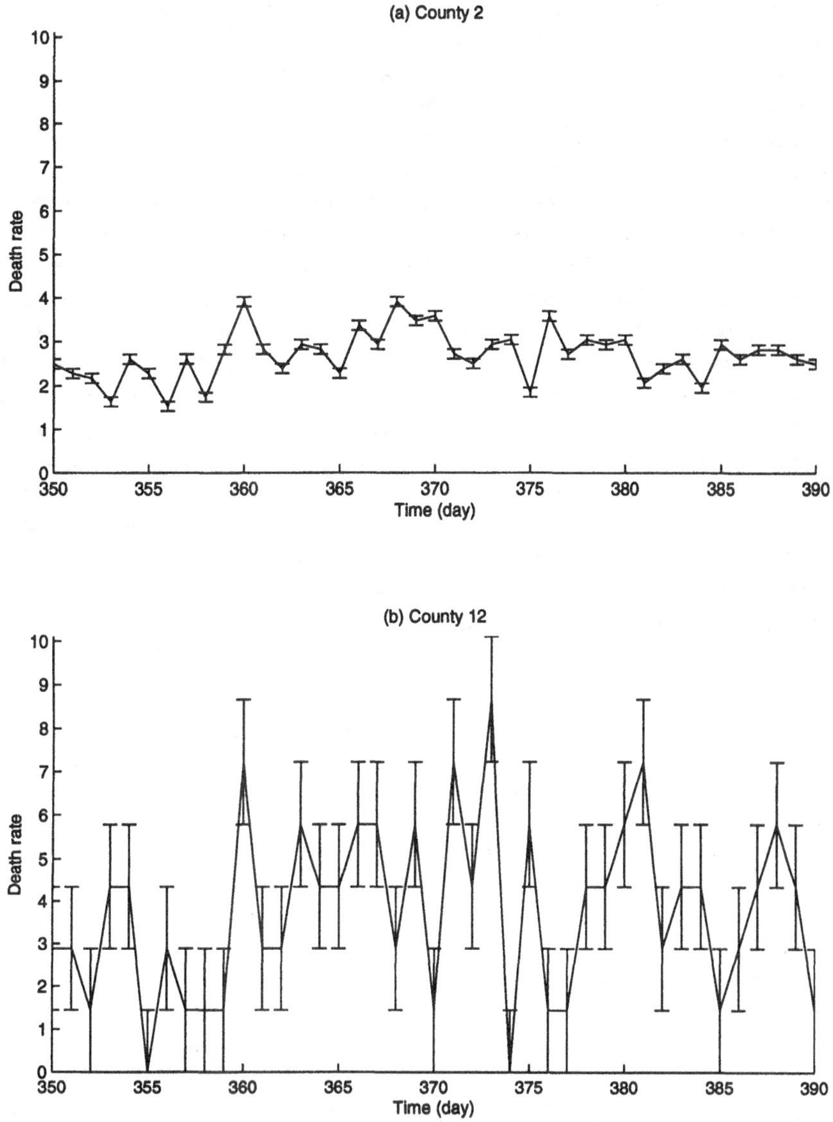

Figure 9.8. Soft (interval) death-rate data for County 2 and County 12 (in number of deaths per 100,000 cases per day). The error bars represent the uncertainty in the data.

raise some interesting issues, particularly with respect to confounding factors such as socio-economic condition, pollution, *etc.* Also, the list of possible risk factors may include age, gender, preexisting health conditions, *etc.* Nevertheless, on the basis of the information available, no considerable confounding was identified (no confounding variables were found to be closely associated with both the health effect and the risk factor; *etc.*). This decision seems to be justified by the fact that—as it will be shown in Example 9.15 below—the incorporation of PM_{10} distribution as a possible confounder has little effect on the quantitative assessment of the temperature exposure–mortality association. As a result, condition (*iii.*) was assumed valid at this stage of the analysis; the consideration of confounders is examined in Example 9.15. Regarding the main spatiotemporal condition of Postulate 9.1, its application requires death-rate predictions based on: (a) soft death-rate data only; and (b) on the combination of soft death-rate and hard temperature-exposure data. Both predictions (a) and (b) are obtained with the help of BME techniques. Death-rate predictions were first obtained using soft death-rate data which were available in the form of intervals; *i.e.* $D(\boldsymbol{s}, t) \in (l, u)$, where l is a lower rate value and u is an upper value. Examples of such interval soft data are shown in Figure 9.8. Let $f_D(D_k)$ denote the posterior pdf of the death rates $D_k = D(\boldsymbol{s}_k, t_k)$ which are predicted at space/time points $(\boldsymbol{s}_k, t_k)$ using death-rate data at neighboring counties as well as at the same county but during different days. For illustration, the pdf obtained for County 7 during the day $t_k = t_7 = 415$ is shown in Figure 9.9. On the basis of the $f_D(D_7)$ of Figure 9.9, the most probable death-rate prediction (mode) $\hat{D}_{7|D} = 2.78$ (deaths per 100,000 people per day) was found, as well as other desirable estimates. Next, using both the interval death-rate data and the hard temperature-exposure data, the vector BME approach provided the posterior pdf $f_{DX}(D_7)$ for County 7 during the same day $t_k = t_7 = 415$ (shown in Fig. 9.9).

The $f_{DX}(D_7)$ is clearly an improvement over $f_D(D_7)$. It leads, *e.g.*, to the mode death-rate prediction of $\hat{D}_{7|DX} = 3.35$ (deaths per 100,000 people per day) that is closer to the actual death rate $D_7 = 3.77$ than the $\hat{D}_{7|D}$ previously calculated (D_7 was assumed unknown during the prediction process). According to the PEP criterion, these results support an association between colder temperatures and death rates at the population level. To proceed further with the study of the temperature exposure–death-rate association, death-rate predictions $\hat{D}_{k|D}$ and $\hat{D}_{k|DX}$ were produced at all counties in the state for all days of the 1996 winter season 5. As before, while the $\hat{D}_{k|D}$ predictions were obtained at each space/time point using only death-rate data, the $\hat{D}_{k|DX}$ predictions were calculated using death-rate data as well as temperature data. On the basis of these predictions, the death-rate prediction errors $e_{k|D} = |D_k - \hat{D}_{k|D}|$ and $e_{k|DX} = |D_k - \hat{D}_{k|DX}|$ were calculated at each space/time point $(\boldsymbol{s}_k, t_k)$. The association between exposure to colder temperature and death rate was then investigated by computing the PEP parameter β_{DX} of Equation 9.48 above (in %), where E_{DX} and E_D, respectively, are the arithmetic averages of the $e_{k|DX}$ and $e_{k|D}$ values over a spatiotemporal domain consisting of all counties and a time window of 30 days. According to the PEP criterion, negative β_{DX}-values would support the cold temperature exposure–death-rate association (whereas zero values would not support such an association).

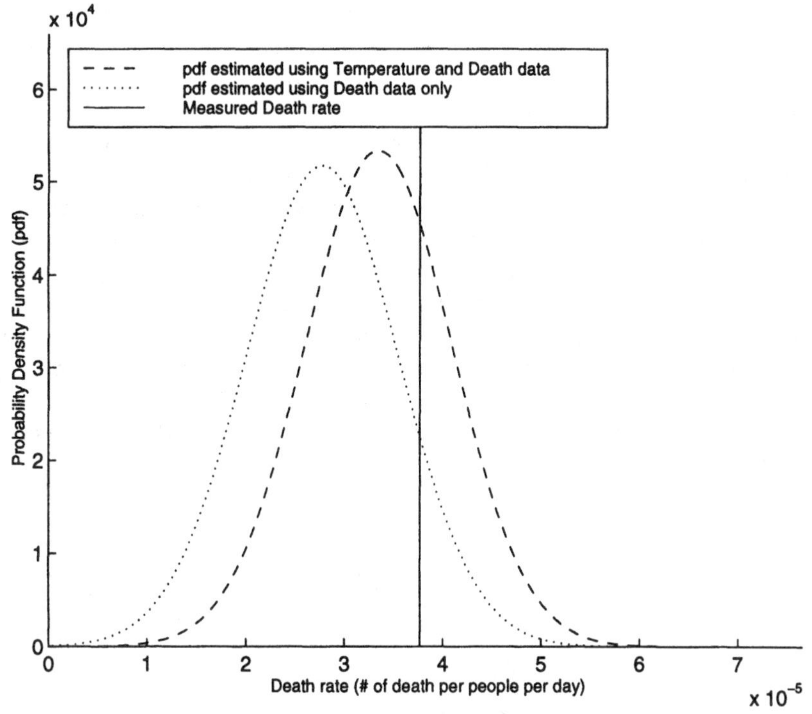

Figure 9.9. Posterior pdf of the death rate for County 7 at day 415 obtained using: (a) death-rate data (dotted line) and (b) death-rate and temperature data (dashed line). The vertical solid line represents the actual recorded death rate (in number of deaths per 100,000 cases per day).

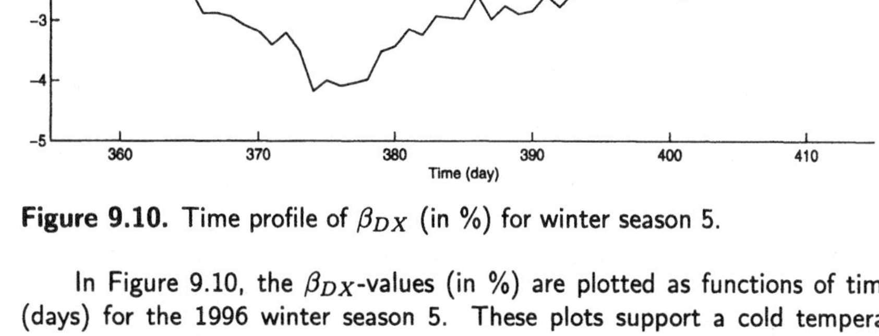

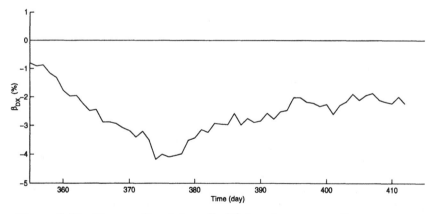

Figure 9.10. Time profile of β_{DX} (in %) for winter season 5.

In Figure 9.10, the β_{DX}-values (in %) are plotted as functions of time (days) for the 1996 winter season 5. These plots support a cold temperature exposure–death-rate association during the season of interest (notice that the β_{DX} is consistently negative). Moreover, the β_{DX} magnitude is indicative of the strength of this association as a function of time. As is shown in Figure 9.10, the temperature exposure–death-rate association is of varying magnitude during winter season 5.

Finally, in Figure 9.11 the spatial temperature exposure–death-rate association at the population level is represented in terms of the parameter β_{DX} map (in %). In this case, however, E_{DX} and E_D denote temporal averages calculated for each one of the North Carolina counties using all $e_{k\,|\,DX}$ and $e_{k\,|\,D}$ values obtained during the season. The spatial distribution of β_{DX} is everywhere negative. Notice that during winter season 5, while the temperature exposure–death-rate association is relatively weaker at the eastern part of the state and along the coastline (a fact which is probably due to the moderating effect of the ocean on the cold temperature distribution), it is stronger at the western side of the state (*i.e.*, along the mountain ranges of the Great Smoky Mountains National Park).

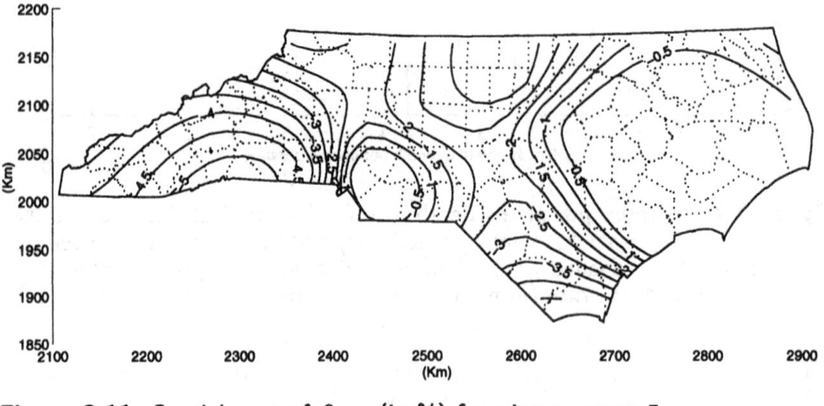

Figure 9.11. Spatial map of β_{DX} (in %) for winter season 5.

PEP provides a useful framework of spatiotemporal analysis of the exposure–mortality association, and its extension to more than one environmental variable is straightforward. If, *e.g.*, a possible confounder Z has been identified, we can use the PEP parameter

$$\beta_{DXZ} = (E_{DXZ} - E_{DZ})/E_D \tag{9.49}$$

(in %), where E_{DXZ} is the mortality prediction error based on death-rate D, exposure X, and confounder Z data, and E_{DZ} is the mortality prediction error based on death-rate D and confounder Z data. The β_{DXZ} parameter incorporates Z in the vector BME analysis of Postulate 9.1. Then, by comparing β_{DXZ} *vs.* β_{DX}, one could assess the importance of the confounder Z in the exposure–effect association under investigation. The approach is illustrated in Example 9.15. The same approach may also serve to evaluate competing association theories by means of prediction accuracy at a set of crucial observation points.

EXAMPLE 9.15: Particulate matter has been associated with public health risks by a number of authors (*e.g.*, Anderson *et al.*, 1992; USEPA, 1996; 1997). These risks are associated mainly with particulate matter of aerodynamic particle size of 10μm or smaller. The PM_{10} is a possible confounding factor, since it may have a causal association with death rate, while being correlated with temperature data. The PM_{10} data were obtained from the Aerometric Information Retrieval System (AIRS) of the USEPA (U.S. Environmental Protection Agency). The data used in this study were collected at 47 monitoring stations located throughout the State of North Carolina (Fig. 9.12). Also shown in Figure 9.12 are the contour lines of PM_{10} concentration (in $\mu g/m^3$), *e.g.*, for August 31, 1995. In order to assess the confounding effect of PM_{10}, we define the PEP parameter $\beta_{DXP} = (E_{DXP} - E_{DP})/E_D$, where E_{DXP} is the mortality prediction error using temperature exposure, PM_{10}, and death-rate data, while E_{DP} is the prediction error using only death-rate and PM_{10} data. As before, the $D(s, t)$, $X(s, t)$, and $P(s, t)$ represent death-rate, cold temperature exposure, and PM_{10} distributions, respectively.

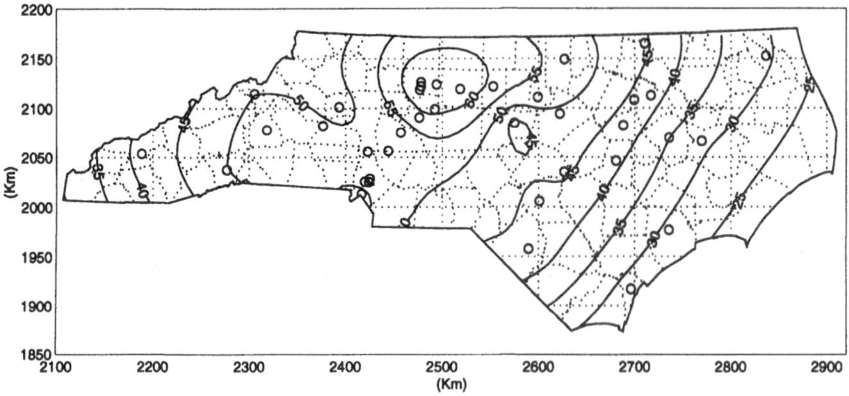

Figure 9.12. Locations of the 47 North Carolina monitoring stations for PM_{10} data (circles). Also shown are the contour lines of PM_{10} concentration (in $\mu g/m^3$) for August 31, 1995.

Including PM_{10} in the analysis resulted in a total of six covariance and cross-covariance models (as compared to only three covariance models used in Example 9.14). All six covariance and cross-covariance models were spatially isotropic and temporally stationary (Christakos and Serre, 2000a). The physical variables (temperature exposure and PM_{10} distributions) exhibited considerably larger spatial correlation ranges than the death-rate distribution. The spatiotemporal distributions of exposure to cold temperature X and of PM_{10} concentrations P were found to be negatively correlated (*i.e.*, as the temperature T drops, the exposure to cold temperature X increases and the concentration P decreases). Moreover, exposure to colder temperatures and higher levels of PM_{10} were correlated with higher death rates. The corresponding β_{DXP}-values are plotted in Figure 9.13 as a function of time. Also plotted for comparison is the PEP parameter β_{DX} of Figure 9.10 (which did not account for the confounding effect of PM_{10}). It is clear from these plots that accounting for PM_{10} results only in a slightly different strength in the reported cold temperature–mortality association for winter season 5. The spatial map of parameter β_{DXP} for the same season is plotted in Figure 9.14; this figure is similar to that obtained for β_{DX} in the previous section (Fig. 9.11). Both Figures 9.13 and 9.14 seem to support the claim that the PM_{10} space/time distribution does not have a significant confounding effect on the association between cold temperature exposure and mortality association. □

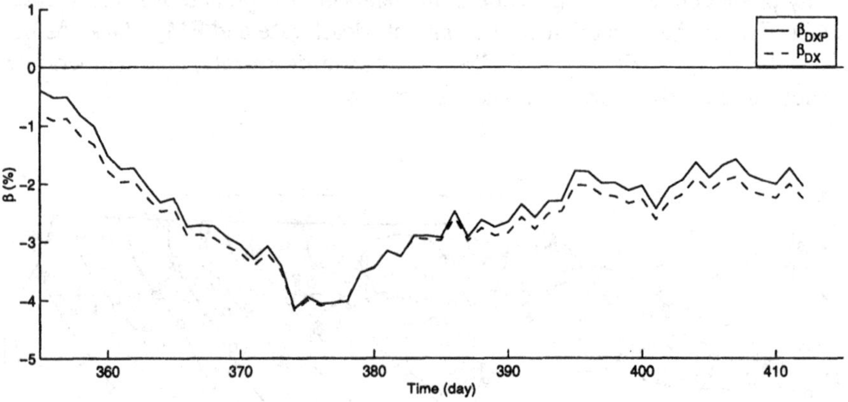

Figure 9.13. Time profile of β_{DXP} (in %) for winter season 5 (solid line). Also shown is the time profile for the parameter β_{DX} of Figure 9.11 (dotted line).

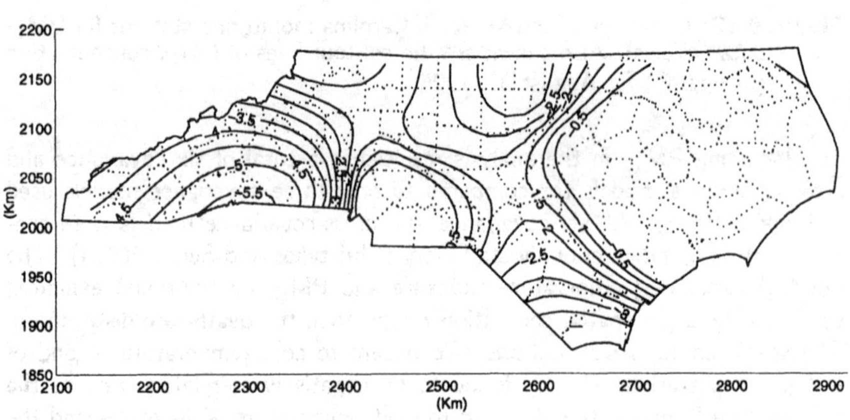

Figure 9.14. Spatial map of β_{DXP} (in %) for winter season 5.

The main goal of Examples 9.14 and 9.15 above was to demonstrate the significant advantages of the rigorous spatiotemporal BME modeling of human-exposure systems and discuss the practical usefulness of a novel stochastic criterion in assessing exposure–health-effect associations. In conclusion, BME analysis can provide important new insights into human-exposure phenomena, possibly leading to improved environmental exposure–health-effect assessments. Human-exposure problems, viewed as stochastic spatiotemporal systems, provide models which may challenge certain assumptions of traditional exposure analysis, and could shed light on some environmental pollution–population damage associations now coming under study.

Bringing Plato and Odysseus Together

As is clear from our discussion so far, the modern geostatistics paradigm is an open system that integrates (*i.*) epistemic rules, (*ii.*) physical knowledge, and (*iii.*) control variables and multiple objectives which depend on the application being considered. In other words, BME analysis takes place in a spiral form within which knowledge bases are developed recursively; it is their integration that guarantees the openness of the system and enables it to evolve within the limits (physical, economic, *etc.*) specified by the application under consideration.

All this points toward the development of a "reality checklist" regarding the necessary steps that must be taken and the appropriate decisions that need to be made by the modern geostatistician confronted with real-world problems. Such a checklist is summarized below:

Step 1. Obtaining a deeper ontologic and epistemic understanding of the phenomenon of interest, the resources, and the procedures available, including:

(a) the spatiotemporal geometry of the study domain (Euclidean *vs.* non-Euclidean, intrinsic *vs.* extrinsic, coordinate system, metric, *etc.*);

(b) the main physical characteristics of the natural variables involved (spatial homogeneity, anisotropy, temporal stationarity, additive or non-additive, small- *vs.* large-scale variability, *etc.*);

(c) the knowledge bases available (physical laws, space/time statistics, hard data, soft data, *etc.*); and

(d) the operationally defined measurement and sampling procedures (sample shapes and sizes, sampling networks, *etc.*).

Step 2. Deciding what kind of S/TRF best represents the natural variables. This decision should involve the consideration of a number of issues, as follows:

(a) ordinary or generalized models, multiple-point statistics;

(b) permissible correlation functions (covariance, variogram, *etc.*); and

(c) discrete or continuous representations.

Step 3. Choosing the BME mapping scheme. This is a multifold task including issues like the following:

(a) single-point or multipoint;

(b) point or functional; and

(c) single-variable or multivariable.

Step 4. Using feedback to coordinate the developments in (a)–(c) of Step 3 above with the application-specific goals and objectives.

The modern geostatistician's reality checklist is based on a combination of two perspectives: theoretical and practical. Following Whitehead (1969), we shall call the *theoretical* perspective the "Reason of Plato," and the *practical* perspective, the "Reason of Odysseus." Plato was the great philosopher of ancient Greece who championed the theoretical approach to scientific problems. Odysseus, Homer's mythic hero, on the other hand, was always capable of coming up with smart solutions to all kinds of practical problems he and his crew faced during their long journey. This metaphor serves our case well. Indeed, the aim of modern spatiotemporal geostatistics is to *integrate* effectively the powerful theoretical features of the "Reason of Plato" with the practical thinking of the "Reason of Odysseus." If, as it has been said, "Plato shared his perspective with the gods and Odysseus with the foxes," then the modern geostatistician shares his or her perspective with both!

10

SINGLE-POINT ANALYTICAL FORMULATIONS

"Knowing ignorance is strength. Ignoring knowledge is sickness." Lao Tsu

The Basic Single-Point BME Equations

In the preceding chapters we presented the main concepts of BME and we studied several scientific applications of modern geostatistics. When we talk about any sort of scientific discipline we are essentially talking about two main components:

1. an organized body of physical knowledge (*ontological* component), and
2. a distinctive methodology for obtaining and processing knowledge of the subject matter of the science (*epistemic* component).

The epistemic viewpoint a discipline adopts is a part of its very characterization of its scientific content. It is by virtue of the epistemic component that we can say, *e.g.*, that modern physics is importantly different from scholastic physics. In the case of modern spatiotemporal geostatistics, this viewpoint has led to the development of the BME concepts and methods, which have considerable advantages over many traditional geostatistical approaches. While searching for solutions to real-world problems, BME analysis forces us to determine explicitly the available physical knowledge, and to develop logically plausible rules for processing this knowledge. All these issues are incorporated up front in the mapping process, and nothing is swept under the carpet. What distinguishes the "knowledge-based" approach from the "axiomatic" approach is the role played by the $\mathcal{G}$ and $\mathcal{S}$ bases of past experience. While the "axiomatic" approach is based on deduction from a set of basic principles, in the "knowledge-based" approach the challenge in space/time analysis is to make use of the $\mathcal{G}$ and $\mathcal{S}$ bases in the most effective way.

In this chapter we will continue our study of the *mathematical* features of the single-point BME model (multipoint analysis is discussed in Chapter 11). For the purposes of this study, the *basic BME equations* of single-point spatiotemporal mapping developed in the previous sections are summarized in Table 10.1 (the equation numbers used in previous chapters are also indicated in Table 10.1; the parameters and operators A, B, D, and Ξ_S were defined in Table 6.1 on p. 133).

Table 10.1. The basic BME equations.

Equation*	Eq. no.
$\overline{h_\alpha}(\boldsymbol{p}_{map}) = \int d\boldsymbol{\chi}_{map}\, g_\alpha(\boldsymbol{\chi}_{map}) \exp\{\mu_0 + \mathcal{Y}_G[\boldsymbol{\chi}_{map};\, \boldsymbol{p}_{map}]\}$, $\alpha = 0, 1, \ldots, N_c$	5.9
$f_{\mathcal{K}}(\boldsymbol{\chi}_k) = A^{-1}\mathcal{Y}_S[\boldsymbol{\chi}_{map};\, \boldsymbol{p}_k] = A^{-1}B\int_D d\Xi_S(\boldsymbol{\chi}_{soft}) \exp[\mu_0 + \mathcal{Y}_G]$	6.17
$\frac{\partial}{\partial\chi_k}\mathcal{Y}_S\big\vert_{\chi_k=\hat{\chi}_k} = B\int_D d\Xi_S(\boldsymbol{\chi}_{soft})\{\exp[\mathcal{Y}_g]\,\frac{\partial}{\partial\chi_k}\mathcal{Y}_G\}_{\chi_k=\hat{\chi}_k} = 0$	7.10
$\boldsymbol{\chi}_{map} = (\chi_1, \ldots, \chi_m, \chi_k)$	

*Equations appear on p. 107, 132, and 137, respectively.

In the following sections we consider several interesting analytical formulations of the basic BME equations in Table 10.1. These formulations will be associated with the various knowledge bases considered in previous chapters (in fact, some of these formulations have been used in applications discussed in previous chapters).

Ordinary Covariance and Variogram—Hard and Soft Data

We start with a fundamental proposition. Notice that, as mentioned in the previous chapter (Comment 9.1, p. 178), when dealing with vector or matrix multiplications, the vectors involved are considered as column vectors.

PROPOSITION 10.1: Let $\boldsymbol{\chi}_{hard}$ be a vector of hard data at points $\boldsymbol{p}_i$ ($i = 1, 2, \ldots, m_h$) and let $\boldsymbol{\chi}_{soft}$ be a vector of soft data of various possible forms (see Table 6.1, p. 133) at points $\boldsymbol{p}_i$ ($i = m_h + 1, \ldots, m$). General knowledge includes the mean and the (centered) ordinary covariance. Then, the BME posterior pdf is given by Equation 6.17 (p. 132), with

$$\mathcal{Y}_G[\boldsymbol{\chi}_{map};\, \boldsymbol{p}_{map}] = -\tfrac{1}{2}(\boldsymbol{\chi}_{map} - \overline{\boldsymbol{x}_{map}})^T\, \boldsymbol{c}_{map}^{-1}\,(\boldsymbol{\chi}_{map} - \overline{\boldsymbol{x}_{map}}) \tag{10.1}$$

and

$$\mu_0 = -\log Z = -\log\left[(2\pi)^{(m+1)/2} / \,|\,\boldsymbol{c}_{map}^{-1}\,|^{1/2}\right] \tag{10.2}$$

where

$$\overline{x_{map}} = (\,\overline{x_1}, \ldots, \overline{x_m}, \overline{x_k}\,) \tag{10.3}$$

is the mean vector for points p_i ($i = 1, \ldots, m$ and k), and

$$c_{map} = \overline{(\,x_{map} - \overline{x_{map}}\,)\,(x_{map} - \overline{x_{map}}\,)^T} \tag{10.4}$$

is the (centered) covariance matrix between all these points.

Proof: Essentially, we need to determine the form of the operator $\mathcal{Y}_G$ in Equation 6.17 (p. 132). Given the general knowledge above, the $\mathcal{Y}_G$ is given by

$$\mathcal{Y}_G[\boldsymbol{\chi}_{map};\, \boldsymbol{p}_{map}] = \sum_{i=1}^{m,k} \left[\mu_i \chi_i + \mu_{ii}\,(\chi_i - \overline{x_i})^2\right]$$
$$+ \sum_{i \neq j=1}^{m,k} \mu_{ij}(\chi_i - \overline{x_i})\,(\chi_j - \overline{x_j}) \tag{10.5}$$

In the following calculations it will be convenient to define the coefficients

$$\lambda_{ij} = -2\mu_{ij} \tag{10.6}$$

In view of Equation 10.6, Equation 5.6 (p. 106) becomes

$$f_G(\boldsymbol{\chi}_{map}) = Z^{-1} \exp\left[\sum_{i=1}^{m,k} \mu_i \chi_i - \tfrac{1}{2} \sum_{i,j=1}^{m,k} \lambda_{ij}\,(\chi_i - \overline{x_i})\,(\chi_j - \overline{x_j})\right]$$
$$= Z^{-1} \exp\left[\boldsymbol{\mu}^T \boldsymbol{\chi}_{map} - \tfrac{1}{2}\,(\boldsymbol{\chi}_{map} - \overline{\boldsymbol{x}_{map}})^T \boldsymbol{\lambda}\,(\boldsymbol{\chi}_{map} - \overline{\boldsymbol{x}_{map}})\right] \tag{10.7}$$

where $\boldsymbol{\mu} = [\,\mu_1 \ldots \mu_m\, \mu_k]^T$, and

$$\boldsymbol{\lambda} = \begin{bmatrix} \lambda_{11} & \lambda_{12} & \cdots & \lambda_{1m} & \lambda_{1k} \\ \vdots & & & & \\ \lambda_{m1} & \lambda_{m2} & \cdots & \lambda_{mm} & \lambda_{mk} \\ \lambda_{k1} & \lambda_{k2} & \cdots & \lambda_{km} & \lambda_{kk} \end{bmatrix} \tag{10.8}$$

By letting $\boldsymbol{\psi}_{map} = \boldsymbol{\chi}_{map} - \overline{\boldsymbol{x}_{map}}$, the partition function can be written as

$$Z = \exp\left[\,\boldsymbol{\mu}^T \overline{\boldsymbol{x}_{map}}\right] \int d\boldsymbol{\psi}_{map} \exp\left[\,\boldsymbol{\mu}^T \boldsymbol{\psi}_{map} - \tfrac{1}{2}\,\boldsymbol{\psi}_{map}^T \boldsymbol{\lambda}\, \boldsymbol{\psi}_{map}\right]$$
$$= (2\pi)^{(m+1)/2}\, |\boldsymbol{\lambda}|^{-1/2} \exp\left[\,\tfrac{1}{2}\boldsymbol{\mu}^T (\boldsymbol{\lambda}^{-1} \boldsymbol{\mu} + 2\overline{\boldsymbol{x}_{map}})\right] \tag{10.9}$$

Furthermore,

$$\overline{x_i} = Z^{-1} \partial Z / \partial \mu_i = \sum_{j=1}^{m,k} \lambda_{ij}^{-1}\, \mu_j + \overline{x_i} \tag{10.10}$$

where λ_{ij}^{-1} is the ij-th element of the inverse matrix $\boldsymbol{\lambda}^{-1}$ and, since $\lambda_{ij}^{-1} > 0$, we find

$$\mu_j = 0 \tag{10.11}$$

for $j = 1, 2, \ldots, m, k$. Also,

$$c_{ij} = \overline{[x_i - \overline{x_i}]\,[x_j - \overline{x_j}]} = \int d\boldsymbol{\chi}_{map}\,(\chi_i - \overline{x_i})\,(\chi_j - \overline{x_j})\,f_{\mathcal{G}}(\boldsymbol{\chi}_{map})$$

$$= (2\pi)^{-(m+1)/2}\,|\boldsymbol{\lambda}|^{\,1/2} \int d\boldsymbol{\chi}_{map}\,(\chi_i - \overline{x_i})\,(\chi_j - \overline{x_j})$$

$$\exp\left[-\tfrac{1}{2}(\boldsymbol{\chi}_{map} - \overline{\boldsymbol{x}_{map}})^T \boldsymbol{\lambda} (\boldsymbol{\chi}_{map} - \overline{\boldsymbol{x}_{map}})\right] \qquad (10.12)$$

From Equation 10.12 and well-known properties of multivariate Gaussian laws it follows that

$$\boldsymbol{\lambda} = \boldsymbol{c}_{map}^{-1} \qquad (10.13)$$

In light of Equations 10.11 and 10.13, Equation 10.9 reduces to Equation 10.2, and Equation 10.5 leads to Equation 10.1. □

The following corollary is a straightforward consequence of Proposition 10.1 above, with particularly useful applications in practice.

COROLLARY 10.1: Given the general and specificatory knowledge bases described in Proposition 10.1, the BME estimate $\hat{\chi}_k = \chi_{k,mode}$ is the solution of the equation

$$\sum_{i=1}^{m_h} c_{ik}^{-1}(\chi_i - \overline{x_i}) + \sum_{i=m_h+1}^{m} c_{ik}^{-1}\left[\overline{x_i}(\hat{\chi}_k) - \overline{x_i}\right] + c_{kk}^{-1}(\hat{\chi}_k - \overline{x_k}) = 0 \qquad (10.14)$$

where

$$\overline{x_i}(\hat{\chi}_k) = \frac{B\int_D d\Xi_S\,\chi_i \exp[\mathcal{Y}_{\mathcal{G}}]_{\chi_k=\hat{\chi}_k}}{B\int_D d\Xi_S \exp[\mathcal{Y}_{\mathcal{G}}]_{\chi_k=\hat{\chi}_k}} \qquad (10.15)$$

where the $\mathcal{Y}_{\mathcal{G}}$ is the operator of Equation 10.1.

Proof: The BMEmode estimate is the solution of Equation 7.10 (p. 137), where $\mathcal{Y}_{\mathcal{G}}$ is given by Equation 10.1, *i.e.*,

$$B\int_D d\Xi_S \left\{ \exp[\mathcal{Y}_{\mathcal{G}}]\,\frac{\partial}{\partial \chi_k}\left[\sum_{i=1}^{m,k}\sum_{j=1}^{m,k}(\chi_i - \overline{x_i})\,c_{ij}^{-1}(\chi_j - \overline{x_j})\right]\right\}_{\chi_k=\hat{\chi}_k} = 0$$

or

$$\sum_{i=1}^{m,k} c_{ik}^{-1}\,B\int_D d\Xi_S \left\{\exp[\mathcal{Y}_{\mathcal{G}}]\,(\chi_i - \overline{x_i})\right\}_{\chi_k=\hat{\chi}_k} = 0$$

which, in light of Equation 10.15, leads to Equation 10.14. □

EXAMPLE 10.1: Example 7.2 (p. 138) is a special case of Corollary 10.1. Indeed, Equation 7.14 (p. 140) could have been obtained directly from Equation 10.14 above. This shows the great simplifications in calculations provided by analytical results such as Proposition 10.1 and Corollary 10.1. □

The following proposition deals with the situation in which the space/time variability is expressed in terms of non-centered covariances.

PROPOSITION 10.2: The specificatory knowledge is as described in Proposition 10.1. The general knowledge consists of the non-centered ordinary covariance in space/time. The BMEmode estimate $\hat{\chi}_k$ is the solution of the equation

$$\sum_{i=1}^{m_h} C_{ik}^{-1}\chi_i + \sum_{i=m_h+1}^{m} C_{ik}^{-1}\,\overline{x_i}\,(\hat{\chi}_k) + C_{kk}^{-1}\,\hat{\chi}_k = 0 \qquad (10.16)$$

C_{ik}^{-1} and C_{kk}^{-1} are the ik-th and k-th elements, respectively, of the inverse matrix C_{map}^{-1}, where

$$C_{map} = \overline{x_{map}\,x_{map}^T} \qquad (10.17)$$

is the matrix of the spatiotemporal non-centered covariances C_{ij} between all points p_i and p_j $(i,\, j = 1,\, \ldots,\, m,\, k)$; and the parameter $\overline{x_i}(\hat{\chi}_k)$ is of the form of Equation 10.15 with

$$\mathcal{Y}_G[\boldsymbol{\chi}_{map};\, \boldsymbol{p}_{map}] = -\tfrac{1}{2}\boldsymbol{\chi}_{map}^T\, C_{map}^{-1}\, \boldsymbol{\chi}_{map} \qquad (10.18)$$

Proof: Working along the lines of the proof of Corollary 10.1 above, we find that Equation 7.10 (p. 137) can be written as

$$\sum_{i,j=1}^{m,k} \mu_{ij}\, B \int_D d\Xi_S \{\exp[\mathcal{Y}_G]\, \frac{\partial}{\partial \chi_k}\, g_{ij}(\chi_i, \chi_j)\}_{\chi_k=\hat{\chi}_k} = 0 \qquad (10.19)$$

where $\mu_{ij} = -\frac{1}{2}C_{ij}^{-1}$ (C_{ij}^{-1} is the ij-th element of the inverse matrix C_{map}^{-1}), $g_{ij}(\chi_i, \chi_j) = \chi_i\,\chi_j$ $(i,\, j = 1,\, \ldots,\, m$ and $k)$, and

$$\mathcal{Y}_G[\boldsymbol{\chi}_{map};\, \boldsymbol{p}_{map}] = -\frac{1}{2}\sum_{i,j=1}^{m,k} C_{ij}^{-1}\chi_i\,\chi_j$$

Equation 10.19 can be simplified as follows:

$$B \int_D d\Xi_S \left[-\sum_{j=1}^{m} C_{ik}^{-1}\chi_j - C_{kk}^{-1}\hat{\chi}_k\right] \exp[\mathcal{Y}_G]_{\chi_k=\hat{\chi}_k} = 0 \qquad (10.20)$$

which—taking into account Equations 10.15 and 10.18—can be written in the form of Equation 10.16. □

In certain practical applications the spatiotemporal variogram can be calculated more efficiently than certain other second-order statistical moments. In such cases the following proposition is useful, for it provides the appropriate BME formulation in terms of variogram functions.

PROPOSITION 10.3: The specificatory knowledge is as described in Proposition 10.1. The general knowledge consists of the variogram between all pairs of points in space/time. The BMEmode estimate is the solution of equation

$$\hat{\chi}_k \sum_{i=1}^{m} \gamma_{ik}^{-1} - \sum_{i=1}^{m_h} \gamma_{ik}^{-1} \chi_i - \sum_{i=m_h+1}^{m} \gamma_{ik}^{-1} \overline{x_i}(\hat{\chi}_k) = 0 \qquad (10.21)$$

where γ_{ik}^{-1} is the ik-th element of the inverse matrix $\boldsymbol{\gamma}_{map}^{-1}$ such that

$$\boldsymbol{\gamma}_{map} = \begin{bmatrix} \gamma_{1k} & \dots & 0 & 0 \\ \vdots & & & \\ 0 & \dots & \gamma_{mk} & 0 \\ 0 & \dots & 0 & 0 \end{bmatrix} \qquad (10.22)$$

is the variogram matrix between the points $\boldsymbol{p}_i$ $(i = 1, \dots, m)$ and $\boldsymbol{p}_k$; and $\overline{x_i}(\hat{\chi}_k)$ is of the form of Equation 10.15 above with

$$\mathcal{Y}_{\mathcal{G}}[\boldsymbol{\chi}_{map};\, \boldsymbol{p}_{map}] = -\tfrac{1}{4} \sum_{i=1}^{m} \gamma_{ik}^{-1} (\chi_i - \chi_k)^2 \qquad (10.23)$$

Proof: The prior pdf is

$$f_{\mathcal{G}}(\boldsymbol{\chi}_{map}) = Z^{-1} \exp[\mathcal{Y}_{\mathcal{G}}] = Z^{-1} \exp[-\tfrac{1}{2}\, \delta \boldsymbol{\chi}_{map}^{T} \boldsymbol{\lambda} \delta \boldsymbol{\chi}_{map}] \qquad (10.24)$$

where $Z^{-1} = (2\pi)^{-(m+1)/2}\, |\boldsymbol{\lambda}|^{1/2}$, $\delta \boldsymbol{\chi}_{map} = (\chi_1 - \chi_k, \dots, \chi_m - \chi_k\ 0)$, and $\boldsymbol{\lambda} = \{\lambda_{ii} = \lambda_i\ (i = 1, \dots, m, k);\ \lambda_{ij} = 0\ (i \neq j)\}$. Since

$$\begin{aligned} c_{\delta(ik,jk)} &= \overline{\delta X_{ik}\, \delta X_{jk}} = \overline{[X(\boldsymbol{p}_i) - X(\boldsymbol{p}_k)][X(\boldsymbol{p}_j) - X(\boldsymbol{p}_k)]} \\ &= 2\gamma_x(\boldsymbol{p}_i, \boldsymbol{p}_j, \boldsymbol{p}_k) \end{aligned} \qquad (10.25)$$

it follows from Equation 10.24 that $\boldsymbol{\lambda}^{-1} = 2\boldsymbol{\gamma}_{map}$, where the $\boldsymbol{\gamma}_{map}$ is given by Equation 10.22 above. Equation 7.10 (p. 137) can now be written as

$$\sum_{i=1}^{m} \gamma_{ik}^{-1} B \int_D d\Xi_S \left\{ \exp[\mathcal{Y}_{\mathcal{G}}] \frac{\partial}{\partial \chi_k} (\chi_i - \chi_k)^2 \right\}_{\chi_k = \hat{\chi}_k} = 0$$

or

$$\sum_{i=m_h+1}^{m} \gamma_{ik}^{-1} B \int_D d\Xi_S\, \chi_i \exp[\mathcal{Y}_{\mathcal{G}}]_{\chi_k = \hat{\chi}_k} +$$

$$\left[\sum_{i=1}^{m_h} \gamma_{ik}^{-1} \chi_i - \hat{\chi}_k \sum_{i=1}^{m} \gamma_{ik}^{-1} \right] B \int_D d\Xi_S \exp[\mathcal{Y}_{\mathcal{G}}]_{\chi_k = \hat{\chi}_k} = 0 \qquad (10.26)$$

with $\mathcal{Y}_{\mathcal{G}}$ as in Equation 10.23, which leads to Equation 10.21. □

COMMENT 10.1: *The result suggested by Proposition 10.3 can easily be extended to incorporate cross-variograms $\gamma_x(\boldsymbol{p}_i, \boldsymbol{p}_j, \boldsymbol{p}_k)$ between more than one random field.*

Equations 10.14, 10.16, and 10.21 tell us how to combine physical knowledge about the space/time variation of the natural variable, hard, and soft data to arrive at the most probable estimate. The BME estimate obtained from the analytical formulations above is, in general, a nonlinear function of the data. This is in contrast to the classical kriging estimators, which are restricted to linear combinations of hard data and do not possess a systematic mechanism to incorporate the variety of knowledge bases available in physical applications (probabilistic data, scientific laws, *etc.*). As already mentioned, in addition to the BMEmode estimates provided by the above analytical expressions (Corollary 10.1 and Propositions 10.2 and 10.3), other BME estimates can be obtained from the general BME posterior pdf derived in the fundamental Proposition 10.1. A common case is the BMEmean estimate (see Chapter 7, "Other BME Estimates," p. 147) which is used in the particulate matter study discussed next.

Particulate Matter Distributions in North Carolina

Particulate matter (PM) is the general term used for a mixture of solid particles and liquid droplets found in the air (*e.g.*, see USEPA, 1997). Particles may come in a wide range of sizes; "fine" particles have a diameter $d < 2.5\,\mu$m (*e.g.*, $PM_{2.5}$) and "coarser" particles are such that $d > 2.5\,\mu$m (*e.g.*, PM_{10}). Particles originate from many different stationary and mobile sources as well as from natural sources. PM_{10} is generally emitted from sources such as vehicles traveling on unpaved roads, materials handling, and crushing and grinding operations, as well as windblown dust. Some particles are emitted directly from their sources (*e.g.*, smokestacks and cars). Also, gases such as SO_2, NO_x, and VOC interact with other compounds in the air to form fine particles. Their chemical and physical compositions vary depending on location, time, weather, *etc.* PM has been linked to public health risks by a number of authors (Milne *et al.*, 1982; Anderson *et al.*, 1992; Dockery and Pope, 1994; USEPA, 1996; Janssen *et al.*, 1999). These risks are associated mainly with PM of aerodynamic particle sizes of 10 μm or smaller. PM particles can accumulate in the respiratory system and are associated with numerous health effects. Exposure to coarse particles is primarily associated with the aggravation of respiratory conditions, such as asthma; fine particles are most closely associated with increased hospital admissions and emergency room visits for heart and lung disease, increased respiratory symptoms and disease, decreased lung function, and even premature death. Sensitive groups that appear to be at greatest risk to such effects include the elderly, individuals with cardiopulmonary disease, and children. In addition to health problems, PM is a major cause of reduced visibility in many parts of the United States. Airborne particles also can cause damage to paints and building materials.

In the case of the State of North Carolina, the specificatory knowledge base $\mathcal{S}$ includes measurements of ambient PM_{10} which are available from the EPA (Environmental Protection Agency) Aerometric Information Retrieval System (AIRS). Averaged (24-hr) air ambient concentrations of PM_{10} were measured at 47 monitoring stations distributed throughout North Carolina during the years 1995 and 1996. The PM_{10} data were collected at these monitoring stations every 6 days (thus, while these two years had 729 calendar days, there were only 122 measurement days, starting January 3, 1995). The PM_{10} measurements of the EPA data base that were considered accurate were classified as hard data χ_{hard}. Furthermore, the classification of uncertain measurements as soft data χ_{soft} was based on expert opinion (*e.g.*, expertise gained in similar situations, intuition, understanding of the atmospheric processes, and error biases). This implies that, (*a*) the missing PM_{10} values were replaced by soft intervals (Eq. 3.32, p. 85) and (*b*) at some other points, the shape of the soft probabilistic data (Eq. 3.33) that were used represented the distribution of measurement errors around the reported uncertain values of PM_{10} concentration. At each missing data point, the lower bound of the interval was assumed equal to zero and the upper bound was assumed equal to the maximum PM_{10} concentration measured within the local neighborhood surrounding the missing data point. Using local neighborhoods leads to physically meaningful bounds, as well as smaller (and, thus, more informative) interval data.

For illustration, in Figure 10.1 we show the PM_{10} measurements collected every six days at monitoring station no. 13 during the year 1995 and part of 1996 (Christakos and Serre, 2000b). The time axis is labeled in calendar days, with day 1 corresponding to January 1, 1995. The first measurement day was January 3, 1995; hard (exact) data, as well as soft (uncertain) data of the interval and the probabilistic types are shown in Figure 10.1. The general knowledge base included the space/time covariance of the particulate matter distribution. The experimental covariance was first calculated from the available PM_{10} data for the years 1995 and 1996. A theoretical model was then obtained by fitting the following model to the experimental values

$$c_x(r,\tau) = \sum_{i=1}^{2} c_i \exp[-r/a_{r,i} - \tau/a_{\tau,i}] \tag{10.27}$$

where $c_1 = 45$, $c_2 = 50$ [both in $(\mu g/m^3)^2$], $a_{r,1} = 20$, $a_{r,2} = 1,000$ (in km), and $a_{\tau,1} = 1$, $a_{\tau,2} = 5$ (in days). The covariance model (Eq. 10.27) is plotted in Figure 10.2. Equation 10.27 is the sum of two distinct exponential covariance functions which are separable with respect to space and time. While the first exponential term addresses short-range interactions with respect to both space and time, the second term addresses long-range interactions (this behavior is in agreement with the pattern of curvature shown in Fig. 10.2). The short-range interaction term has parameters consistent with a metropolitan scale of 20 km

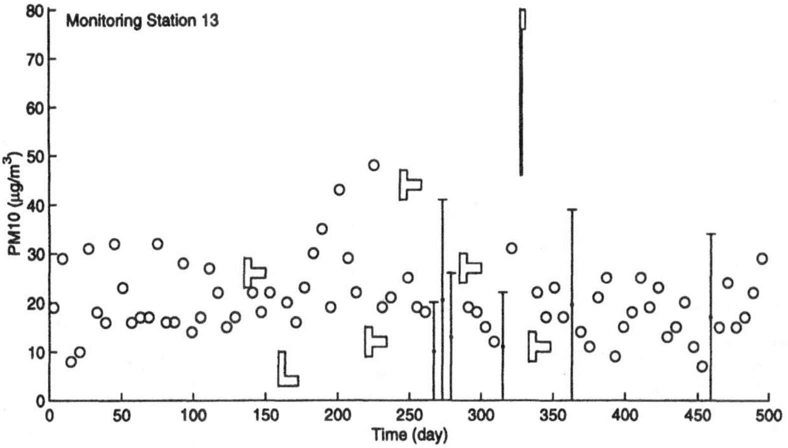

Figure 10.1. PM_{10} concentrations (in $\mu g/m^3$) at monitoring station no. 13. Circles denote hard data, bars denote soft data (interval type), and pulse-shaped curves denote soft data (probabilistic type).

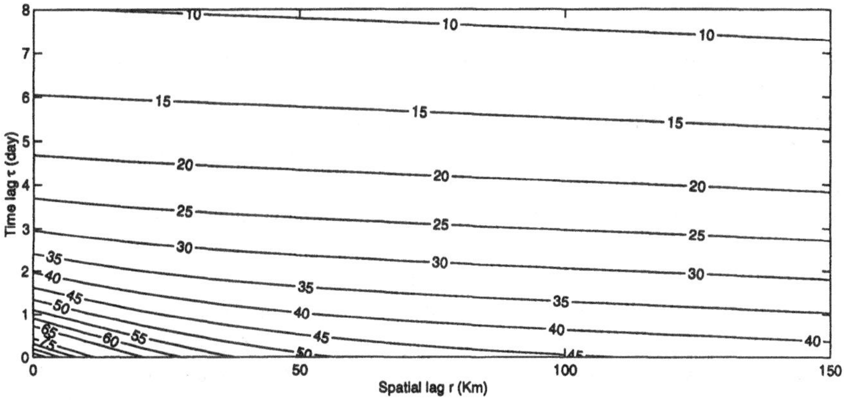

Figure 10.2. Space/time covariance [in $(\mu g/m^3)^2$] of the PM_{10} concentrations shown as a function of the time lag τ and the spatial lag r.

and 1 day. Likewise, the long-range interaction term has parameters consistent with a regional scale of 1,000 km and 5 days. Thus, short-range interactions are consistent with the spatiotemporal scale of emissions from point (smokestacks) or line sources (roads), whereas long-range interactions are consistent with the scale of mass-transport and dispersion over regional domains as affected by weather patterns and other meteorological considerations (North Carolina is bounded by mountains on the west with a prevailing weather pattern of west–to–east, except when influenced by tropical storms and hurricanes).

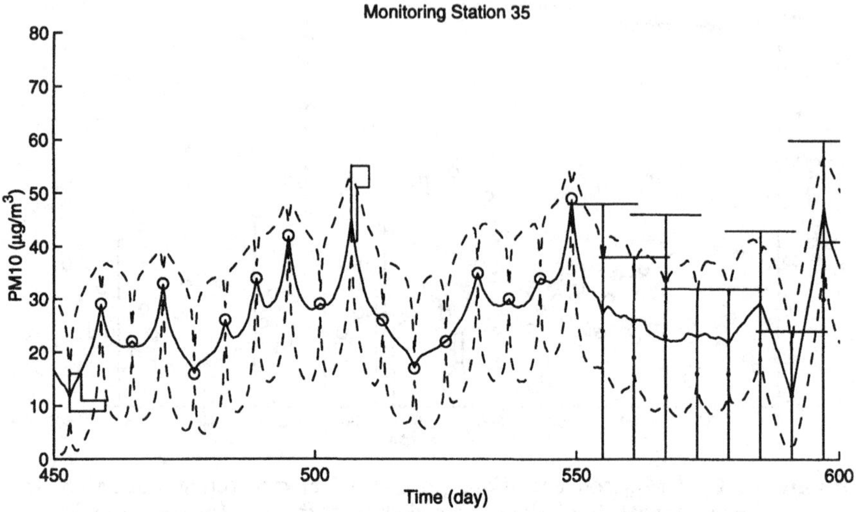

Figure 10.3. BMEmean PM_{10} profile (in $\mu g/m^3$) at monitoring station no. 35. The estimated profile is shown as a plain line and the 90% confidence intervals are denoted by a dashed line. Circles represent hard data, bars denote soft data (interval type), and pulse-shaped curves represent soft data (probabilistic type).

On the basis of hard and soft PM_{10} data (collected every six days at each monitoring station, as well as at a set of neighboring stations) the BMEmean estimate of the PM_{10} concentration was calculated as a function of time. For illustration, in Figure 10.3 we show the corresponding BMEmean PM_{10} profile at station no. 35; also shown are the corresponding 90% confidence intervals. The PM_{10} profile shows considerable temporal variation. The PM_{10} estimates honor the hard data and are consistent with the soft data available, interval and probabilistic. In Figure 10.3, the incorporation of soft data by the BME approach allows PM_{10} estimation at times where this would be otherwise impossible due to the lack of hard data. Note that the estimated PM_{10} profile is characterized by "valleys" and "dips." This happens because, while the estimates are equal to the measured values at the hard data points, in between these points the estimates are affected by the stochastic mean. When two data points are above the mean, *e.g.*, the interpolated profile between these points will generally take the shape of a "valley" (leaning towards the mean); on the other hand, when two data points are below the mean, the interpolated profile may include a "dip." The upper and lower bounds of the 90% confidence interval are also shown in Figure 10.3 as functions of time. The shape of these bounds implies that the confidence interval values are zero at the hard data points and increase gradually in the time period between the data points.

Furthermore, using hard and soft PM_{10} data collected on August 25, 1995, as well as during the preceding and following days, we calculated the BMEmean estimate for PM_{10} at the nodes of a regular grid covering North Carolina. On the basis of these estimated values, we constructed the contour map of PM_{10} for August 25, 1995, shown in Figure 10.4a. The PM_{10} concentration for the following two days (*i.e.*, August 31 and September 6) are shown in Figures 10.4b and 10.4c, respectively. Note that August 31 (Fig. 10.4b) was chosen because it experienced the highest spatially averaged PM_{10} value during the 1995 period (August 25 and September 6 were simply the preceding and following measuring days). Also, note the considerable spatial variability of PM_{10} concentration depicted by each map, as well as the temporal PM_{10} variation between the three maps. The level of PM_{10} concentration reached a maximum that exceeded the 60 $\mu g/m^3$ on August 31 (at the north-central region of the state) and then decreased to below 40 $\mu g/m^3$ on September 6.

These maps are based on an integrated analysis of space and time which accounts for important cross-correlations and dependencies between PM_{10} concentrations at various spatial locations and time instants. In Figure 10.5 we plot the BME error standard deviation $\sigma_{k\,|\,\mathcal{K}}$ of the PM_{10} map obtained on August 31, 1995. Note that the uncertainty distribution is affected by two factors: (*i.*) the amount of information (*i.e.*, data points available in the neighborhood surrounding each estimation point), and (*ii.*) the quality of information available (*i.e.*, the uncertainty level of the soft data). For example, the estimation error ranges from 0 $\mu g/m^3$ at monitoring stations where hard data are available, to higher values (up to 7.5 $\mu g/m^3$) at regions away from these stations. In addition, a high level of uncertainty occurs when the BME estimates use soft data characterized by large intervals. The analysis above emphasizes two essential points: (*a*) in the presence of considerable space/time variability, the PM_{10} estimates obtained by a mapping method may be inadequate or misleading, and need to be considered together with an accuracy measure such as the error standard deviation (in some cases, a more appropriate uncertainty characterization could involve the full posterior pdf); and (*b*) other forms of knowledge may need to be processed in order to reduce estimation uncertainty. These kinds of uncertainty maps are very valuable in environmental risk analysis: they can demonstrate the limitations of an existing network of monitoring stations, identify areas where additional observations are needed, optimize the design of future networks, provide uncertainty indicators for use in health risk management systems, *etc.* In conclusion, numerical implementations of the BME method produce accurate pollutant estimates for use in spatiotemporal mapping applications, and realistic measures of the relevant uncertainties for use in decision making and environmental risk assessment applications (a discussion of these applications may be found, *e.g.*, in Dab *et al.*, 1996 or in Itô and Thurston, 1996).

We continue our discussion by addressing the *rear mirror metaphor*. In many cases, estimating the future on the basis of past experience is like trying to drive a car looking only in the rear mirror: there is no problem if the road is straight, but serious problems could arise if there are sharp bends. Therefore, in addition to past experience one often needs to use soft data regarding future

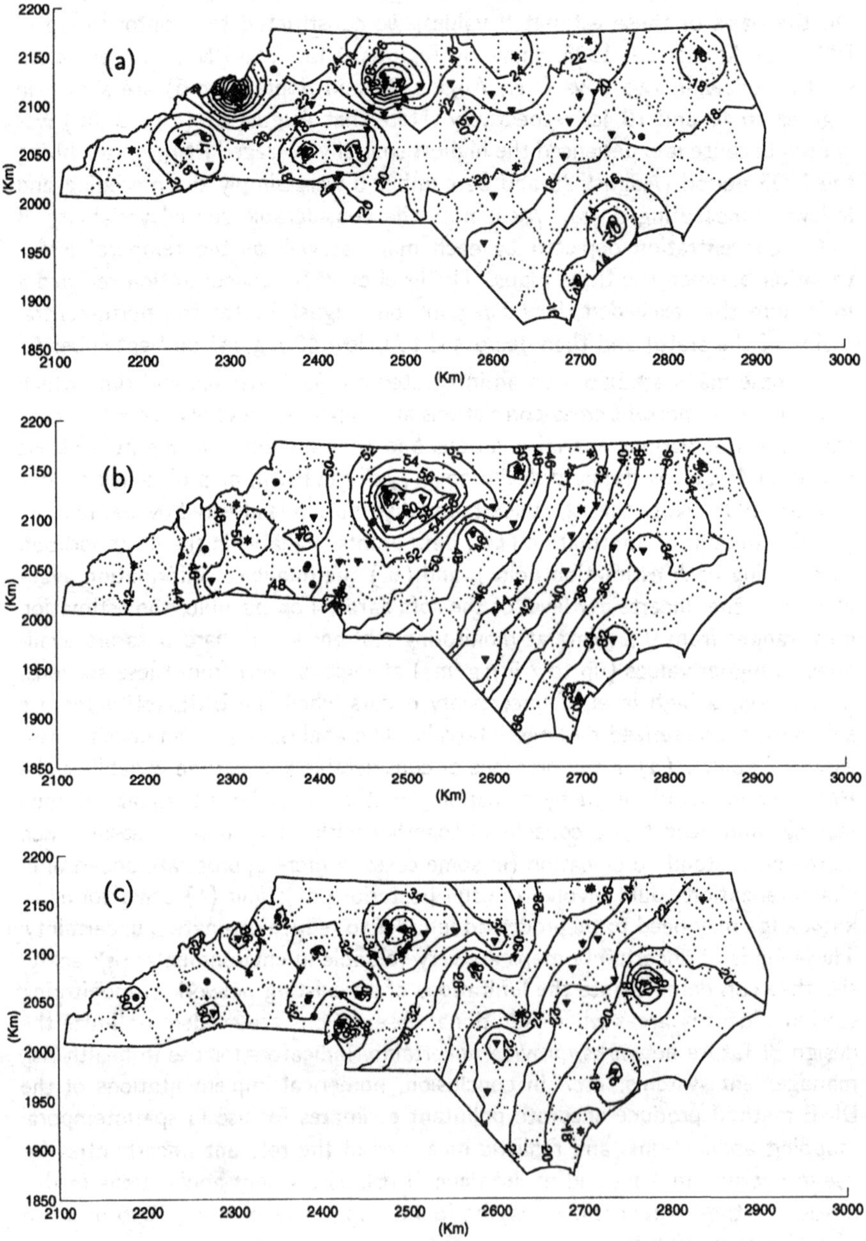

Figure 10.4. Spatial map of PM_{10} concentrations (in $\mu g/m^3$) obtained using BMEmean estimates on (a) August 25, 1995, (b) August 31, 1995, and (c) September 6, 1995. Monitoring stations where hard, interval, and probabilistic data are available in space/time are depicted by triangles, circles, and hexagons, respectively.

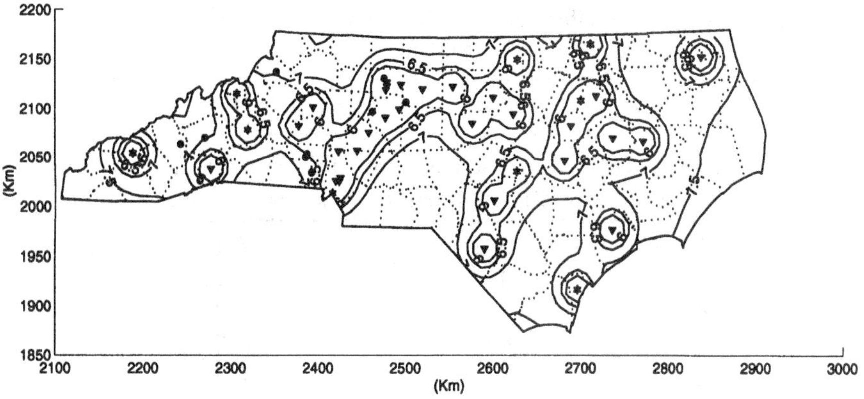

Figure 10.5. Spatial map of BME estimation error standard deviation for PM_{10} concentration (in $\mu g/m^3$) associated with the PM_{10} map of Figure 10.4b.

estimation points. As we saw in Chapter 6 (p. 132), BME's versatility provides it with the means to analyze soft data at the estimation points. There are various methods for encoding soft data at the estimation points. Given, *e.g.*, hard measurements of a physical variable X at points $\boldsymbol{p}_{hard}$, a technique (*e.g.*, polynomial fitting or model simulation) can be used to derive X-values at the estimation points $\boldsymbol{p}_k$. The new X-values, which are uncertain, can be used to generate soft data at these points (*e.g.*, probability functions having these values as means).

In order to provide a numerical illustration of the effect of soft data at the estimation points, Christakos and Serre (2000b) examined the following situation. Consider a set of points throughout North Carolina where PM_{10} values are available (but considered unknown for the purposes of the analysis). These values were then estimated: (a) by assuming that no soft data are available at the estimation points, and (b) by using probability soft data at the estimation points. The estimation errors (in $\mu g/m^3$) were calculated for both approaches (a) and (b). The results are shown in Table 10.2. Clearly, approach (b) provides a better estimation accuracy than approach (a). Moreover, in Figure 10.6 we plot the difference Δe in estimation errors (a)–(b) throughout North Carolina averaged over a three-day period (August 25, August 31, and September 6 of 1995). Note that the Δe values are consistently positive over the entire state (ranging from about 1.0 $\mu g/m^3$ to about 10.0 $\mu g/m^3$). The Δe map, therefore, demonstrates that the incorporation of soft data at the estimation points (whenever available) can dramatically improve the quality of PM_{10} estimation.

Since space/time estimation is improved by using soft data, it should be interesting to evaluate the numerical work involved when incorporating soft

Table 10.2. The effect of soft PM_{10} data at a set of estimation points on August 31, 1995.

	Estimation error ($\mu g/m^3$)	
Monitoring station no.	(a) Without soft data at the estimation points	(b) With soft data at the estimation points
40	0.75	0.12
44	0.73	0.13
46	1.33	0.22
39	4.95	0.67
1	3.20	0.54

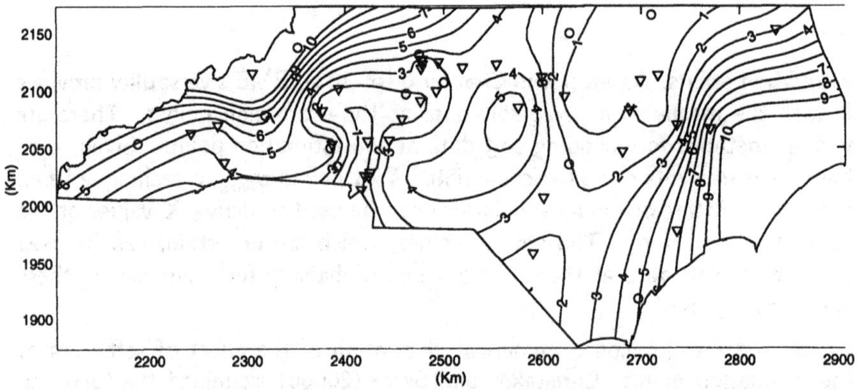

Figure 10.6. Map of the Δe values (in $\mu g/m^3$) over North Carolina averaged during a three-day period. This map demonstrates the improvement gained by considering soft data at the estimation points.

data into the mapping process. An assessment of this work is presented in the following example.

EXAMPLE 10.2: In Figure 10.7 we show the numerical work (CPU time) required by the BME technique on an HP-9000 computer for a typical case (SANLIB99, 1999). The CPU time is plotted as a function of the number of soft (interval) data points (for 2, 8, and 32 hard data points). It is evident from Figure 10.7 that the CPU time remains small ($e.g.$, less than 0.15 sec for up to about 7 soft data points), which makes BME a numerically efficient method for spatiotemporal analysis. For larger numbers of soft data points one should use a Monte Carlo method. □

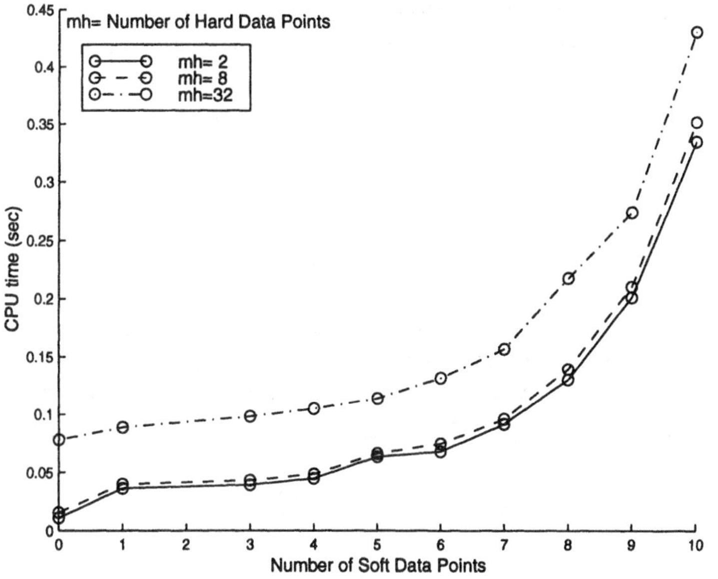

Figure 10.7. CPU time (sec) of BME estimation on an HP-9000/C160 workstation.

Naturally, as improvements are made in the computer programming of the BME techniques, the complexity of the numerical algorithms and the associated computational costs will continue to decrease.

Generalized Covariance—Hard and Soft Data

Readers unfamiliar with the S/TRF-ν/μ theory (Christakos, 1991b, 1992) may skip this section. We will start by proving a fundamental proposition, which deals with situations in which the spatiotemporal pattern of the natural variable under consideration is characterized by spatially nonhomogeneous and temporally nonstationary variations.

PROPOSITION 10.4: The specificatory knowledge is as described in Proposition 10.1. The general knowledge consists of the generalized spatiotemporal covariance of orders ν/μ between any pair of points in $\boldsymbol{p}_{map} = \{\boldsymbol{p}_i;\ i = 1, \dots\ m,\ k\}$. Then, the BME posterior pdf is given by Equation 6.17 (p. 132) with

$$\mathcal{Y}_G[\boldsymbol{\chi}_{map};\ \boldsymbol{p}_{map}] = -\tfrac{1}{2}\,\boldsymbol{Q}^T(\boldsymbol{\chi}_{map})\,c_Q^{-1}(\boldsymbol{\kappa}_{map})\,\boldsymbol{Q}\,(\boldsymbol{\chi}_{map}) \tag{10.28}$$

and

$$\mu_0 = -\log\left[(2\pi)^{\theta/2}/\,|\,c_Q^{-1}\,|^{1/2}\right] \tag{10.29}$$

where θ is the number of the spatiotemporal increments considered, $\boldsymbol{Q}$ is the vector operator associated with the S/TRF-ν/μ , and κ_{map} is the generalized covariance matrix between the space/time points $\boldsymbol{p}_i$ such that

$$c_Q\,(\kappa_{map}) = \overline{\boldsymbol{Q}(\boldsymbol{x}_{map})\,\boldsymbol{Q}^T\,(\boldsymbol{x}_{map})} \tag{10.30}$$

Proof: The $\mathcal{Y}_G$-operator is given by

$$\mathcal{Y}_G = \sum\nolimits_{i,j=1}^{\theta} \mu_{ij}\,\psi_i\,\psi_j \tag{10.31}$$

where $\psi_i = Q_i(\boldsymbol{\chi}_{map})$, $i = 1, 2, \ldots, \theta$, are spatiotemporal increments determined on the basis of the theory of S/TRF-ν/μ (Christakos, 1992), and $\boldsymbol{\psi}(\boldsymbol{\chi}_{map}) = (\psi_1, \ldots, \psi_\theta)$. In view of Equation 10.31, and letting $\mu_{ij} = -\frac{1}{2}\lambda_{ij}$, $\overline{g_i}(\boldsymbol{\chi}_{map}) = \overline{Q_i^2(\boldsymbol{x}_{map})}$, Equations 5.6 and 5.7 (p. 106–107) lead to

$$f_G\,(\boldsymbol{\chi}_{map}) = Z^{-1}\exp\left[-\tfrac{1}{2}\,\boldsymbol{Q}^T\boldsymbol{\lambda}\boldsymbol{Q}\,\right] \tag{10.32}$$

where $\boldsymbol{\lambda} = \{\lambda_{ij}\}$, $i, j = 1, \ldots, \theta$. The partition function is now written as

$$Z = \int d\boldsymbol{\chi}_{map}\,\exp\left[-\tfrac{1}{2}\,\boldsymbol{Q}^T\boldsymbol{\lambda}\boldsymbol{Q}\,\right] = (2\pi)^{\theta/2}\,|\boldsymbol{\lambda}|^{-1/2} \tag{10.33}$$

Since Equation 10.32 is of a Gaussian form and $c_{y,ij} = c_{Q,\,ij}(\boldsymbol{\kappa}_{map})$, we immediately get

$$\boldsymbol{\lambda} = c_{y,\,\theta}^{-1} = c_Q^{-1}(\boldsymbol{\kappa}_{map}) \tag{10.34}$$

In light of Equations 10.33 and 10.34 and taking into account the relation $Z^{-1} = \exp\,\mu_0$, the $\mathcal{Y}_G$ and μ_0 can be expressed by Equations 10.28 and 10.29, respectively. □

In order to illustrate the application of Proposition 10.4, let us consider the following example.

EXAMPLE 10.3: A hard datum χ_1 is available at point $\boldsymbol{p}_1$ and a soft datum χ_2 at point $\boldsymbol{p}_2$ in space/time. We seek the BME estimate at the point $\boldsymbol{p}_k$. Assume that the operator associated with the underlying S/TRF-1/1 model is given by

$$Q\,[\boldsymbol{\chi}_{map}] = \chi_k - \tfrac{1}{2}\sum\nolimits_{i=1}^{2}\chi_i \tag{10.35}$$

in which case the prior statistic is written as

$$\begin{aligned}\overline{g}\,(\boldsymbol{\chi}_{map}) &= \overline{[x_k - \tfrac{1}{2}\sum\nolimits_{i=1}^{2} x_i]^2}\\ &= \tfrac{1}{4}[\kappa_{11} + \kappa_{22}] + \kappa_{kk} - \kappa_{1k} - \kappa_{2k} + \tfrac{1}{2}\kappa_{12} = c_Q(\boldsymbol{\kappa}_{map})\end{aligned} \tag{10.36}$$

where κ_{map} is the generalized covariance matrix. Then,

$$\mathcal{Y}_{\mathcal{G}} = -\tfrac{1}{2}\lambda Q^2 \tag{10.37}$$

and, hence,

$$f_{\mathcal{G}}(\boldsymbol{\chi}_{map}) = Z^{-1}\exp\{-\tfrac{1}{2}\lambda Q^2\} \tag{10.38}$$

From Equation 10.38 we get $\lambda = c_Q^{-1}(\boldsymbol{\kappa}_{map})$ and $Z = \left[2\pi\, c_Q\,(\boldsymbol{\kappa}_{map})\right]^{1/2}$. The posterior pdf is given by

$$f_{\mathcal{K}}(\chi_k) = B(AZ)^{-1}\int_D d\Xi_S(\boldsymbol{\chi}_{soft})\exp[\mathcal{Y}_{\mathcal{G}}]$$

$$= B(AZ)^{-1}\int_D d\Xi_S(\chi_2)\exp\{-\tfrac{1}{2}\lambda Q^2\} \tag{10.39}$$

where the A, B, $\boldsymbol{D}$ and Ξ_S depend on the form of the soft data available (Chapter 6). If, *e.g.*, the soft data are of the interval form, Equation 10.39 reduces to

$$f_{\mathcal{K}}(\chi_k) = (AZ)^{-1}\int_I d\chi_2\,\exp\left\{-\tfrac{1}{2}\lambda\left[\chi_k - \tfrac{1}{2}\sum\nolimits_{i=1}^{2}\chi_i\right]^2\right\} \tag{10.40}$$

where the $\mathcal{Y}_{\mathcal{G}}$ is given by Equation 10.37. In this case the BME equation becomes

$$2\hat{\chi}_k - \overline{x_2}(\hat{\chi}_k) - \chi_1 = 0 \tag{10.41}$$

where $\overline{x_2}(\hat{\chi}_k) = \left\{A^{-1}\int_I d\chi_2\,\chi_2\,\exp[\mathcal{Y}_{\mathcal{G}}]\right\}_{\chi_k=\hat{\chi}_k}$. Equation 10.41 can then be solved with respect to the BME estimate $\hat{\chi}_k$. □

The choice of the S/TRF operator Q involved in the BME calculations above should be made in a way that is mathematically rigorous as well as internally consistent (the Q-operator and the various physical theories and laws governing the natural variables involved in a specific application must be interrelated and corroborative). It is possible, *e.g.*, that Q represents the finite difference scheme obtained from the discretization of the differential equation law governing a physical phenomenon. In many cases, the form of the S/TRF operator Q may change from one space/time neighborhood to another (Christakos, 1992).

Some Non-Gaussian Analytical Expressions

The preceding analytical results are concerned about general knowledge bases $\mathcal{G}$ that involve second-order space/time moments (*e.g.*, covariance or variogram functions). This essentially implies that the $\mathcal{G}$-based operator $\mathcal{Y}_{\mathcal{G}}$ has one of the familiar quadratic expressions, and the resulting prior pdf $f_{\mathcal{G}}$ of Equation 5.6 (p. 106) has a *Gaussian* form. If the $\mathcal{G}$-base includes higher order space/time moments, $f_{\mathcal{G}}$ shows a deviation from the Gaussian shape and Equation 5.6 can be decomposed as follows

$$f_{\mathcal{G}}(\boldsymbol{\chi}_{map};\, \boldsymbol{p}_{map}) = Z^{-1} \exp\left\{\mathcal{Y}_{\mathcal{G},0}\left[\boldsymbol{\chi}_{map};\, \boldsymbol{p}_{map}\right] + \delta\, \mathcal{Y}_{\mathcal{G}}\left[\boldsymbol{\chi}_{map};\, \boldsymbol{p}_{map}\right]\right\} \qquad (10.42)$$

where the Gaussian operator $\mathcal{Y}_{\mathcal{G},0}\left[\boldsymbol{\chi}_{map};\, \boldsymbol{p}_{map}\right]$ has the form of Equation 10.1 and $\delta\, \mathcal{Y}_{\mathcal{G}}\left[\boldsymbol{\chi}_{map};\, \boldsymbol{p}_{map}\right]$ is the so-called *non-Gaussian* perturbation. In the case, *e.g.*, that the $\mathcal{G}$-base includes fourth-order moments we find that

$$\delta\, \mathcal{Y}_{\mathcal{G}}\left[\boldsymbol{\chi}_{map};\, \boldsymbol{p}_{map}\right] = \sum\nolimits_{i=1}^{m,\,k} \mu_i\, \chi_i^4 \qquad (10.43)$$

where the coefficients μ_i are calculated from the BME equations. The implication of Equation 10.42 is that the prior pdf of BME analysis is the product of a Gaussian pdf and a non-Gaussian perturbation pdf. In order to derive the shape of $f_{\mathcal{G}}$, one has to calculate the Lagrange multipliers involved in Equations 10.1 and 10.43. As usual, Equation 10.43 should be substituted into the set of BME equations (Eq. 5.9, p. 107) which are solved for the multipliers. These multipliers are then inserted back into Equation 10.42 to obtain the desired shape of the pdf $f_{\mathcal{G}}$. In practice, a variety of computational techniques may be used to solve the BME equations. Hristopulos and Christakos (2000) discuss the implementation of modern Monte Carlo techniques in cases where the pdf $f_{\mathcal{G}}$ shows large deviations from the Gaussian shape.

In light of Equation 10.42, some interesting analytical expressions can be derived. For example, the expectation of a functional $L[\cdot]$ is given by ,

$$\overline{L}\left[\boldsymbol{p}_{map}\right] = \Lambda^{-1} \left\langle L\left[\boldsymbol{x}_{map}\right] \exp\left\{\delta\, \mathcal{Y}_{\mathcal{G}}\left[\boldsymbol{x}_{map};\, \boldsymbol{p}_{map}\right]\right\}\right\rangle \qquad (10.44)$$

where $\Lambda = \left\langle \exp\left\{\delta\, \mathcal{Y}_{\mathcal{G}}\left[\boldsymbol{x}_{map};\, \boldsymbol{p}_{map}\right]\right\}\right\rangle$, and the $\langle\cdot\rangle$ denotes expectation with respect to the Gaussian pdf. Equation 10.44 is a general expression that can provide non-Gaussian space/time moments in terms of Gaussian expectations. In Christakos *et al.* (1999b) and Hristopulos and Christakos (2000), space/time moment expressions in terms of low-order and diagrammatic perturbations are investigated. We illustrate the implementation of Equation 10.44 by means of the following example.

EXAMPLE 10.4: The (non-centered) covariance between two points $\boldsymbol{p}_i$ and $\boldsymbol{p}_j$ can be expressed with the help of Equation 10.44 as follows

$$C_x(\boldsymbol{p}_i,\, \boldsymbol{p}_j) = \Lambda^{-1} \left\langle x_i\, x_j \exp\left\{\delta\, \mathcal{Y}_{\mathcal{G}}\left[\boldsymbol{x}_{map};\, \boldsymbol{p}_{map}\right]\right\}\right\rangle \qquad (10.45)$$

In the case of a zero-mean homogeneous/stationary random field with $\delta\, \mathcal{Y}_{\mathcal{G}}$ given by Equation 10.43, Equation 10.44 leads to

$$C_x(\boldsymbol{p}_i,\, \boldsymbol{p}_j) \approx \left\langle x_i\, x_j \left\{1 + \delta\, \mathcal{Y}_{\mathcal{G}}\left[\boldsymbol{x}_{map};\, \boldsymbol{p}_{map}\right]\right\}\right\rangle \left\langle 1 + \delta\, \mathcal{Y}_{\mathcal{G}}\left[\boldsymbol{x}_{map};\, \boldsymbol{p}_{map}\right]\right\rangle^{-1}$$

$$\approx \left\langle x_i\, x_j \left[1 + \sum\nolimits_{i=1}^{m,\,k} \mu_i\, x_i^4\right]\right\rangle \left\langle 1 + \sum\nolimits_{i=1}^{m,\,k} \mu_i\, x_i^4 \right\rangle^{-1} \qquad (10.46)$$

After some algebraic manipulations involving the properties of Gaussian expectations, Equation 10.46 yields

$$C_x(\boldsymbol{p}_i, \boldsymbol{p}_j) \approx \langle x_i\, x_j \rangle + 12 \mu\, \sigma_x^2 \sum_{\lambda=1}^{m,k} \langle x_i\, x_\lambda \rangle \langle x_\lambda\, x_j \rangle \tag{10.47}$$

where $\sigma_x^2 = \langle x^2 \rangle$ and the μ is calculated from the BME equations. Equation 10.47 offers a good approximation in the case of a small deviation of the pdf f_G from the Gaussian shape (*i.e.*, the non-Gaussian perturbation is sufficiently weak). Similarly, the fourth-order moment is given by

$$\overline{x_i^4} = 3\sigma_x^4 + 6\,\mu \sum_{\lambda=1}^{m,k} \left[\sigma_x^8 + 3 \langle x_i\, x_\lambda \rangle^4 + 12\, \sigma_x^4 \langle x_i\, x_\lambda \rangle^4 \right] \tag{10.48}$$

Higher order moments are derived in a similar manner. □

Approximate but in many cases useful analytical expressions are obtained in terms of perturbation expansions. In the following example we compare two analytical approximations and a numerical method.

EXAMPLE 10.5: In the univariate case of the non-Gaussian term (Eq. 10.43), the following leading-order *perturbation approximation* for the variance is derived, $\sigma^2_{PER-1} = \sigma_0^2(1+12\,\mu_4\,\sigma_0^4)$. If we use diagrammatic analysis, the leading-order *diagrammatic approximation* for the variance is $\sigma^2_{DIA-1} = \sigma_0^2/[1 - 12\,\mu_4\,\sigma_0^4]$. The low-order perturbation σ^2_{PER-1} and the diagrammatic σ^2_{DIA-1} approximations are compared in Figure 10.8, along with estimates of

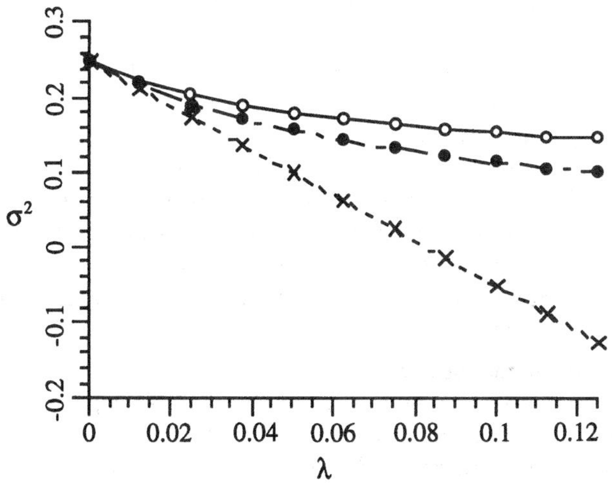

Figure 10.8. Plot of the variance estimates *vs.* the parameter $\lambda = -\mu_4\,\sigma_0^4$. Numerical estimates are indicated by open circles, diagrammatic approximations by solid circles, and the leading-order perturbations are shown by ×.

the variance obtained by a numerical *quadrature integration* scheme for $\sigma_0^2 = 0.25$. The variance is plotted *vs.* the dimensionless non-Gaussian perturbation parameter $\lambda = -\mu_4 \sigma_0^4$. The estimate σ_{PER-1}^2 is accurate for $\lambda \leq 0.04$. The estimate σ_{DIA-1}^2 is more accurate than σ_{PER-1}^2 for $\lambda > 0.04$. In fact, the diagrammatic approach is more accurate than lower-order perturbation for multivariate distributions as well. □

The above approach can be used as well in terms of the posterior pdf $f_{\mathcal{K}}(\chi_k;\ \boldsymbol{p}_k)$. Such an approach is particularly useful in the calculation of expectation functionals with respect to the posterior pdf. As we saw in previous chapters, such functionals arise in a variety of applications, including BME estimation (Chapter 7, p. 147), decision making (Chapter 8, p. 163; Chapter 9, p. 174), and systems analysis (Chapter 9, p. 181).

Theory, Practice, and Computers

The crux of our discussion so far is that BME analysis, just as any scientific approach, requires reasoning at two levels: (*i.*) at the *theoretical* level (which includes the mathematical formulations and proofs presented in the preceding sections); and (*ii.*) at the *practical* level (which involves computational formulations, cost, efficiency, workable schemes, *etc.*). At the practical level (*ii.*), the efficient implementations of the analytical BME formulations above are made possible with the help of computers which are capable of collating knowledge from a number of sources before plotting out the result as a map. SANLIB99 (1999) is the latest version of a continuously updated research library of modern geostatistics computer programs which can work on any UNIX workstation network.

Computerized versions of the BME equations take advantage of two distinct elements of the computer—its ability to store vast amounts of various forms of knowledge, and its ability to process this knowledge in obedience to the strict logical procedures of BME analysis. On the basis of theory and data, the basic BME equations possess significant generalization power. This generalization may occur at several levels of data availability, extending from abundance to near absence. At the former level, cautious generalizations can take place. At the latter level, the generalizations are considerably riskier and take on the nature of hypotheses.

Our experience so far has been that BME is, indeed, a very good approach capable of dealing successfully with a variety of practical mapping situations. Certainly, there is plenty of room for improvement. Improving, *e.g.*, the efficiency with which physical knowledge is acquired will lead to more rapidly produced spatiotemporal maps. This will, in turn, speed up the process of interpretation, understanding, and when necessary, revision.

11

MULTIPOINT ANALYTICAL FORMULATIONS

"Science begins with myths and with the criticism of myths."
K. Popper

The Basic Multipoint BME Equations

The modern spatiotemporal geostatistics model is built upon three fundamental postulates that may be termed the *spatiotemporal physical geometry relationships*, the *epistemic (observer–observed) paradigm*, and the *ontological (context-dependent) knowledge bases*. Before our image of the BME concept becomes clouded with more analytical and computational formulations in the remaining sections, it is worth summarizing its main stages. In a nutshell, the main stages of the BME conception of scientific mapping are as follows:

1. We set up the basic BME equations using whatever sources of knowledge (general and specificatory) are available.
2. We solve the BME equations and see if they lead to experimentally verified predictions. If the predictions are verified, this means that our knowledge bases that led to these predictions were sufficiently accurate and detailed enough for our purposes.
3. If the predictions are not verified, this means that there may be unknown physical influences which are relevant and which should be sought at the ontological level.

Working along the lines of scientific reasoning proposed by modern spatiotemporal geostatistics, in this chapter we derive *analytical* results for various multipoint BME mapping scenarios. The basic BME equations of multipoint spatiotemporal mapping are summarized in Table 11.1 (equation numbers used in previous chapters are included in the table; parameters and the operators A, B, D, and Ξ_S are defined in Table 6.1 on p. 133). These equations offer a theoretical construction which encompasses the mathematical and physical features that geostatisticians seek. Due to space limitations, theoretical formulations for only a few selected general and specificatory knowledge bases are considered below (some of these formulations have been used in applications discussed in previous chapters). There are infinite possibilities, however, limited only by the availability of physical knowledge bases in practice. As emphasized throughout the book, the theoretical richness of the BME construction is a powerful development in modern spatiotemporal geostatistics.

Table 11.1. The basic BME equations.

Equation*	Eq. no.
$\overline{h_\alpha}(\boldsymbol{p}_{map}) = \int d\boldsymbol{\chi}_{map}\, g_\alpha(\boldsymbol{\chi}_{map}) \exp\left[\mu_0 + \mathcal{Y}_{\mathcal{G}}(\boldsymbol{\chi}_{map};\, \boldsymbol{p}_{map})\right]$, $\alpha = 0, 1, \ldots, N_c$	9.27
$f_{\mathcal{K}}(\boldsymbol{\chi}_k) = A^{-1}\mathcal{Y}_{\mathcal{S}}[\boldsymbol{\chi}_{map};\, \boldsymbol{p}_k] = A^{-1}B \int_D d\Xi_{\mathcal{S}}(\boldsymbol{\chi}_{soft}) \exp[\mu_0 + \mathcal{Y}_{\mathcal{G}}]$	9.32
$\frac{\partial}{\partial \chi_{k\ell}} \mathcal{Y}_{\mathcal{S}}\rvert_{\boldsymbol{\chi}_k = \hat{\boldsymbol{\chi}}_k} = B \int_D d\Xi_{\mathcal{S}}(\boldsymbol{\chi}_{soft}) \left\{\exp[\mathcal{Y}_{\mathcal{G}}] \frac{\partial}{\partial \chi_{k\ell}} \mathcal{Y}_{\mathcal{G}}\right\}_{\boldsymbol{\chi}_k = \hat{\boldsymbol{\chi}}_k} = 0$, $\ell = 0, 1, \ldots, \rho$	9.33
$\boldsymbol{\chi}_{map} = (\chi_1, \ldots, \chi_m, \chi_{k_1}, \ldots, \chi_{k_\rho})$	

*Equations appear on p. 175–176.

Ordinary Covariance—Hard and Soft Data

We start with the following fundamental proposition [the proof is very similar to that of Proposition 10.1 (p. 198) and is not included here].

PROPOSITION 11.1: Let $\boldsymbol{\chi}_{hard}$ be a vector of hard data at points $\boldsymbol{p}_i$ ($i = 1, 2, \ldots, m_h$) and $\boldsymbol{\chi}_{soft}$ be a vector of soft data (of various possible forms; see Table 6.1) at points $\boldsymbol{p}_i$ ($i = m_h + 1, \ldots, m$). General knowledge includes the mean and the (centered) ordinary covariance. Then, the BME posterior pdf is given by Equation 9.32 (p. 176) with

$$\mathcal{Y}_{\mathcal{G}}[\boldsymbol{\chi}_{map};\, \boldsymbol{p}_{map}] = -\tfrac{1}{2}(\boldsymbol{\chi}_{map} - \overline{\boldsymbol{x}_{map}})^T \boldsymbol{c}_{map}^{-1}(\boldsymbol{\chi}_{map} - \overline{\boldsymbol{x}_{map}}) \tag{11.1}$$

and

$$\mu_0 = -\log\left[(2\pi)^{(m+1)/2} / \lvert \boldsymbol{c}_{map}^{-1} \rvert^{1/2}\right] \tag{11.2}$$

where now

$$\overline{\boldsymbol{x}_{map}} = (\overline{x_1}, \ldots \overline{x_m}, \overline{x_{k_1}}, \ldots, \overline{x_{k_\rho}}) \tag{11.3}$$

is the mean vector for points $\boldsymbol{p}_i$, $i = 1, \ldots, m, k_1, \ldots, k_\rho$ (notice the difference compared to Eq. 10.3, p. 199), and

$$\boldsymbol{c}_{map} = \overline{(\boldsymbol{x}_{map} - \overline{\boldsymbol{x}_{map}})(\boldsymbol{x}_{map} - \overline{\boldsymbol{x}_{map}})^T} \tag{11.4}$$

is the centered covariance matrix between all the points. The BMEmode

vector estimates $\hat{\boldsymbol{\chi}}_k$ are solutions of the equations

$$\sum_{i=1}^{m_h} c_{ik_\ell}^{-1}(\chi_i - \overline{x_i}) + \sum_{i=m_h+1}^{m} c_{ik_\ell}^{-1}\left[\overline{x_i}(\hat{\boldsymbol{\chi}}_k) - \overline{x_i}\right] + \sum_{i=k_1}^{k_\rho} c_{ik_\ell}^{-1}(\hat{\chi}_i - \overline{x_i}) = 0 \tag{11.5}$$

$(\ell = 1, \ldots, \rho)$ where

$$\overline{x_i}(\hat{\boldsymbol{\chi}}_k) = \frac{B\int_D d\Xi_S \chi_i \exp[\mathcal{Y}_G]_{\chi_k=\hat{\chi}_k}}{B\int_D d\Xi_S \exp[\mathcal{Y}_G]_{\chi_k=\hat{\chi}_k}} \tag{11.6}$$

The following proposition is a straightforward extension in the multipoint case of the single-point situation treated in Proposition 10.2 (p. 201).

PROPOSITION 11.2: Assume that hard data are given at points $\boldsymbol{p}_i$ $(i = 1, 2, \ldots, m_h)$ and soft data of various possible forms (Table 6.1) at points $\boldsymbol{p}_i$ $(i = m_h + 1, \ldots, m)$. General knowledge is the (non-centered) ordinary covariance. The multipoint BMEmode estimate $\hat{\boldsymbol{\chi}}_k = (\hat{\chi}_{k_1}, \ldots \hat{\chi}_{k_\rho})$ is a solution of the set of equations

$$\sum_{i=1}^{m_h} C_{ik_\ell}^{-1}\chi_i + \sum_{i=m_h+1}^{m} C_{ik_\ell}^{-1}\overline{x_i}(\hat{\boldsymbol{\chi}}_k) + \sum_{i=k_1}^{k_\rho} C_{ik_\ell}^{-1}\hat{\chi}_i = 0, \quad \ell = 1, \ldots, \rho \tag{11.7}$$

$C_{ik_\ell}^{-1}$ is the ik_ℓ-th element of the inverse matrix C_{map}^{-1}, where C_{map} is the matrix of the non-centered covariances C_{ij} between all the points $\boldsymbol{p}_i$ and $\boldsymbol{p}_j$ $(i, j = 1, \ldots, m, k_1 \ldots, k_\rho)$; and the parameter $\overline{x_i}\,(\hat{\boldsymbol{\chi}}_k)$ is of the form of Equation 11.6 above.

Proof: Equation 10.18 (p. 201) can be extended to the multipoint case where now $\boldsymbol{\chi}_{map} = (\chi_1, \ldots, \chi_m, \chi_{k_1}, \ldots, \chi_{k_\rho})$ and

$$\boldsymbol{C}_{map} = \overline{\boldsymbol{x}_{map}\boldsymbol{x}_{map}^T} = \begin{bmatrix} C_{11} & \ldots & C_{1m} & C_{1k_1} & \ldots & C_{1k_\rho} \\ \vdots & & & & & \\ C_{m1} & \ldots & C_{mm} & C_{mk_1} & \ldots & C_{mk_\rho} \\ C_{k_11} & \ldots & C_{k_1m} & C_{k_1k_1} & \ldots & C_{k_1k_\rho} \\ \vdots & & & & & \\ C_{k_\rho1} & \ldots & C_{k_\rho m} & C_{k_\rho k_1} & \ldots & C_{k_\rho k_\rho} \end{bmatrix} \tag{11.8}$$

is the non-centered covariance matrix between all points $\boldsymbol{p}_i (i = 1, \ldots, m, k_1, \ldots, k_\rho)$. Equation 9.32 (p. 176) is valid with

$$\mathcal{Y}_G[\boldsymbol{\chi}_{map};\, \boldsymbol{p}_{map}] = \sum_{i,j=1}^{m,\, k_1, \ldots, k_\rho} \mu_{ij}\, g_{ij}(\chi_i, \chi_j) \tag{11.9}$$

where $\mu_{ij} = -\frac{1}{2}C_{ij}^{-1}$ (C_{ij}^{-1} is the ij-th element of the inverse matrix $\boldsymbol{C}_{map}^{-1}$), $g_{ij}(\chi_i, \chi_j) = \chi_i\chi_j$ $(i, j = 1, \ldots, m, k_1, \ldots, k_\rho)$. After some manipulations

$$\mathcal{Y}_G = -\frac{1}{2}\sum_{i,j=1}^{m} C_{ij}^{-1}\chi_i\chi_j - \sum_{i=1}^{m}\sum_{j=k_1}^{k_\rho} C_{ij}^{-1}\chi_i\chi_j - \frac{1}{2}\sum_{i,j=k_1}^{k_\rho} C_{ij}^{-1}\chi_i\chi_j \tag{11.10}$$

and

$$\partial \mathcal{Y}_{G}/\partial \chi_{k_\ell}|_{\chi_k=\hat{\chi}_k} = -\sum_{j=k_1}^{k_\rho} C_{k_\ell j}^{-1}\hat{\chi}_j - \sum_{j=1}^{m} C_{k_\ell j}^{-1}\chi_j, \quad (\ell = 1, \ldots, \rho) \tag{11.11}$$

Hence, the BME equations can be written as

$$\left[\sum_{i=k_1}^{k_\rho} C_{ik_\ell}^{-1}\hat{\chi}_i + \sum_{i=1}^{m_h} C_{ik_\ell}^{-1}\chi_i\right] B \int_D d\Xi_s(\boldsymbol{\chi}_{soft}) \exp\left[\mathcal{Y}_G(\boldsymbol{\chi}_{map})\right]_{\chi_k=\hat{\chi}_k}$$

$$+\sum_{i=m_h+1}^{m} C_{ik_\ell}^{-1} B \int_D d\Xi_s(\boldsymbol{\chi}_{soft})\chi_i \exp\left[\mathcal{Y}_G(\boldsymbol{\chi}_{map})\right]_{\chi_k=\hat{\chi}_k} = 0,$$

$$\ell = 1, \ldots \rho$$

which leads to Equation 11.7. □

In the following developments, it is convenient to define the partitioned matrices

$$\mathbf{c}_{map} = \begin{bmatrix} \mathbf{c}_{hs,hs} & \mathbf{c}_{hs,k} \\ \mathbf{c}_{k,hs} & \mathbf{c}_{k,k} \end{bmatrix} = \begin{bmatrix} \mathbf{c}_{h,h} & \mathbf{c}_{h,s} & \mathbf{c}_{h,k} \\ \mathbf{c}_{s,h} & \mathbf{c}_{s,s} & \mathbf{c}_{s,k} \\ \mathbf{c}_{k,h} & \mathbf{c}_{k,s} & \mathbf{c}_{k,k} \end{bmatrix} \tag{11.12}$$

and

$$\mathbf{c}_{kh,kh} = \begin{bmatrix} \mathbf{c}_{k,k} & \mathbf{c}_{k,h} \\ \mathbf{c}_{h,k} & \mathbf{c}_{h,h} \end{bmatrix} \tag{11.13}$$

where the subscripts h, s, and k denote hard points, soft points, and estimation points, respectively. Computationally efficient formulations of the posterior (or integration) pdf can be derived in some situations, as described by the following two propositions (Serre and Christakos, 1999a).

PROPOSITION 11.3: Assume that hard data are given at points $\boldsymbol{p}_i$ ($i = 1, 2, \ldots, m_h$) and soft data of the interval type (Eq. 3.32, p. 85) at points $\boldsymbol{p}_i$ ($i = m_h + 1, \ldots, m$). General knowledge is the (centered) ordinary covariance. The posterior pdf is as follows

$$f_{\mathcal{K}}(\boldsymbol{\chi}_k) =$$

$$A^{-1}\phi(\boldsymbol{\chi}_k; \boldsymbol{B}_{k|h}\boldsymbol{\chi}_{hard}, \mathbf{c}_{k|h})\int_{\boldsymbol{l}-\boldsymbol{B}_{s|kh}\boldsymbol{\chi}_{kh}}^{\boldsymbol{u}-\boldsymbol{B}_{s|kh}\boldsymbol{\chi}_{kh}} d\boldsymbol{\chi}_{soft}\,\phi(\boldsymbol{\chi}_{soft}; \boldsymbol{O}, \mathbf{c}_{s|kh}) \tag{11.14}$$

where $\boldsymbol{\chi}_{kh} = (\boldsymbol{\chi}_k, \boldsymbol{\chi}_{hard})$, $\boldsymbol{B}_{k|h} = \mathbf{c}_{k,h}\mathbf{c}_{h,h}^{-1}$, $\mathbf{c}_{k|h} = \mathbf{c}_{k,k} - \boldsymbol{B}_{k|h}\mathbf{c}_{h,k}$, $\boldsymbol{B}_{s|kh} = \mathbf{c}_{s,kh}\mathbf{c}_{kh,kh}^{-1}$, $\mathbf{c}_{s|kh} = \mathbf{c}_{s,s} - \boldsymbol{B}_{s|kh}\mathbf{c}_{kh,s}$, $\boldsymbol{l} = (l_{m_h+1}, \ldots, l_m)$, and $\boldsymbol{u} = (u_{m_h+1}, \ldots, u_m)$; the $\phi(\boldsymbol{\chi}; \overline{\boldsymbol{x}}, \mathbf{c})$ denotes a Gaussian distribution with mean vector $\overline{\boldsymbol{x}}$ and covariance matrix $\mathbf{c}$; and $A = \int_{\boldsymbol{l}-\boldsymbol{B}_{s|h}\boldsymbol{\chi}_{hard}}^{\boldsymbol{u}-\boldsymbol{B}_{s|h}\boldsymbol{\chi}_{hard}} d\boldsymbol{\chi}_{soft}\,\phi(\boldsymbol{\chi}_{soft}; \boldsymbol{O}, \mathbf{c}_{s|h})$.

Note that the multiple integral in Equation 11.14 has the form of a multivariate Gaussian probability, which is very useful in numerical implementations (see, *e.g.*, Genz, 1992).

PROPOSITION 11.4: Assume that hard data are given at points p_i ($i = 1, 2, \ldots, m_h$) and soft data of the probabilistic type (Eq. 3.33) at points p_i ($i = m_h + 1, \ldots, m$). General knowledge is the (centered) ordinary covariance. The posterior pdf is

$$f_{\mathcal{K}}(\boldsymbol{\chi}_k) =$$

$$A^{-1}\,\phi(\boldsymbol{\chi}_k;\ \boldsymbol{B}_{k|h}\,\boldsymbol{\chi}_{hard},\ \boldsymbol{c}_{k|h}) \int d\boldsymbol{\chi}_{soft}\, f_S(\boldsymbol{\chi}_{soft})\phi(\boldsymbol{\chi}_{soft};\ \boldsymbol{B}_{s|kh}\,\boldsymbol{\chi}_{kh},\ \boldsymbol{c}_{s|kh}) \tag{11.15}$$

where $A = \int d\boldsymbol{\chi}_{soft}\, f_S(\boldsymbol{\chi}_{soft})\,\phi(\boldsymbol{\chi}_{soft};\ \boldsymbol{B}_{s|h}\,\boldsymbol{\chi}_{hard},\ \boldsymbol{c}_{s|h})$.

The BME posterior pdf's (Eqs. 11.14 and 11.15) may be interpreted as a *natural synthesis* of the two knowledge bases, *i.e.*, the general (covariance) and the specificatory (soft data). In the case of single-point analysis ($\rho = 1$), important parameters in spatiotemporal mapping are the conditional mean

$$\hat{\chi}_k = A^{-1} \int d\boldsymbol{\chi}_{soft}\, f_S(\boldsymbol{\chi}_{soft})\, \boldsymbol{B}_{k|hs}\,\boldsymbol{\chi}_{hs}\,\phi(\boldsymbol{\chi}_{soft};\ \boldsymbol{B}_{s|h}\,\boldsymbol{\chi}_{hard},\ \boldsymbol{c}_{s|h}) \tag{11.16}$$

and the variance

$$\sigma_k^2 = \boldsymbol{c}_{k|hs} + A^{-1} \int d\boldsymbol{\chi}_{soft}\, f_S(\boldsymbol{\chi}_{soft})\,(\boldsymbol{B}_{k|hs}\,\boldsymbol{\chi}_{data} - \hat{\chi}_k)^2\,\phi(\boldsymbol{\chi}_{soft};$$

$$\boldsymbol{B}_{s|h}\,\boldsymbol{\chi}_{hard},\ \boldsymbol{c}_{s|h}) \tag{11.17}$$

where $\boldsymbol{B}_{k|hs} = \boldsymbol{c}_{k,hs}\,\boldsymbol{c}_{hs,hs}^{-1}$, $\boldsymbol{c}_{k|hs} = \boldsymbol{c}_{k,k} - \boldsymbol{B}_{k|hs}\,\boldsymbol{c}_{hs,k}$, $\boldsymbol{B}_{s|h} = \boldsymbol{c}_{s,h}\,\boldsymbol{c}_{h,h}^{-1}$, and $\boldsymbol{c}_{s|h} = \boldsymbol{c}_{s,s} - \boldsymbol{B}_{s|h}\,\boldsymbol{c}_{h,s}$, and the A is defined as in Proposition 11.4.

COMMENT 11.1: *The computational BME formulations presented above account for various specificatory knowledge bases (*e.g.*, combinations of hard and soft data), thus leading to a non-Gaussian posterior pdf, in general. Furthermore, the BME estimates are nonlinear. When only hard data are used,* i.e., *$m_h = m$, the multiple integral and the subscript s is dropped in Equations 11.14–11.17. In this case, the mean and the variance of the posterior pdf become, respectively, $\bar{x}_{k|\mathcal{K}} = \boldsymbol{c}_{k,hs}\,\boldsymbol{c}_{hs,hs}^{-1}\,\boldsymbol{\chi}_{hard}$ and $\sigma_{k|\mathcal{K}}^2 = \boldsymbol{c}_{k,k} - \boldsymbol{c}_{k,hs}\,\boldsymbol{c}_{hs,hs}^{-1}\,\boldsymbol{c}_{hs,k}$. These quantities coincide with the simple kriging (SK) estimate and its error variance, respectively. This result shows that BME analysis can offer a unified framework which contains the existing kriging methods as its limiting cases. More specifically, the implication of the unified framework in numerical calculations is that when $\mathcal{G} = \{$mean and covariance$\}$ and $\mathcal{S} = \{$hard data$\}$, BME will provide the same estimates as the kriging techniques (for a detailed discussion see Chapter 12). When higher order space/time moments, physical laws, soft data,* etc. *are added, the kriging methods cannot be used. The BME techniques, however, can still be implemented, leading to useful results.*

Ordinary Variogram—Hard and Soft Data

Next we consider the common case in which the ordinary variogram function is available together with a set of hard and soft data.

PROPOSITION 11.5: Assume that hard data exist at points $\boldsymbol{p}_i$ $(i = 1, 2, \ldots, m_h)$ and soft data of various possible forms (Chapter 3) at points $\boldsymbol{p}_i$ $(i = m_h + 1, \ldots, m)$. General knowledge is expressed in terms of the variogram. The multipoint BMEmode estimate $\hat{\boldsymbol{\chi}}_k$ is the solution of the set of equations $(\ell = 1, \ldots, \rho)$

$$\sum\nolimits_{j=k_1}^{k_\rho} \gamma_{k_\ell j}^{-1} (\hat{\chi}_{k_\ell} - \chi_j) + \tfrac{1}{2} \sum\nolimits_{j=1}^{m_h} \gamma_{k_\ell j}^{-1} (\hat{\chi}_{k_\ell} - \chi_j) + \tfrac{1}{2} \sum\nolimits_{j=m_h+1}^{m} \gamma_{k_\ell j}^{-1} \overline{x_j} (\hat{\chi}_{k_\ell}) = 0 \tag{11.18}$$

where γ_{ij}^{-1} is the ij-th element of the inverse matrix $\boldsymbol{\gamma}_{map}^{-1}$ such that

$$\boldsymbol{\gamma}_{map} = \begin{bmatrix} \boldsymbol{\gamma}_1 & O \\ O & \boldsymbol{\gamma}_2 \end{bmatrix} \tag{11.19}$$

is a variogram matrix; $\boldsymbol{\gamma}_1$ is an $(m \times \rho) \times (m \times \rho)$ diagonal matrix of variograms between the points $\boldsymbol{p}_i$ $(i = 1, \ldots, m)$ and $\boldsymbol{p}_{k_\ell}$ $(\ell = 1, \ldots, \rho)$; $\boldsymbol{\gamma}_2$ is a $\rho \times \rho$ diagonal matrix of variograms between the points $\boldsymbol{p}_{k_\ell}$ themselves; and

$$\overline{x_j}(\hat{\chi}_{k_\ell}) = \frac{B \int_D d\Xi_S (\hat{\chi}_{k_\ell} - \chi_j) \exp[\mathcal{Y}_G]_{\boldsymbol{\chi}_k = \hat{\boldsymbol{\chi}}_k}}{B \int_D d\Xi_S \exp[\mathcal{Y}_G]_{\boldsymbol{\chi}_k = \hat{\boldsymbol{\chi}}_k}} \tag{11.20}$$

with

$$\mathcal{Y}_G = -\tfrac{1}{4} \sum\nolimits_{i=k_1}^{k_\rho} \left[\sum\nolimits_{j=1}^{m} \gamma_{ij}^{-1} (\chi_i - \chi_j)^2 + \sum\nolimits_{j=k_1}^{k_\rho} \gamma_{ij}^{-1} (\chi_i - \chi_j)^2 \right] \tag{11.21}$$

Proof: The prior pdf is of the form

$$f_G(\boldsymbol{\chi}_{map}) = Z^{-1} \exp[\mathcal{Y}_G] \tag{11.22}$$

where

$$\mathcal{Y}_G = -\tfrac{1}{2} \sum\nolimits_{i=k_1}^{k_\rho} \left[\sum\nolimits_{j=1}^{m} \lambda_{ij} (\chi_i - \chi_j)^2 + \sum\nolimits_{j=k_1}^{k_\rho} \lambda_{ij} (\chi_i - \chi_j)^2 \right] \tag{11.23}$$

Since the $c_{\delta(ik,jk)} = 2\gamma_x(\boldsymbol{p}_i, \boldsymbol{p}_j, \boldsymbol{p}_k)$ is given by Equation 10.25 (p. 202), it follows that

$$\boldsymbol{\lambda}^{-1} = 2\boldsymbol{\gamma}_{map} \tag{11.24}$$

where

$$\left. \begin{array}{l} \boldsymbol{\lambda} = diag\,[\boldsymbol{\lambda}_{k_\ell m}] \\ \boldsymbol{\lambda}_{k_\ell m} = diag\,[\lambda_{k_\ell j}],\ j = 1, \ldots, m,\ k_1, \ldots, k_\rho \text{ and } \ell = 1, \ldots, \rho \end{array} \right\} \tag{11.25}$$

In light of the aforementioned analysis, the BME equation is given by Equation 9.29 (p. 175), where the operator $\mathcal{Y}_G$ is of the form of Equation 11.21 and

$$\frac{\partial}{\partial \chi_{k_\ell}} \mathcal{Y}_{\mathcal{G}} = -\tfrac{1}{2}\sum\nolimits_{j=1}^{m} \gamma_{k_\ell j}^{-1}(\chi_{k_\ell}-\chi_j) - \sum\nolimits_{j=k_1}^{k_\rho} \gamma_{k_\ell j}^{-1}(\chi_{k_\ell}-\chi_j), \quad \ell = 1, \ldots, \rho \tag{11.26}$$

By substituting the last equation into the BME equation (Eq. 9.29), we find

$$\Big[\sum\nolimits_{j=k_1}^{k_\rho} \gamma_{k_\ell j}^{-1}(\hat{\chi}_{k_\ell}-\chi_j) + \tfrac{1}{2}\sum\nolimits_{j=1}^{m_h} \gamma_{k_\ell j}^{-1}(\hat{\chi}_{k_\ell}-\chi_j)\Big] B \int_D d\Xi_S \exp[\mathcal{Y}_{\mathcal{G}}]_{\boldsymbol{\chi}_k=\hat{\boldsymbol{\chi}}_k}$$
$$+\tfrac{1}{2}\sum\nolimits_{j=m_h+1}^{m} \gamma_{k_\ell j}^{-1} B \int_D d\Xi_S(\hat{\chi}_{k_\ell}-\chi_j)\exp[\mathcal{Y}_{\mathcal{G}}]_{\boldsymbol{\chi}_k=\hat{\boldsymbol{\chi}}_k} = 0, \quad \ell = 1, \ldots, \rho \tag{11.27}$$

which can also be written as Equation 11.18. □

Several of the single-point results of Chapter 10 may be derived as special cases of the analysis of the present chapter. The BME equation, Equation 10.21 (p. 202), *e.g.*, is a special case of Equation 11.18 above; *etc.*

Other Combinations

Analytical results for applications involving combinations of vectorial and multipoint scenarios can also be derived. For illustration, consider the following example.

EXAMPLE 11.1: Assume that the specificatory knowledge about the natural fields $X(\boldsymbol{p})$ and $Y(\boldsymbol{p})$ are the hard data $\boldsymbol{\chi}_{hard}$ and $\boldsymbol{\psi}_{hard}$, as well as the soft data $\boldsymbol{\chi}_{soft}$ and $\boldsymbol{\psi}_{soft}$ (of any of the various forms considered in Chapter 3). The general knowledge consists of the ordinary auto- and cross-covariance matrices

$$\boldsymbol{C}_x = \overline{\boldsymbol{x}_{map}\boldsymbol{x}_{map}^T}, \; \boldsymbol{C}_y = \overline{\boldsymbol{y}_{data}\boldsymbol{y}_{data}^T}, \text{ and } \boldsymbol{C}_{xy} = \overline{\boldsymbol{x}_{map}\boldsymbol{y}_{data}^T} \tag{11.28}$$

Under these circumstances, the BME estimates $\hat{\chi}_{k_\ell}$ will be the solutions of the following set of BME equations

$$\sum\nolimits_{j=1}^{m_h} (\boldsymbol{C}_x^{-1})_{k_\ell,j}\,\chi_j + \sum\nolimits_{j=k_1}^{k_\rho} (\boldsymbol{C}_x^{-1})_{k_\ell,j}\,\hat{\chi}_j + \sum\nolimits_{j=1}^{n_h} (\boldsymbol{C}_{xy}^{-1})_{k_\ell,j}\,\psi_j$$
$$+\tau^{-1}\sum\nolimits_{j=m_h+1}^{m} (\boldsymbol{C}_x^{-1})_{k_\ell,j} B \int d\Xi_S(\boldsymbol{\chi}_{soft}, \boldsymbol{\psi}_{soft})\,\chi_j \exp[\mathcal{Y}_{\mathcal{G}}]_{\boldsymbol{\chi}_k=\hat{\boldsymbol{\chi}}_k}$$
$$+\tau^{-1}\sum\nolimits_{j=n_h+1}^{n} (\boldsymbol{C}_{xy}^{-1})_{k_\ell,j} B \int d\Xi_S(\boldsymbol{\chi}_{soft}, \boldsymbol{\psi}_{soft})\,\psi_j \exp[\mathcal{Y}_{\mathcal{G}}]_{\boldsymbol{\chi}_k=\hat{\boldsymbol{\chi}}_k} = 0 \tag{11.29}$$

where $\ell = 1, \ldots, \rho$, $(\boldsymbol{C}_x^{-1})_{i,j}$ denotes the ij-th element of the inverse covariance matrix, $\tau = B\int d\Xi_S \exp[\mathcal{Y}_{\mathcal{G}}]_{\boldsymbol{\chi}_k=\hat{\boldsymbol{\chi}}_k}$, and $\mathcal{Y}_{\mathcal{G}} = -\tfrac{1}{2}\boldsymbol{\chi}_{map}^T\boldsymbol{C}_x^{-1}\boldsymbol{\chi}_{map} - \tfrac{1}{2}\boldsymbol{\psi}_{data}^T\boldsymbol{C}_y^{-1}\boldsymbol{\psi}_{data} - \boldsymbol{\chi}_{map}^T\boldsymbol{C}_{xy}^{-1}\boldsymbol{\psi}_{data}$. □

At this point we have in our possession a number of models which are largely analytically lucid. Certainly, the analytical formulations become more elaborate as the number of knowledge sources that BME takes into consideration increases. The reader, however, should remember: "There's no free lunch!" It is, after all, a matter of choice. If we decide to limit our analysis to a few low-order statistical moments and a set of hard data, BME will have no difficulty generating simple, linear estimators. If, however, we believe that the available knowledge bases cannot be ignored, we have no choice but to use the more elaborate BME formulations. Modern spatiotemporal geostatistics allows such a choice, something that is not possible with most classical techniques.

Spatiotemporal Covariance and Variogram Models

The BME framework is very general, and one has considerable freedom in the choice of the covariance and variogram models (ordinary and generalized). Indeed, the covariance and variogram models used in the BME equations can be separable or nonseparable functions, they may be associated to homogeneous/stationary or nonhomogeneous/nonstationary random fields, *etc.*

Separable models

Separable covariances and variograms (ordinary or generalized), which are obtained by combining permissible spatial and temporal models, offer useful solutions in a variety of applications. In particular, a wide variety of space/time separable covariance models are obtained by means of the product

$$c_x(\boldsymbol{h},\, \tau) = c_s(\boldsymbol{h})\, c_t(\tau) \tag{11.30}$$

where $c_s(\boldsymbol{h})$ and $c_t(\tau)$ are valid spatial and temporal models.

EXAMPLE 11.2: The Gaussian model

$$c_x(\boldsymbol{h},\, \tau) = c_0 \exp(-\boldsymbol{h}^2 - v^2\tau^2) \tag{11.31}$$

and the exponential model

$$c_x(\boldsymbol{h},\, \tau) = c_0 \exp(-\,|\,\boldsymbol{h}\,| - v\tau) \tag{11.32}$$

are among the most popular separable models (see also p. 64–65 in this volume). Another interesting separable model is given by

$$c_x(r,\, \tau) = r^{-1} \exp(-ar) \exp(-b\tau) \tag{11.33}$$

where $r = |\,\boldsymbol{h}\,|$ and $\tau = t - t'$ (the spectral density of the above model decreases as either the spatial frequency k or the temporal frequency ω increases). In an effort to study rainfall fields in $R^2 \times T$, Rodriguez-Iturbe and Mejia (1974) used the model

$$c_x(r,\, \tau) = a\, r\, K_1(a\, r) \exp(-b\, \tau) \tag{11.34}$$

where K_1 is the modified Bessel function of the 2nd kind. □

Several other examples of separable space/time covariances and variograms can be found in Whittle (1954), Christakos (1992), Johnson and Dudgeon (1993), Christakos and Hristopulos (1998), and references therein.

Nonseparable models

Nonseparable spatiotemporal covariances for *homogeneous/stationary* random fields have been proposed by a number of researchers in various scientific disciplines. Let us discuss a few examples.

EXAMPLE 11.3: Heine (1955) developed the following nonseparable covariance model in $R^1 \times T$,

$$c_x(r,\tau) = \tfrac{1}{2}\sigma^2 \Big\{ \exp(-a\,r)\, Erfc\left[(2a\,\tau - c\,r)/2\sqrt{c\,\tau}\,\right] + \exp(a\,r)\, Erfc\left[(2a\,\tau + c\,r)/2\sqrt{c\,\tau}\,\right]\Big\} \qquad (11.35)$$

where a, c, and σ^2 are suitable coefficients associated with a parabolic-type partial differential equation; and $Erfc$ is the complementary error function. □

EXAMPLE 11.4: Jones and Zhang (1997) suggested a set of covariances in $R^n \times T$ having the spectral density

$$\tilde{c}_x(k,\omega) = \sigma^2 \left[(k^2 + a^2)^{2p} + c^2\omega^2\right]^{-1} \qquad (11.36)$$

Equation 11.36 represents homogeneous/stationary random fields. In $R^2 \times T$, for $p = 2$, Equation 11.36 leads to a covariance that can be calculated as follows

$$c_x(r,\tau) = 2a^2\sigma^2 \int_0^\infty \frac{k\,\exp[-(k^2+a^2)^2\,\tau/c]}{(k^2+a^2)^2}\, J_0(k\,r)\, d\,k \qquad (11.37)$$

where J_0 is the Bessel function of the 1st kind of order 0. □

While the above covariance models represent homogeneous/stationary random fields only, Christakos (1992) and Christakos and Hristopulos (1998) presented several classes of nonseparable models which are valid for homogeneous/stationary fields as well as models for *nonhomogeneous/nonstationary* fields. The examples below include nonseparable covariance models derived on the basis of stochastic partial differential equations, nonlinear systems, suitably chosen spectral densities, dynamic rules, fractal processes, *etc.*

EXAMPLE 11.5: Nonseparable spatiotemporal covariance models can be associated with the *stochastic partial differential equation*

$$\partial X(\boldsymbol{p})/\partial t = \mathcal{L}_s\left[X(\boldsymbol{p})\right] \qquad (11.38)$$

where $\mathcal{L}_s$ is a linear spatial differential operator. These models include the following (Christakos and Hristopulos, 1998)

$$c_x(\boldsymbol{s},t;\,\boldsymbol{s}',t') = \begin{cases} \sum_{n,m} c_{nm}\,\chi_{1n}(\boldsymbol{s})\,\chi_{1m}(\boldsymbol{s}')\,\chi_{2n}(t)\,\chi_{2m}(t') \\ \sum_{n,m} A_n A_m\, c_{\chi(n,m)}(\boldsymbol{s},\boldsymbol{s}')\,\chi_{2n}(t)\chi_{2m}(t') \\ \sum_{n,m} \overline{A_n A_m\,\chi_{1n}(\boldsymbol{s})\,\chi_{1m}(\boldsymbol{s}')}\,\chi_{2n}(t)\,\chi_{2m}(t') \end{cases} \tag{11.39}$$

where χ_{1n} and χ_{2n} represent eigenfunctions (modes) of Equation 11.38, the coefficients c_{nm} represent correlations of the mode coefficients, *i.e.*, $c_{nm} = \overline{A_n A_m}$, the two-point function $c_{\chi(n,m)}(\boldsymbol{s},\boldsymbol{s}')$ denotes the correlation $\overline{\chi_{1n}(\boldsymbol{s})\,\chi_{1m}(\boldsymbol{s}')}$, and A_n are random variables to be determined from the initial and boundary conditions. Randomness in the three models of Equation 11.39 can be introduced, respectively, by: (*i.*) the initial or boundary conditions leading to random coefficients A_n ; (*ii.*) the differential operator $\mathcal{L}_s$ leading to random eigenfunctions $\chi_{1n}(\boldsymbol{s})$; and (*iii.*) by both of the above. Models (Eq. 11.39) may be homogeneous/nonstationary due to a number of reasons including the boundary and initial conditions. Similarly, on the basis of the noisy *Burgers equation*, an interesting nonseparable covariance model in $R^1 \times T$ is given by

$$c_x(r,\tau) = r^{-\alpha} f(\tau/r^z) \tag{11.40}$$

(large r, τ), where $f(w = \tau/r^z) = \frac{\sigma^2}{4D\sqrt{\pi\,D\,w}}\ \exp[-1/4\,Dw]$. □

EXAMPLE 11.6: A diffusion-inspired covariance model is associated with the spectral density $\tilde{c}_x(k,\omega) = 2\,\alpha\,\sigma^2/\left[\,\omega^2 + (\alpha\,k^2)^2\,\right]$, which satisfies *Bochner's theorem*. By calculating the inverse transform we find the nonseparable space/time covariance model

$$c_x(r,\tau) = \sigma^2 \exp\left[-r^2/4\alpha\,\tau\right] \Big/ (4\alpha\pi\,\tau)^{n/2} \tag{11.41}$$

in $R^n \times T$. The nonseparable variogram below that is used in several applications in the $R^1 \times T$ domain has been obtained from the application of Bochner's theorem

$$\gamma_x(r,\tau) = \sigma^2\{1 - \exp\left[-\sqrt{r^2/a^2 + \tau^2/b^2}\right]\} \tag{11.42}$$

The a and b are coefficients corresponding to the spatial and temporal scales (see, *e.g.*, Christakos, 1992); this model represents homogeneous/stationary random fields. □

EXAMPLE 11.7: A rich class of nonseparable covariance models in $R^n \times T$ is generated from the *spectral density*

$$\tilde{c}_x(\boldsymbol{k},\omega) = \delta(\omega + \boldsymbol{k}\cdot\boldsymbol{v})\,\tilde{c}_s(\boldsymbol{k}) \tag{11.43}$$

where $\boldsymbol{v}$ is a vector of known coefficients and $\tilde{c}_s(\boldsymbol{k})$ is the spectral density of a purely spatial covariance $c_x(\boldsymbol{r})$. Thus, based on the already available

$c_x(\boldsymbol{r})$ models, a variety of space/time covariances can be derived, such as the following two examples: the covariance

$$c_x(\boldsymbol{h},\tau) = \sigma^2 \exp\left[-\,|\,\boldsymbol{h} - \boldsymbol{v}\,\tau\,|\,/a\right] \tag{11.44}$$

associated with $\tilde{c}_s(\boldsymbol{k}) = \sigma^2\,(2a)^n\,\pi^{n/2}/\left[\Gamma(2-n/2)\,(1+a^2\,k^2)^{(n+1)/2}\right]$, where $k = |\,\boldsymbol{k}\,|$ and Γ is the gamma function; and the covariance

$$c_x(\boldsymbol{h},\tau) = \sigma^2 \exp\left[-\,(\boldsymbol{h} - \boldsymbol{v}\,\tau)^2/a^2\right] \tag{11.45}$$

associated with the density $\tilde{c}_s(\boldsymbol{k}) = \sigma^2(a\sqrt{\pi}\,)^n\,\exp\left[\,-\,a^2\,k^2/4\right]$. □

EXAMPLE 11.8: A certain class of random fields includes models of growth and pattern formation in which the spatiotemporal evolution is governed by a set of *dynamic rules*. In such a physical context, it is worth mentioning a nonseparable covariance model that satisfies the dynamic scaling form

$$c_x(r,\tau) = r^{-1}\,g(r^z/\tau) \tag{11.46}$$

where $z = 1.82$, and $g(x) \sim x^{\alpha}$ $(x << 1)$, $\sim x^{-b}$ $(x >> 1)$ with exponents $\alpha \cong 1.4$ and $b \cong 0.6$ (for details, see Christakos and Hristopulos, 1998). □

EXAMPLE 11.9: A class of *fractal*-related nonseparable covariance models in space/time are available, such as

$$c_x(r,\tau;\ u_c, w_c) = \sigma^2 \hat{f}_z(\tau/r^{\beta};\ u_c)\hat{f}_\alpha(r;\ w_c) \tag{11.47}$$

where:

$\hat{f}_\nu(x;\,y) = f_\nu(x;\,y)/f_\nu(0;\,y)$ and $f_\nu(x;\,y) = \frac{1}{\Gamma(-\nu)}\int_0^y du\; e^{-u\,x}\,u^{-(\nu+1)}$, with $\nu = z,\ \alpha$; $x = \tau/r^{\beta},\ r$, and $y = u_c,\ w_c$. The function $\hat{f}_z\,(\tau/r^{\beta};\,u_c)$ has an unusual dependence on the space and time lags through τ/r^{β}. For large τ, the ratio τ/r^{β} is close to zero (for r sufficiently large), and the value of the function $\hat{f}_z$ is close to 1. With regard to $\hat{f}_z$, two pairs of space/time points are equidistant if $\tau_1/r_1^{\beta} = \tau_2/r_2^{\beta}$. Hence, the equation for equidistant space/time contours is $\tau/r^{\beta} = c$. This dependence is physically quite different than that implied by, *e.g.*, a Gaussian space/time covariance function (in the latter case, equidistant lags satisfy the equation $r^2/\xi_r^2 + \tau^2/\xi_\tau^2 = c$). □

EXAMPLE 11.10: A variety of nonseparable generalized spatiotemporal covariance models can be defined on the basis of the S/TRF-ν/μ theory. One such model is as follows:

$$\kappa_x(r,\tau) = \sum\nolimits_{\rho=0}^{2\nu+1}\sum\nolimits_{\zeta=0}^{2\mu+1}(-1)^{s(\rho)+s(\zeta)}\,a_{\rho/\zeta}\,r^{\rho}\,\tau^{\zeta} \tag{11.48}$$

where $a_{\rho/\zeta}$ are coefficients such that the permissibility conditions are satisfied. The $s(\rho)$, $s(\zeta)$ are sign functions given by $s(x) = \frac{1}{2}(x - \delta_{x,2l+1})$, where $x = \rho,\ \zeta$ (Christakos and Hristopulos, 1998). □

EXAMPLE 11.11: A useful class of nonhomogeneous/nonstationary covariances is determined by the *decomposition relationship* (Christakos, 1992)

$$c_x(s, t;\ s', t') = \kappa_x(r, \tau) + p_{\nu/\mu}(s, t)\, p_{\nu/\mu}(s', t') \tag{11.49}$$

where $p_{\nu/\mu}(s, t)$ and $p_{\nu/\mu}(s', t')$ are suitable polynomials in space and time. Furthermore, models of homogeneous/stationary covariances c_y can be derived from

$$c_y(h, \tau) = U_Q\, \kappa_x \tag{11.50}$$

where U_Q is a linear space/time differential operator. An interesting generalized spatiotemporal covariance derived from Equation 11.50 is as follows

$$\begin{aligned}\kappa_x(r,\tau) &= \alpha_0\, \delta(r)\, \delta(\tau) + \delta(r) \sum_{\zeta=0}^{\mu} a_\zeta\, (-1)^{\zeta+1} \tau^{2\zeta+1} \\ &+ \delta(\tau) \sum_{\rho=0}^{\nu} b_\rho (-1)^{\rho+1} r^{2\rho+1} + \sum_{\rho=0}^{\nu} \sum_{\zeta=0}^{\mu} d_{\rho/\zeta}\, (-1)^{\rho+\zeta} r^{2\rho+1}\, \tau^{2\zeta+1} \\ &+ \delta_{n,2}\, r^{2\nu} \log r \sum_{\zeta=0}^{\mu} (-1)^{\zeta}\, c_\zeta\, \tau^{2\zeta+1}\end{aligned} \tag{11.51}$$

where the coefficients α_0, a_ζ, b_ρ, c_ζ, and $d_{\rho/\zeta}$ must satisfy certain relationships derived from the permissibility conditions. The first three terms in Equation 11.51 represent space/time nuggets; the fourth term is purely polynomial. The last term which is logarithmic in the space lag is obtained only in 2-D. □

EXAMPLE 11.12: Yet another interesting set of nonseparable covariance models in $R^n \times T$ can be defined from separable covariances. In general, a sum of separable covariances is a nonseparable covariance; a model belonging to this class is (see also Eq. 10.27, p. 204)

$$c_x(r,\tau) = \sum_{i=1}^{N} c_i \exp\left[-r/\xi_i - \tau/b_i\right] \tag{11.52}$$

where b_i, c_i, and ξ_i are suitable coefficients. Note that a superposition of separable terms enables one to take into account correlations that are not captured by a single separable term. □

And Still the Garden Grows!

We conclude this chapter by expressing the view that the great charm of BME analysis lies in its almost unlimited versatility and generality. Any possible combination of scalar or vectorial natural processes, single-point or multipoint maps, Euclidean or non-Euclidean spaces, homogeneous or nonhomogeneous spatial patterns, stationary or nonstationary temporal trends, linear or nonlinear predictors, *etc.* arising in practical problems can be examined starting from essentially the same few basic BME equations. In addition, most of the existing classical techniques fit naturally into the BME framework, within which they acquire additional strength and significance. And still the garden grows!

12

POPULAR METHODS IN THE LIGHT OF MODERN SPATIOTEMPORAL GEOSTATISTICS

> ***"In science one must search for ideas. If there are no ideas, there is no science. A knowledge of facts is only valuable in so far as facts conceal ideas: facts without ideas are just the sweepings of the brain and the memory."*** V.G. Belinskii

The Generalization Power of BME

In the detective story "Adventure of the Sussex Vampire," Sherlock Holmes makes the following remark:

> One forms provisional theories and waits for time and fuller knowledge to explore them. A bad habit, Mr. Ferguson, but human nature is weak.

In terms of the BME epistemic paradigm, the first part of Holmes' remark, "One forms provisional theories...," clearly refers to general knowledge at the prior stage, while the second part, "...waits for time and fuller knowledge to explore them," refers to specificatory knowledge at the integration (posterior) stage. Building on such a logical distinction between general and specificatory knowledge, one of the basic BME postulates is that two basic concepts—*informativeness* and *probabiliorism*—are both desirable features of space/time analysis, which refer, though, to different stages of the epistemic process and are associated with different goals of scientific investigation. While informativeness is associated with the goal of building a *model* that expresses the available general knowledge base $\mathcal{G}$ at the prior stage and is maximally informative (in a well-defined mathematical sense), probabiliorism is associated with the aim of producing at the integration (posterior) stage space/time *estimates* (map $\hat{\chi}_{map}$) that are highly probable in light of the specificatory knowledge base $\mathcal{S}$ being considered. A symbolic representation of the BME process may be written:

$$\mathcal{G}\text{ base} \xrightarrow{\textit{Informativeness}} f_{\mathcal{G}} \xrightarrow{\mathcal{S}\text{ base}} f_{\mathcal{K}} \xrightarrow{\textit{Probabiliorism}} \hat{\chi}_{map}$$

The integration of the $\mathcal{S}$ base into the prior probability model $f_{\mathcal{G}}$ yields the new (posterior) model $f_{\mathcal{K}}$. Hence, the mathematical form of $f_{\mathcal{K}}$ depends on that of $\mathcal{S}$ and $f_{\mathcal{G}}$. The space/time map $\hat{\chi}_{map}$ may also depend on the knowledge-based conditionals considered in the probability model (*i.e.*, Bayesian, truth-functional, *etc.* conditionals). A central feature of the BME approach is its considerable *generality*. In sciences, the desire for generality constantly calls for relentless and coordinated exploration of their foundations, and for greater freedom as well. Modern geostatistics is not merely a collection of hard-data processing techniques. Indeed, as discussed in the previous chapters, many applications involve plenty of physical knowledge that must be taken into consideration by the mapping method. Such a consideration requires sounder foundations and new additions to the mathematical structure of geostatistics.

Certainly, there are situations (*e.g.*, in the early stages of development of a scientific field) in which only limited sets of observations are available, and in such instances, the use of popular data-processing techniques may be justified. The present chapter examines certain of these data-processing techniques in the light of modern spatiotemporal geostatistics. As happens with any novel theory that seeks to successfully replace old ones, the techniques of classical geostatistics are special cases of the considerably more general theory of modern geostatistics. This is a process that shows up many times in the development of sciences: obtaining certain already-known results as logical or mathematical consequences of newly stated principles. Comparative studies of BME approaches and those of classical geostatistics are also discussed in this chapter, and the powerful features of BME are demonstrated through synthetic examples as well as real-world applications. Modern spatiotemporal geostatistics includes a variety of mathematical models and stochastic techniques. As will be shown in this chapter, a unified framework is provided by the generalized spatiotemporal random field theory. A large number of popular models, including coarse-grained random fields, wavelet random fields, and fractal random fields, can be derived as special cases of the generalized spatiotemporal random field theory.

Minimum Mean Squared Error Estimators

Assume that the specificatory knowledge consists of only a set of hard data χ_{hard} about the natural variable $X(\boldsymbol{p})$ at the space/time points $\boldsymbol{p}_i$ ($i = 1, \ldots, m$). It is a well-known result (*e.g.*, Cramer and Leadbetter, 1967) that the best of all spatiotemporal *minimum mean squared error* (MMSE) estimators of $X(\boldsymbol{p}_k)$ at a point $\boldsymbol{p}_k$ $(k \neq i)$ is the conditional mean

$$\hat{\chi}_{k,\,MMSE} = \overline{X(\boldsymbol{p}_k) \mid \boldsymbol{\chi}_{hard}} \tag{12.1}$$

The derivation of specific analytical expressions for the estimator (Eq. 12.1) depends on the probability law of the S/TRF. Consider, *e.g.*, the case of a *Gaussian* S/TRF. Then, Equation 12.1 reduces to a linear estimator, which is optimal among all MMSE estimators. Typically, such linear MMSE estimators are expressed in terms of the hard data, *i.e.*,

$$\hat{\chi}_{k,\,MMSE} = \boldsymbol{\lambda}^T \boldsymbol{\chi}_{hard} \tag{12.2}$$

where $\boldsymbol{\lambda}$ is a vector of weights associated with the data points and involving the space/time mean and covariance functions (see, *e.g.*, ordinary and simple kriging; Cressie, 1991).

From the BME perspective, if the general knowledge is limited to the mean and covariance functions and the specificatory knowledge includes only hard data, *i.e.*,

$$\left.\begin{array}{l} \mathcal{G} = \text{spatiotemporal mean and covariance functions} \\ \mathcal{S} = \text{hard data} \end{array}\right\} \tag{12.3}$$

the posterior pdf is Gaussian. The Gaussian pdf is symmetric and the BMEmode estimate is, by definition, the conditional mean, *i.e.*, it is the same as the MMSE estimate (the BME estimator in this case is linear, which is also the case of the MMSE estimator). By summarizing the above discussion we can write that

$$\hat{\chi}_{k,\,MMSE} = \overline{X(\boldsymbol{p}_k) \mid \boldsymbol{\chi}_{hard}} \overset{Gaussian}{=} \left\{ \begin{array}{l} \boldsymbol{\lambda}^T \boldsymbol{\chi}_{hard} \\ \hat{\chi}_{k,\,BME} \end{array} \right. \tag{12.4}$$

A direct consequence of the preceding analysis is the following proposition.

PROPOSITION 12.1: For a Gaussian S/TRF, the MMSE estimate obtained on the basis of the physical knowledge (Eq. 12.3) coincides with the BMEmode estimate derived using the same physical knowledge.

For *non-Gaussian* S/TRF, the two estimation approaches generally give different results. Most MMSE techniques still are based on the first two moments, although they do not offer an adequate characterization of the non-Gaussian law. BME, on the other hand, recognizes the need to involve higher order moments in the analysis and does it in a rigorous fashion.

COMMENT 12.1: *An interesting consequence of Proposition 12.1 is that while MMSE by itself may lack a meaningful epistemic explanation (rather, its use is justified solely on the basis of a suitably favorable track record), it can, nevertheless, acquire such an explanation within the context of the general BME analysis.*

A comparative study of the BME and MMSE approaches is possible in terms of three essential concepts of scientific reasoning and methodology: observational nesting, falsifiability, and fertility degree. The *nesting* of a scientific approach (Newton-Smith, 1981) refers to its ability to include the successes of its predecessors. Indeed, an essential feature of the BME theory is that it is formulated in a way that preserves most of the referents of earlier theories, which are its limiting cases.

EXAMPLE 12.1: As we saw above, MMSE estimation is a special case of the considerably more general BME mapping approach. In the following section, we will see that popular geostatistical estimators such as kriging are merely special cases of BME analysis; these special cases are obtained under restrictive conditions on the form of the estimator and on the physical knowledge bases that can be used. □

The *falsifiability* of a scientific approach (Popper, 1962) measures the extent to which it involves hypotheses and models that can be falsified by empirical (experimental or observational) evidence. The enterprise of science, as the falsificationist sees it, consists of the proposal of highly falsifiable hypotheses, followed by deliberate and tenacious attempts to falsify them. A newly proposed approach will be considered worthy of the consideration of scientists if it is more falsifiable than its rival. The superiority of BME over the MMSE method is, thus, demonstrated on the basis of the relative merits of the competing methods.

EXAMPLE 12.2: By not taking into consideration the physical laws operative on a natural process or mechanism, the MMSE estimate could be falsified on the basis of empirical evidence consistent with the laws. BME, on the other hand, takes into account the physical laws and allows a complete probabilistic characterization of the situation in terms of its posterior pdf. As a consequence, BME has considerably higher chances of withstanding tests that falsify MMSE analysis. □

The *fertility degree* of a scientific approach (Chalmers, 1994) measures the extent to which the approach contains within it objective opportunities for critical thinking and development, or the extent to which it opens up new lines of investigation. In this sense, as is obvious from our discussion so far, the fertility degree of the BME approach is considerably larger than that of the MMSE approach.

EXAMPLE 12.3: BME provides a sound epistemic framework for critical thinking and expands the study domain to include the observer as well as the observed. Non-Gaussian laws are automatically incorporated. Unlike MMSE analysis (which capitalizes on empirical data-fitting techniques), BME capitalizes on the powerful theories and laws of natural sciences and incorporates various forms of knowledge in a rigorous and systematic manner. Also, BME has global prediction features (whereas MMSE predictors are most appropriate for interpolation purposes; Stein, 1999), it allows multipoint mapping, *etc.* □

The following sections present analytical results and numerical applications in which various sorts of geostatistical kriging are compared to BME-based techniques, and by means of which the BME theory is amply confirmed.

Kriging Estimators

As we saw above, the BME theory preserves the referents of earlier geostatistical results which are derived as its limiting cases. This powerful feature of BME is demonstrated in this section by showing that it can easily reproduce certain well-known kriging results. Of course, BME can lead to improved maps in situations in which additional sources of knowledge become available (higher order moments, multiple-point statistics, scientific theories, *etc.*).

Simple and ordinary kriging

We start with some basic results that are valid in the restrictive case in which our knowledge consists of hard data and low-order space/time statistical moments.

PROPOSITION 12.2: When the general knowledge is limited to the mean and (centered) covariance and the specificatory knowledge includes only hard data, the BMEmode estimate is

$$\hat{\chi}_k = \overline{x_k} - \sum_{i=1}^{m} \frac{c_{ik}^{-1}}{c_{kk}^{-1}} (\chi_i - \overline{x_i}) \tag{12.5}$$

where, as before, $\overline{x_i}$ denotes the mean value and c_{ik}^{-1} is the ik-th element of the inverse (centered) covariance matrix c_{map}^{-1}. Equation 12.5 coincides with the simple kriging estimate.

Proof: Under the conditions described in the proposition, the BME equation reduces to

$$\frac{\partial}{\partial \chi_k} \mathcal{Y}_g \Big|_{\chi_k = \hat{\chi}_k} = 0 \tag{12.6}$$

or

$$\sum_{i=1}^{m,k} c_{ii}^{-1} \left[\frac{\partial g_{ii}(\chi_i)}{\partial \chi_k} \right]_{\chi_k = \hat{\chi}_k} + \sum_{i=1}^{m,k} \sum_{i \neq j=1}^{m,k} c_{ij}^{-1} \left[\frac{\partial g_{ij}(\chi_i, \chi_j)}{\partial \chi_k} \right]_{\chi_k = \hat{\chi}_k} = 0 \tag{12.7}$$

where $g_{ii}(\chi_i) = (\chi_i - \overline{x_i})^2$ and $g_{ij}(\chi_i, \chi_j) = (\chi_i - \overline{x_i})(\chi_j - \overline{x_j})$. After derivation, Equation 12.7 reduces to Equation 12.5. □

Working along the same lines, the following proposition can easily be proven.

PROPOSITION 12.3: When the general knowledge is limited to the variogram and the specificatory knowledge includes only hard data, the BMEmode estimate is given by

$$\hat{\chi}_k = \frac{\sum_{i=1}^{m} \gamma_{ki}^{-1}}{\sum_{j=1}^{m} \gamma_{kj}^{-1}} \chi_i \tag{12.8}$$

where γ_{ki}^{-1} is the ki-th element of the inverse variogram matrix $\boldsymbol{\gamma}_{map}^{-1}$. Equation 12.8 coincides with the ordinary kriging estimate.

As should be expected from the analysis of the preceding section, Equations 12.5 and 12.8 coincide with the kriging estimators that rely on the same general and specificatory knowledge bases (recall that in the case of the Gaussian pdf the linear kriging estimator is the best of all possible MMSE estimators). Furthermore, Equations 12.5 and 12.8 provide explicit expressions for the simple kriging coefficients $\lambda_i = -c_{ik}^{-1}/c_{kk}^{-1}$ and the ordinary kriging coefficients $\lambda_i = \gamma_{ki}^{-1}/\sum_{j=1}^{m} \gamma_{kj}^{-1}$, respectively (see also the examples that follow). Various studies have shown that the application of BME Equations 12.5 and 12.8 is computationally efficient (Lee and Ellis, 1997a; Christakos, 1998a and b; Serre *et al.*, 1998; Serre and Christakos, 1999a). Note that the BME interpretation of kriging is fundamentally different than the Bayesian interpretation of spatial estimation discussed, *e.g.*, in Kitanidis (1986).

EXAMPLE 12.4: We will present a comparison of BME with simple kriging (SK). Consider the points $\boldsymbol{p}_1 = (\boldsymbol{s}_1, t_1)$ and $\boldsymbol{p}_2 = (\boldsymbol{s}_2, t_2)$ in space/time, where hard data are available. We seek the BME estimate at point $\boldsymbol{p}_k = (\boldsymbol{s}_k, t_k)$. The S/TRF is homogeneous/stationary with constant mean $\overline{x}$ and variance σ_x^2. The spatial distance and the time period between $\boldsymbol{p}_1$ and $\boldsymbol{p}_k$ are the same as between $\boldsymbol{p}_2$ and $\boldsymbol{p}_k$, so that $c_{1k} = c_{2k} = c_{k1} = c_{k2} = c_x$ and $c_{12} = c_{21} = c_x'$. Under these circumstances, the SK estimate is

$$\hat{\chi}_k = \overline{x} + \frac{c_x}{\sigma_x^2 + c_x'} [(\chi_1 - \overline{x}) + (\chi_2 - \overline{x})] \tag{12.9}$$

On the other hand, the BME equation (Eq. 12.5) yields

$$\hat{\chi}_k = \overline{x} - \frac{c_{1k}^{-1}}{c_{kk}^{-1}} (\chi_1 - \overline{x}) - \frac{c_{2k}^{-1}}{c_{kk}^{-1}} (\chi_2 - \overline{x}) \tag{12.10}$$

where $c_{1k}^{-1} = c_{2k}^{-1}$ and c_{kk}^{-1} are given by

$$\left.\begin{aligned} c_{1k}^{-1} &= (c_{21} c_{k2} - \sigma_x^2 c_{k1})/|\mathbf{c}| \\ c_{2k}^{-1} &= (c_{k1} c_{k2} - \sigma_x^2 c_{k2})/|\mathbf{c}| \\ c_{kk}^{-1} &= (\sigma_x^4 - c_{12}^2)/|\mathbf{c}| \end{aligned}\right\} \tag{12.11}$$

with $|\boldsymbol{c}| = c_{1k}(c_{21} c_{k2} - \sigma_x^2 c_{k1}) - c_{2k}(\sigma_x^2 c_{k2} - c_{12} c_{k1}) + \sigma_x^2(\sigma_x^4 - c_{12}^2)$; see also Example 7.2 (p. 138). Since $c_{1k} = c_{2k} = c_{k1} = c_{k2} = c_x$ and $c_{12} = c_{21} = c'_x$, Equation (12.11) gives $c_{1k}^{-1}/c_{kk}^{-1} = c_{2k}^{-1}/c_{kk}^{-1} = -c_x/(\sigma_x^2 + c_x')$. By substituting the latter into Equation 12.10, we find Equation 12.9, thus showing that the BME and SK estimates coincide in this case (which is an expected result). □

Table 12.1. The g_α functions.

α	g_α	$\overline{g_\alpha}$
	Normalization constraint	
0	$g_0 = 1$	$\overline{g_0} = 1$
	Variogram constraints	
1	$g_{1k}(\chi_1, \chi_k) = \frac{1}{2}(\chi_1 - \chi_k)^2 = \frac{1}{2}\psi_{1k}^2$	$\overline{g_{1k}} = \gamma_{1k}$
2	$g_{2k}(\chi_2, \chi_k) = \frac{1}{2}(\chi_2 - \chi_k)^2 = \frac{1}{2}\psi_{2k}^2$	$\overline{g_{2k}} = \gamma_{2k}$

EXAMPLE 12.5: Consider the case of Example 12.4 above, where the variograms γ_{ij} $(i,\ j = 1,\ 2,\ k)$ between the three points are spatially isotropic/temporally stationary. The well-known MMSE estimate provided by space/time ordinary kriging (OK) is

$$\hat{\chi}_k = \tfrac{1}{2}(\chi_1 + \chi_2) \tag{12.12}$$

We will compare the estimate (Eq. 12.12) with the one obtained by BME using the same knowledge base. Since the known statistics are the variograms, the information available concerns the differences $\psi_{ik} = \chi_i - \chi_k$ $(i = 1,\ 2)$. The statistical constraints are shown in Table 12.1. The difference $\psi_{12} = \chi_1 - \chi_2$ is not considered, for it is a linear combination of the previous two differences ψ_{1k} and ψ_{2k}. The $\mathcal{Y}_G$ is

$$\mathcal{Y}_G = \tfrac{1}{2}\,\mu_1\,\psi_{1k}^2 + \tfrac{1}{2}\,\mu_2\,\psi_{2k}^2 \tag{12.13}$$

The Lagrange multipliers μ_1 and μ_2 are found by solving the following system of equations

$$\overline{g_0} = 1 = Z^{-1}\int\!\!\int d\psi_{1k}\,d\psi_{2k}\,\exp[\mathcal{Y}_G] \tag{12.14}$$

and

$$\overline{g_{ik}} = \gamma_{ik} = \tfrac{1}{2}Z^{-1}\int\!\!\int d\psi_{1k}\,d\psi_{2k}\,\psi_{ik}^2\,\exp[\mathcal{Y}_G], \quad i = 1,\, 2 \tag{12.15}$$

The solution of Equations 12.14 and 12.15 gives

$$\mu_i = -\tfrac{1}{2}\,\gamma_{ik}^{-1}, \quad i = 1,\, 2 \tag{12.16}$$

The BME equation reduces to

$$\frac{\partial}{\partial\chi_k}\left[\mu_1\,\psi_{1k}^2 + \mu_2\,\psi_{2k}^2\right]_{\chi_k=\hat{\chi}_k} = -\gamma_{1k}^{-1}(\chi_1 - \hat{\chi}_k) - \gamma_{2k}^{-1}(\chi_2 - \hat{\chi}_k) = 0 \tag{12.17}$$

the solution of which is (note that due to isotropy/stationarity, $\gamma_{1k} = \gamma_{2k} = \gamma$)

$$\hat{\chi}_k = \tfrac{1}{2}(\chi_1 + \chi_2) \tag{12.18}$$

Hence, the BME estimate (Eq. 12.18) is the same as the OK estimate (Eq. 12.12), as was expected. □

To make some numerical comparisons between BME and kriging, a simulation study is examined below. The fact that BME can account for both hard and soft data allows it to produce more accurate numerical results than SK, which relies only on hard data. Remarkably, this is true even when SK is allowed to use all hard data available, while BME is restricted to using only a few hard data points.

EXAMPLE 12.6: Let us revisit Example 8.2 (p. 151). The estimates $\hat{\chi}^{(\ell)}_{BME}(\boldsymbol{p}_k)$ obtained by BME at locations $\boldsymbol{p}_k \in D$, which are the nodes of a dense grid covering the shaded region D in Figure 8.1, can be compared with the estimates $\hat{\chi}^{(\ell)}_{SK}(\boldsymbol{p}_k)$ obtained by space/time SK. For each realization $\chi^{(\ell)}(\boldsymbol{p}_k)$ ($\ell = 1, 2, \ldots, 200$), the estimation errors $e^{(\ell)}_I(\boldsymbol{p}_k) = \hat{\chi}^{(\ell)}_I(\boldsymbol{p}_k) - \chi^{(\ell)}(\boldsymbol{p}_k)$ were computed at all $\boldsymbol{p}_k \in D$ for both $I = BME$ and SK. The difference $\Delta e^{(\ell)}(\boldsymbol{p}_k) = \left| e^{(\ell)}_{BME}(\boldsymbol{p}_k) \right| - \left| e^{(\ell)}_{SK}(\boldsymbol{p}_k) \right|$ was calculated for each realization, and the average over all 200 realizations was obtained at each point $\boldsymbol{p}_k \in D$ by $\Delta e(\boldsymbol{p}_k) = \langle \Delta e^{(\ell)}(\boldsymbol{p}_k) \rangle$, where the averaging operator $\langle \cdot \rangle = 200^{-1} \sum_{\ell=1}^{200} [\cdot]$ is used. The $\Delta e(\boldsymbol{p}_k)$ map of Figure 12.1 is everywhere negative, implying that the BME estimate is stochastically more accurate than the SK estimate at every point $\boldsymbol{p}_k \in D$. The plot of the average error difference $\overline{\Delta e}$ over all points $\boldsymbol{p}_k \in D$ for various values of the time interval Δt (Fig. 12.2) also shows that BME performs considerably better than SK; in fact, BME improves as Δt increases. With each map $\hat{\chi}^{(\ell)}_{BME}(\boldsymbol{p}_k)$ ($\ell = 1, 2, \ldots, 200$), BME associates an accuracy map in terms of the standard deviation of the estimate $\sigma^{(\ell)}_{BME}(\boldsymbol{p}_k)$.

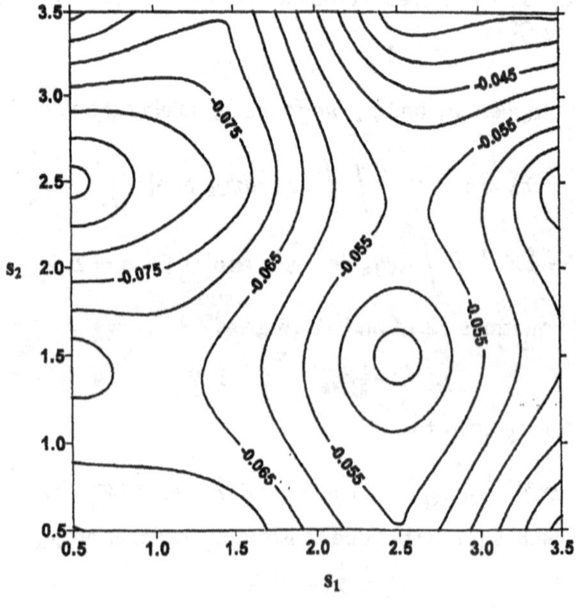

Figure 12.1. Map of the error difference $\Delta e(\boldsymbol{p}_k)$ between BME and SK over D ($\Delta t = 0.5$).

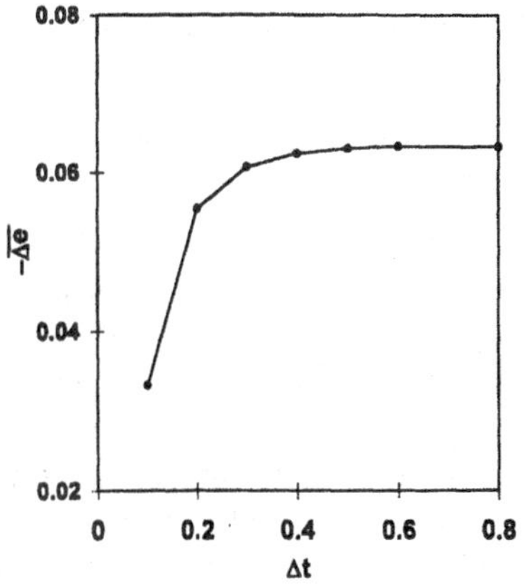

Figure 12.2. Average error difference $\overline{\Delta e}$ (over all points $p_k \in D$) between BME and SK as a function of the time interval Δt.

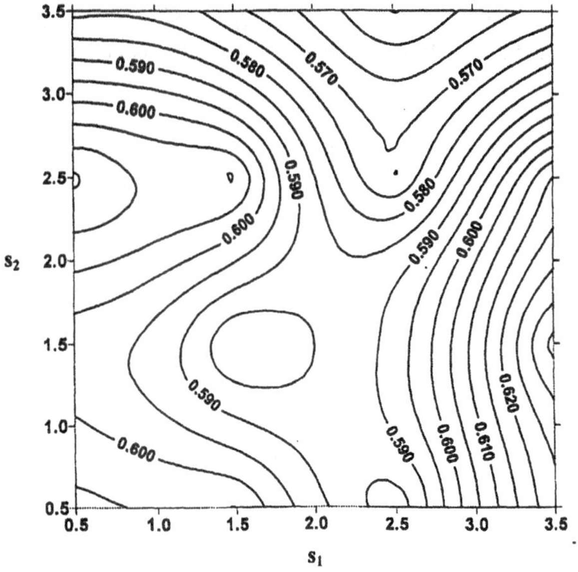

Figure 12.3. Map of the SK estimation error standard deviation $\overline{e_{SK}}(p_k)$; $\Delta t = 0.5$.

In Figure 8.2 (p. 152), the average standard deviation of the estimate over all 200 realizations, expressed by $\sigma_{BME}(p_k) = \langle \sigma^{(\ell)}_{BME}(p_k) \rangle$, was plotted for all $p_k \in D$. In Figure 8.3, the average estimation error standard deviation $\overline{e_{BME}}(p_k)$ was calculated over the 200 realizations, and the (actual) $\overline{e_{BME}}(p_k)$ values were found to be of the same magnitude as the $\sigma_{BME}(p_k)$ values. In Figure 12.3, the average estimation error standard deviation for SK over the 200 realizations, *i.e.*, $\overline{e_{SK}}(\boldsymbol{p}_k)$, is plotted. As was expected, $\overline{e_{SK}}(\boldsymbol{p}_k)$ is everywhere larger than $\overline{e_{BME}}(\boldsymbol{p}_k)$. □

The following example presents some numerical results related to the estimation error pdf derived using BME and SK.

EXAMPLE 12.7: Assume that eight measurements of the natural variable $X(\boldsymbol{p})$, $\boldsymbol{p} = (s,t) \in R^1 \times T$, are available at points $\boldsymbol{p}_1$–$\boldsymbol{p}_8$ on a 1×1 grid centered at (0, 0) (as in Fig. 7.1, p. 141). The $X(\boldsymbol{p})$ has a zero mean and covariance $c_x(h,\tau) = \exp\left[-0.125\,\pi\,(h^2 + \tau^2/0.64)\right]$. There is a soft datum at point $\boldsymbol{p}_9 = (s_9, t_9)$: the $X(\boldsymbol{p}_9)$ is uniformly distributed within an interval ϖ of width 0.4. To see how this kind of soft data can improve estimation accuracy at $\boldsymbol{p}_k$, 1,000 $\chi(\boldsymbol{p}_k)$-realizations were generated and pdfs of the estimation errors $e_I(\boldsymbol{p}_k) = \hat{\chi}_I(\boldsymbol{p}_k) - \chi(\boldsymbol{p}_k)$, ($I$ = *BME, SK*), have been plotted (Fig. 12.4) for various positions of $\boldsymbol{p}_9$. The widths of the $e_{BME}(\boldsymbol{p}_k)$-pdfs are clearly smaller than the width of the $e_{SK}(\boldsymbol{p}_k)$-pdf, implying that $\hat{\chi}_{BME}(\boldsymbol{p}_k)$ is stochastically more accurate than $\hat{\chi}_{SK}(\boldsymbol{p}_k)$. Naturally, the significance of knowledge at point $\boldsymbol{p}_9$ decreases as it moves away from $\boldsymbol{p}_k$. At large distances this knowledge has little contribution to the estimation at point $\boldsymbol{p}_k$; BME and SK essentially rely on the same hard data, and BME becomes practically as accurate as SK. □

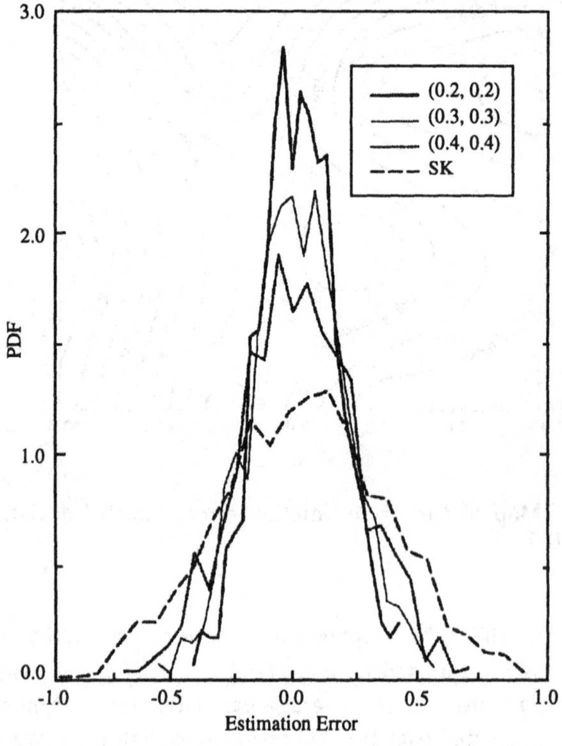

Figure 12.4. Plots of pdf of estimation errors $e_{BME}(\boldsymbol{p}_k)$ and $e_{SK}(\boldsymbol{p}_k)$; $s_9 = t_9 = 0.2$, 0.3, and 0.4.

The next example is a continuation of the geostatistical analysis of the porosity data collected in the West Lyons field of west-central Kansas (Chapter 7, p. 143). This example provides numerical results showing that: (a) the SK and BME maps coincide when the same hard data and low-order moments are assumed (an outcome that is, of course, to be expected from theory); and (b) the BME method produces better maps than SK when additional soft data become available.

EXAMPLE 12.8: Consider the West Lyons porosity data set presented in Chapter 7 and Figure 7.4 (p. 145). A total of 76 data values were available. The general knowledge that was considered included the porosity mean and the covariance model plotted in Figure 7.4. Using all 76 hard data, the SK technique reproduces the map in Figure 7.5, which was the map obtained by the BME method using the same hard data, mean, and covariance model. Furthermore, the SK map obtained by using only 56 hard data (Fig. 12.5) is, not surprisingly, shown to be less accurate than the BME map of Figure 7.6, obtained using the 56 hard data as well as 20 soft interval data of varying widths. □

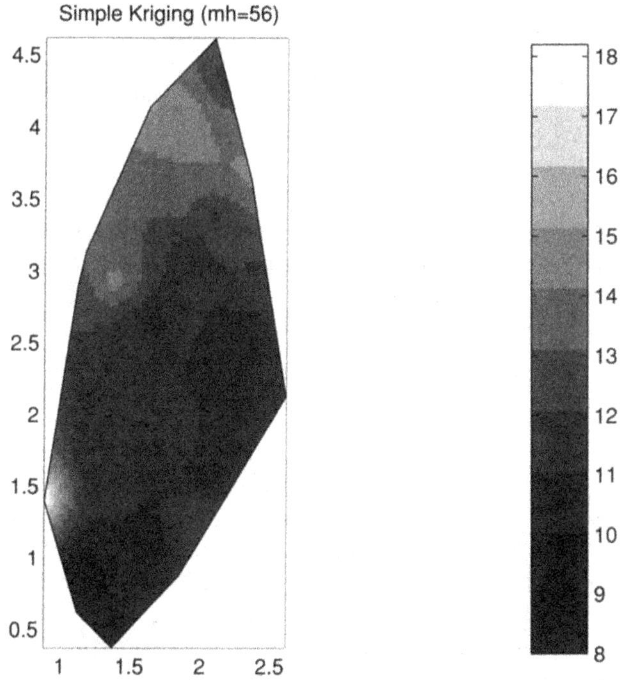

Figure 12.5. The SK porosity map using 56 data (%).

In the following numerical example, soft (probabilistic) data are introduced into the BME analysis, which is then compared to two kriging techniques. BMEmode as well as BMEmean estimates are considered.

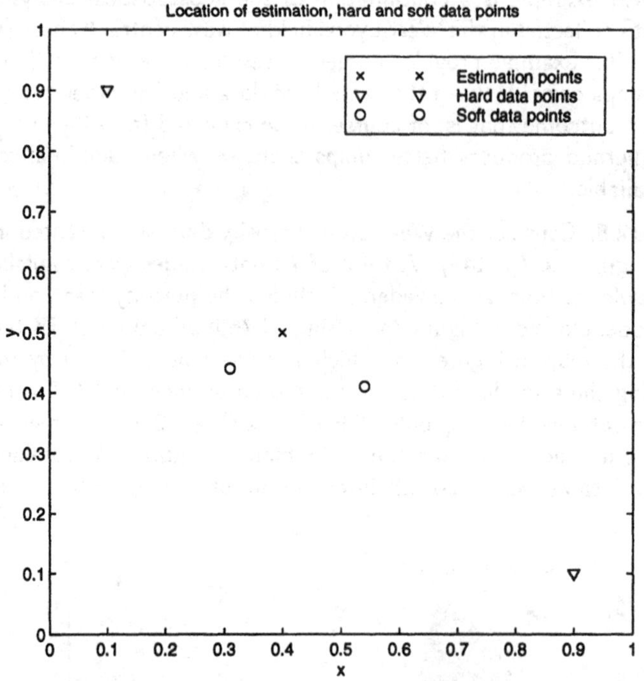

Figure 12.6. Data-point configuration.

EXAMPLE 12.9: Consider the data configuration of Figure 12.6. In addition to the two hard data, two soft data of the probabilistic forms shown in Figure 12.7 (*i.e.*, pdf-1 and pdf-2) are available. In Figures 12.8 and 12.9, we plot simulated estimation error distributions of the BMEmode and the BMEmean estimates obtained using these data. Also, the error distributions resulting from two SK methods are plotted for comparison—SK using only hard data (SKh; see Olea, 1999) and SK with measurement error pdf (SKME; Serre *et al.*, 1998; Serre and Christakos, 1999a). In addition, for each method the mean squared errors E (*i.e.*, the mean of the squared estimation errors) were calculated and their values reported in the legends of Figures 12.8 and 12.9. Again, the BME method provides better estimates than the SK methods. In both figures, the performance of BME is shown to be superior (its BMEmode has the greatest probability of giving an estimation error equal to zero and its BMEmean has the smallest E value). Looking at Figure 12.8, in particular, we first note that SKME provides more accurate estimations than SK, as expected. Indeed, while the mean squared error E for SKh is 0.419, for SKME, it drops to 0.198. This is explained by the fact that SKME incorporates soft (probabilistic) data. What is more interesting is that the BME method produces a mean squared error that is still lower than that of SKME, with a value of only $E = 0.190$ for the BMEmean. Note that $E = 0.231$ for the BMEmode. This should not come

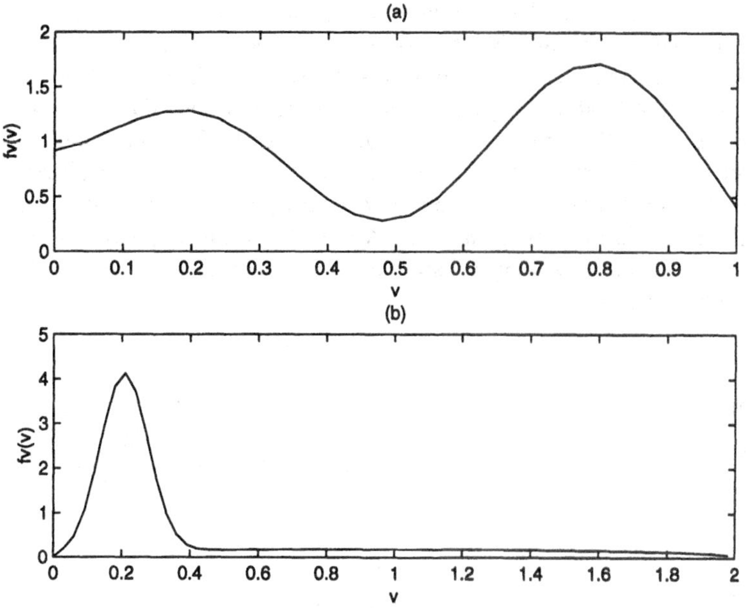

Figure 12.7. Pdfs representing soft data: (a) pdf-1, and (b) pdf-2.

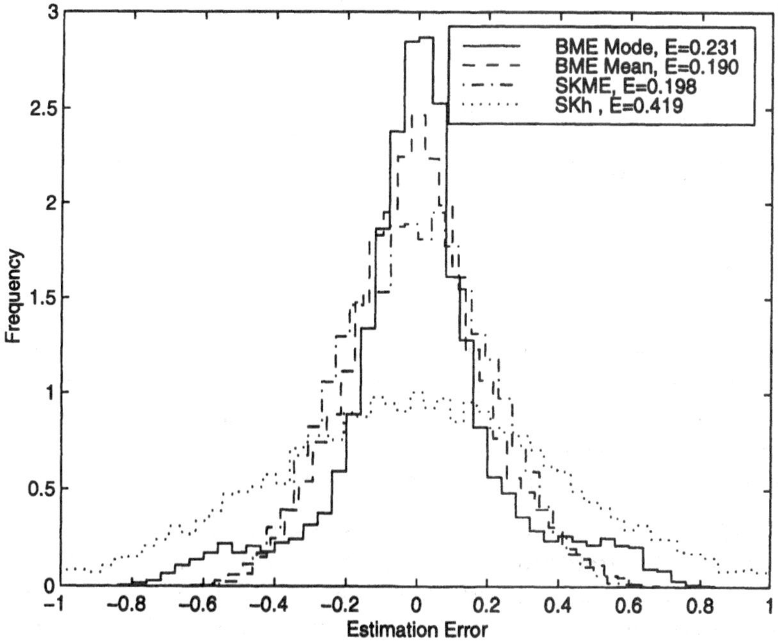

Figure 12.8. Estimation error distributions, BMEmode, BMEmean, SKh, and SKME for pdf-1

as a surprise, since the BMEmode estimator provides the most probable value, while the BMEmean and SKME estimators generate estimates that minimize the mean squared error. In Figure 12.9 the situation is even better for BME. As before, the BMEmode gives the most probable value. Furthermore, the E values for both the BMEmean ($E = 0.032$) and the BMEmode ($E = 0.044$) are substantially smaller than that of SKME ($E = 0.070$). The above results illustrate the fact that by rigorously accounting for the knowledge contained in the probabilistic soft data, the BME method can improve the accuracy of spatiotemporal mapping substantially compared to the traditional kriging methods. □

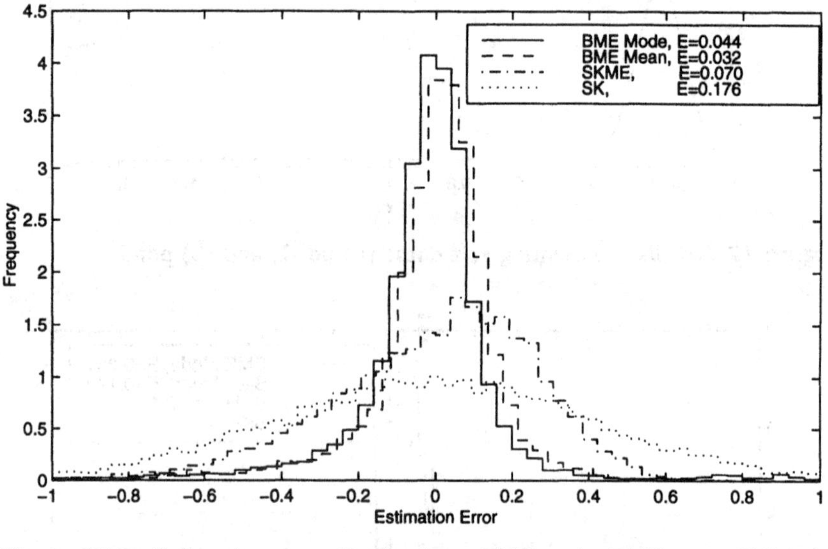

Figure 12.9. Estimation error distributions, BMEmode, BMEmean, SKh, and SKME for pdf-2.

In the following example we revisit Kansas and the Equus Beds aquifer, which was studied with the help of the BME method in Chapter 8 (p. 155). Access to accurate and informative maps is quite valuable in the implementation of strategies to improve the water quality of aquifers, which is the case in the Kansas Equus Beds Recharge Demonstration Project.

EXAMPLE 12.10: Most traditional kriging techniques derive confidence intervals based on the restrictive assumption of a Gaussian posterior pdf (Olea, 1999). As a consequence, space/time estimates that have been produced using kriging are more uncertain than those produced using the BME approach. For comparison purposes, the 90% confidence intervals obtained from the analysis of the Equus Beds aquifer data set (see p. 156–163) were calculated using both the BME and the SK techniques. As discussed in Chapter 8, this data set included both hard and soft (interval and probabilistic) data in space/time. The BME and SK confidence intervals derived at a number of selected wells are plotted as functions of time in Figure 12.10.

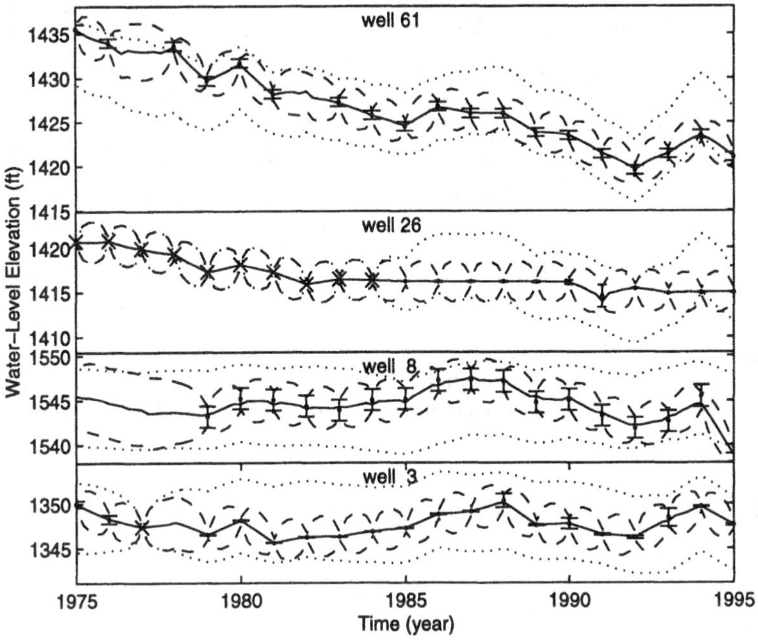

Figure 12.10. BMEmode estimates of water-level elevation at a number of selected wells (shown as solid line) and 90% confidence intervals obtained by BME (dashed line) and SK (dotted line). Hard data are shown by $\times$; soft (interval) data are depicted as error bars.

We should keep in mind that estimation at a specific well uses space/time data from both the same well and from neighboring wells (at the same and at different time periods). One immediately notices that the SK-based 90% confidence intervals shown in Figure 12.10 are much wider and, therefore, less informative than the corresponding BME-based 90% confidence intervals. Also, in well no. 26, the SK method and the BME method give the same confidence intervals of water-level elevations at times in which only hard data happen to be available; at all other time periods during which soft (interval) data are available, the BME technique performs considerably better than the SK technique. Finally, the estimation error standard deviation maps of water-level elevations calculated using the SK technique are plotted in Figure 12.11 for the years 1975 and 1998. As should be expected, the SK maps show larger estimation error standard deviations than the BME maps previously obtained (see the analysis in Chapter 8). □

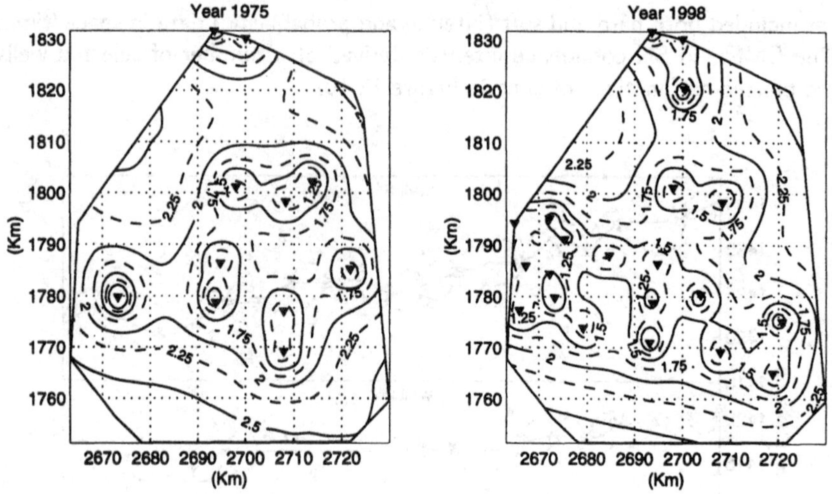

Figure 12.11. Maps of standard deviation error of the SK estimates for the water-level elevations (in ft) for the years 1975 and 1998. Triangles denote observation wells where hard data were available in space/time.

Lognormal kriging

One of the best-known non-Gaussian geostatistical estimators is the *lognormal* SK. An interesting situation is presented in the following proposition, which was proven by Lee and Ellis (1997b).

PROPOSITION 12.4: Assume that the specificatory knowledge consists of hard data χ_{hard} at points p_i ($i = 1, \ldots, m$). Let $X(p)$ be a lognormal field with known mean and (centered) covariance functions. The lognormal simple kriging (LSK) estimate and the lognormal BME (LBME) estimate of $X(p_k)$ at point p_k ($k \neq i$) are given, respectively, by

$$\hat{\chi}_{k,LSK} = \exp\left[\hat{\psi}_{k,SK} + \tfrac{1}{2}\sigma^2_{SK}\right] \tag{12.19}$$

and

$$\hat{\chi}_{k,LBME} = \exp\left[\hat{\psi}_{k,SK} - \sigma^2_{SK}\right] \tag{12.20}$$

where $\hat{\psi}_{k,SK}$ and σ^2_{SK} are the SK estimate and estimation variance of the log-transformed field $Y(p) = \ell og\, X(p)$.

Some interesting applications of Proposition 12.4 in the context of spatial sampling, *etc.*, may be found in Lee and Ellis (1997b)

Nonhomogeneous/nonstationary kriging

The analysis of this section is a special case of Proposition 10.4 (p. 211). Consider only hard data at points p_i ($i = 1, 2, \ldots, m$) and assume that the generalized spatiotemporal covariance of the underlying S/TRF-ν/μ $X(p)$ is

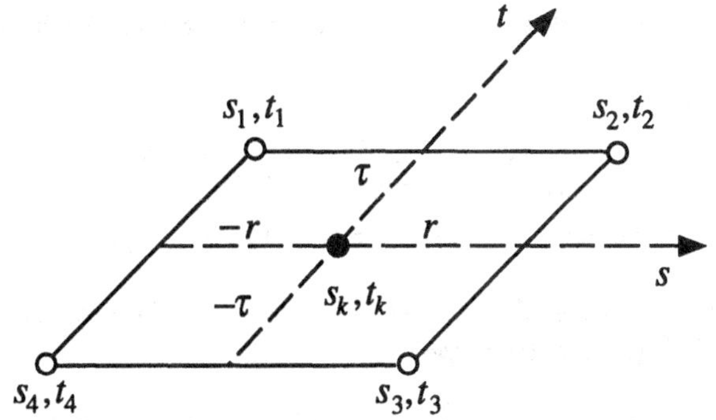

Figure 12.12. The data configuration for Example 12.11.

known. As usual, we seek the BME estimate at the point $\boldsymbol{p}_k \neq \boldsymbol{p}_i$. The BME equation reduces to

$$\frac{\partial}{\partial \chi_k} \mathcal{Y}_{\mathcal{G}} \Big|_{\chi_k = \hat{\chi}_k} = 0 \tag{12.21}$$

where $\mathcal{Y}_{\mathcal{G}}$ is given by Equation 10.28 (p. 211). In this case, therefore, Equation 12.21 provides the BME estimator of the S/TRF-ν/μ $X(\boldsymbol{p})$.

EXAMPLE 12.11: Consider an S/TRF-1/1. Hard data are available at points $\boldsymbol{p}_i = (s_i, t_i)$, $i = 1, 2, 3$, and 4 as in Figure 12.12. For simplicity, let $s_k = t_k = 0$; then, $(s_1, t_1) = (-r, \tau)$, $(s_2, t_2) = (r, \tau)$, $(s_3, t_3) = (r, -\tau)$, and $(s_4, t_4) = (-r, -\tau)$. The space/time kriging estimate is easily found to be

$$\hat{\chi}_k = \tfrac{1}{4} \sum\nolimits_{i=1}^{4} \chi_i \tag{12.22}$$

We seek the BME estimate at the point $\boldsymbol{p}_k = (s_k, t_k)$, which we want to compare with the kriging estimate (Eq. 12.22). To do so we must formulate the BME problem in a way that is consistent with the kriging problem. A random field combination consistent with the analysis above is

$$\psi = Q\left[\boldsymbol{\chi}_{map}\right] = \chi_k - \tfrac{1}{4} \sum\nolimits_{i=1}^{4} \chi_i \tag{12.23}$$

Then, the g_α functions for the problem are $g_0(\boldsymbol{\chi}_{map}) = 1$ and

$$g_1(\boldsymbol{\chi}_{map}) = Q^2[\boldsymbol{\chi}_{map}] = \left[\chi_k - \tfrac{1}{4} \sum\nolimits_{i=1}^{4} \chi_i\right]^2 \tag{12.24}$$

The known prior statistic is

$$\begin{aligned} \overline{g_1(\boldsymbol{x}_{map})} &= \overline{\left[x_k - \tfrac{1}{4} \sum\nolimits_{i=1}^{4} x_i\right]^2} = \tfrac{5}{4}\kappa_x(0,\, 0) + \tfrac{1}{8}\left[\kappa_x(\pm 2r,\, 0)\right. \\ &\quad \left. + \kappa_x(0,\, \pm 2\tau)\right] - \tfrac{1}{2}\kappa_x(\pm r,\, \pm\tau) + \tfrac{1}{16}\left[\kappa_x(-2r,\, \tau)\right. \\ &\quad \left. + \kappa_x(2r,\, -\tau) + \kappa_x(2r,\, 2\tau) + \kappa_x(-2r,\, -2\tau)\right] = -\tfrac{1}{2}\mu_1^{-1} \neq 0 \end{aligned} \tag{12.25}$$

The last equality is a direct consequence of Equations 10.28 and 10.29 (p. 211) with $c_Q = \sigma_y^2 = g_1(\boldsymbol{x}_{map})$. In light of Equation 12.24, the BME equation becomes

$$\mu_1 \frac{\partial}{\partial \chi_k} \left[\chi_k - \tfrac{1}{4} \sum\nolimits_{i=1}^{4} \chi_i\right]^2_{\chi_k = \hat{\chi}_k} = 0 \tag{12.26}$$

which offers the same estimate as kriging (Eq. 12.22). □

EXAMPLE 12.12: Example 12.5 above can be considered in terms of the present analysis, as well. Equation 10.28 yields

$$\mathcal{Y}_G = -\tfrac{1}{2}\left[\overline{Y_1^2}^{\,-1} \psi_1^2 + \overline{Y_2^2}^{\,-1} \psi_2^2\right] = -\tfrac{1}{2}\left[Q^2(\kappa_{1k})^{-1}\psi_1^2 + Q^2(\kappa_{2k})^{-1}\psi_2^2\right] \tag{12.27}$$

where $\overline{Y_i^2} = 2\gamma_{ik} = Q^2(\kappa_{ik}) = -2\kappa_{ik}$, $\overline{Y_i^2}^{\,-1} = \frac{1}{2}\gamma_{ik}^{-1}$, and $\psi_i = Q(\chi_i, \chi_k) = \chi_i - \chi_k$ $(i = 1, 2)$. Hence, the BME equation reduces to

$$\frac{\partial}{\partial \chi_k}\left[\tfrac{1}{2}\gamma_{1k}^{-1}\psi_1^2 + \tfrac{1}{2}\gamma_{2k}^{-1}\psi_2^2\right]_{\chi_k = \hat{\chi}_k}$$

$$= \frac{\partial}{\partial \chi_k}\left[\gamma_{1k}^{-1}(\chi_1 - \chi_k)^2 + \gamma_{2k}^{-1}(\chi_2 - \chi_k)^2\right]_{\chi_k = \hat{\chi}_k} = 0 \tag{12.28}$$

The last equation has the solution

$$\hat{\chi}_k = \frac{\gamma_{2k}\chi_1 + \gamma_{1k}\chi_2}{\gamma_{1k} + \gamma_{2k}} \tag{12.29}$$

Due to isotropy/stationarity, $\gamma_{1k} = \gamma_{2k} = \gamma$ and, hence, Equation 12.29 reduces to the OK estimate (Eq. 12.12). □

Indicator kriging

Next we will provide some numerical comparisons of the BME approach *vs*. the *indicator kriging* (IK) technique (Journel, 1986, 1989; Deutsch and Journel, 1992). The IK technique suffers from certain theoretical and practical problems (see discussion in Olea, 1999; see also p. 133–34 in this volume). Nevertheless, IK was chosen here because it incorporates some kinds of soft data (though not in a systematic and rigorous way as does BME). BME was shown to perform considerably better than IK, with none of IK's theoretical and computational shortcomings.

EXAMPLE 12.13: To compare the BME approach with the IK technique, the following experiment was designed. For a fixed set of 13 spatial points (shown in Fig. 12.13), 500 realizations of a random field $X(\boldsymbol{p})$ were generated. The simulated values follow a multivariate Gaussian law with zero mean and unit variance; the covariance functions used are: (*i.*) the exponential model

$$c_x(r) = \exp(-r/a) \tag{12.30}$$

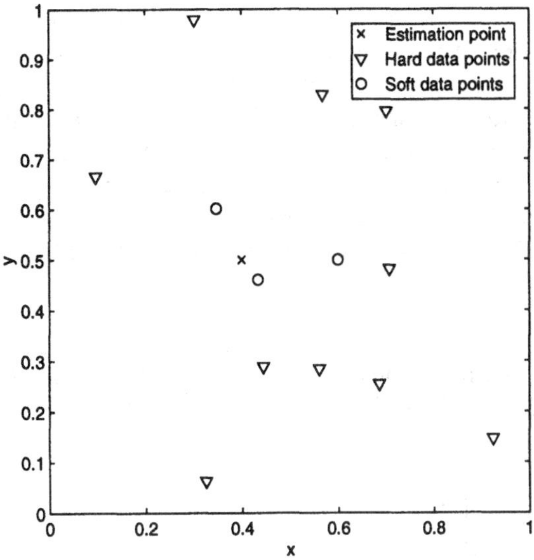

Figure 12.13. Data configuration in R^2. Measurements (hard data) are available at points indicated by ∇; soft data are available at points indicated by open circles. The estimation point is indicated by $\times$.

and (*ii.*) the Gaussian model

$$c_x(\boldsymbol{r}) = \exp\left(-r^2/a^2\right) \tag{12.31}$$

both with $a = 1$. At three points that are kept fixed for all simulations only (soft) interval data are provided. We assume that there are 13 intervals $[\alpha_j, \alpha_{j+1}]$. The limits of these intervals are the threshold values that are considered in IK. The intervals are defined as follows

$$\left.\begin{aligned} \alpha_1 = -\infty, \ \alpha_2 &= F_x^{-1}(0.01) \\ \alpha_{j+2} = F_x^{-1}(0.1\,j), &\quad j = 1, \ldots, 9 \\ \alpha_{12} = F_x^{-1}(0.99), &\quad \alpha_{13} = +\infty \end{aligned}\right\} \tag{12.32}$$

where $F_x^{-1}(p)$ is the p-quantile of the standard Gaussian law. Each point is randomly assigned an interval $\chi(\boldsymbol{p}_i) \in (\alpha_j, \alpha_{j+1}]$. Estimates $\hat{\chi}_k$ are sought at the point $\boldsymbol{p}_k$—the same for all simulations. The ten points with known process values constitute the hard data points, while the three locations with interval data are the soft data points. In order to avoid methodological issues, we assume that all parameters of the multivariate Gaussian law are known in advance for both BME and IK. For the same reason, the same intervals have been chosen for all points. BME does not require that the same intervals be used for all points. Since we assume that complete knowledge about the distribution is available, we can use simple IK to compute the probabilities and indicator covariances from the bivariate Gaussian law. The BME approach directly provides an estimate $\hat{\chi}_k$ at point $\boldsymbol{p}_k$ (*e.g.*, the mode of the posterior pdf) as the solution of the basic BME equation. Since neither the mode nor the mean can be determined reliably for the IK posterior distributions (due to, *e.g.*, non-monotonicity and extreme discretization of the cdf), we used the median of the distributions. Results are shown in Figure 12.14. All estimated values are centered with respect to the known values at $\boldsymbol{p}_k$; thus, each plot gives the corresponding estimation error distribution. It is evident from these plots that the BME approach performs much better than the IK technique.

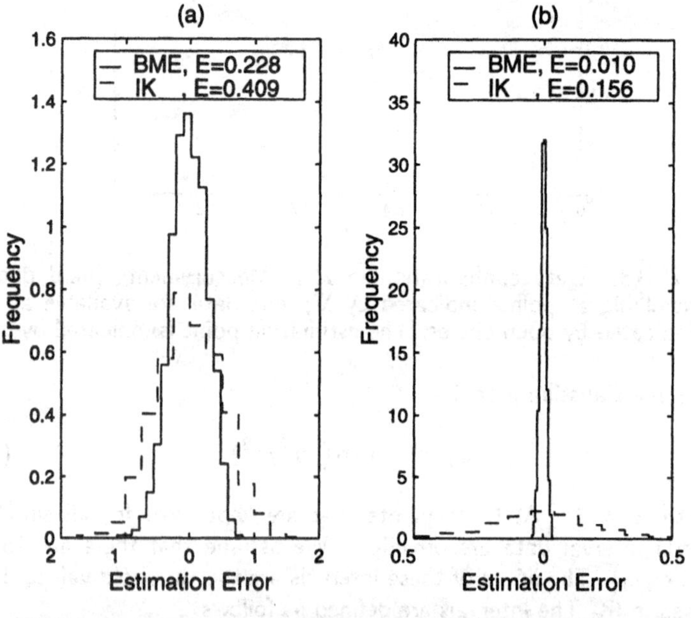

Figure 12.14. Estimation error distributions of BME (continuous line) *vs*. IK (dashed line): (a) exponential covariance, and (b) Gaussian covariance. E denotes the mean of the estimation errors in each case.

A possible reason for the poor performance of IK may be the fact that it uses indicator values to code specificatory knowledge. These indicator values indicate whether a measurement is below or above a threshold value. In order to code (soft) interval data, the thresholds have to correspond to the lower and upper bounds of the interval data. Since the IK technique cannot differentiate between hard and soft data, this constraint on the threshold values must also apply to hard data. In other words, due to the constraints imposed by the coding of the (soft) interval data, the IK technique seems to be "losing" knowledge when coding hard data. It would be interesting to examine whether the noticeably poor performance of IK in the case of the Gaussian covariance (Fig. 12.14b) is related to some problems of this covariance in the context of MMSE estimation, as reported in Stein (1999). In light of results such as those illustrated by Figure 12.14, arguing that IK is synonymous with "practicality" may be as realistic as claiming that Chapel Hill is synonymous with "night life." □

Other sorts of kriging

There exist certain other sorts of kriging commonly used in geostatistical data analysis. Multi-Gaussian kriging (Cressie, 1991) is characterized by the rather strong assumption that a transformation can be established so that the original random field can be transformed into a multivariate Gaussian field (a similar situation was discussed in Example 9.5, p. 174). In multi-Gaussian kriging, while the original field $X(\boldsymbol{p})$ may be characterized by a non-Gaussian multivariate pdf, it is assumed that a transformation $\mathcal{T}[\cdot]$ exists such that the field $Y(\boldsymbol{p}) = \mathcal{T}^{-1}[X(\boldsymbol{p})]$ is multivariate Gaussian. The conditional mean $\overline{y_k \,|\, \boldsymbol{y}_{data}}$ and the variance $Var(y_k \,|\, \boldsymbol{y}_{data})$ can then be estimated, which means that the Gaussian pdf $f_y(\psi_k \,|\, \boldsymbol{\psi}_{data})$ is completely characterized in terms of these two statistics. Finally, the pdf $f_x(\chi_k \,|\, \boldsymbol{\chi}_{data})$ is calculated from the Gaussian pdf. Clearly, in the case in which only hard data are involved, the above pdf coincides with the BME posterior pdf, *i.e.*, $f_{\mathcal{K}}(\chi_k) = f_x(\chi_k \,|\, \boldsymbol{\chi}_{data})$. Therefore, multi-Gaussian kriging is a special case of the general BME analysis.

Additional types of kriging include cokriging and external drift kriging techniques (*e.g.*, Wackernagel, 1995; Batista *et al.*, 1997). Cokriging and external drift kriging are useful in cases in which several secondary natural variables in the mapping of the primary variable must be taken into account. Both techniques share the usual limitations of the MMSE methods; furthermore, they can only use numerical secondary variables, and the primary and secondary variables are assumed to be linearly related (Bardosy *et al.*, 1997). Just as for the previously considered kriging methods, under certain restrictive assumptions, cokriging and external drift kriging can be derived as special cases of the BME model (*e.g.*, cokriging is a special case of vector BME, if up to second-order moments, hard data, and single-point estimation are considered).

Limitations of kriging techniques—Advantages of BME analysis

It may be appropriate to recall briefly some of the *limitations* of the kriging techniques that considerably restrict their theoretical generality as well as their applicability in practice (more details can be found, *e.g.*, in Goovaerts, 1997; Christakos, 1998c; Olea, 1999; and Stein, 1999).

(*a*) It is fairly common practice in the geostatistics literature that the kriging techniques are restricted to the first- and second-order spatial moments, as well as hard data available at a set of neighboring points.

(*b*) Physical laws, high-order space/time moments, and uncertain knowledge of various forms are not taken into consideration by kriging techniques.

Also, the soft data available at estimation points are usually not taken into account.

(*c*) Kriging techniques do not offer multipoint mapping.

(*d*) Most kriging techniques involve linear estimators (see ordinary, simple, intrinsic kriging, *etc.*).

(*e*) Additional constraints on kriging techniques are often imposed on the form of the estimator (unbiasedness, *etc.*).

(*f*) Kriging techniques are mainly interpolative (*e.g.*, extrapolation is not reliable beyond the range of the data).

(*g*) Specialized forms of kriging (indicator kriging, *e.g.*) do not account for the monotonic cdf property, may lead to unfeasible probability values, involve large numbers of kriging systems and variograms (some of them difficult to model), *etc.*

(*h*) Standard practice in geostatistics does not address in a satisfactory manner the circular problem (*i.e.*, covariance or variogram models are estimated empirically from the same data set that is used for kriging).

In contrast, none of these limitations apply to the BME approach. BME, in fact, rigorously takes into consideration many forms of physical knowledge, that improve the accuracy and scientific content of space/time mapping and also provide the means to avoid the circular problem of empirical geostatistics. General knowledge includes scientific laws, multiple-point statistics, and empirical relationships. Soft data at neighboring points or at the estimation points, themselves, are incorporated. Both single-point and multipoint mapping are allowed. Kriging estimators are based on the MMSE criterion that may fail in the case of heavy-tailed random fields with large variances (Painter, 1998). In contrast, BME permits more flexible estimation criteria (*e.g.*, posterior pdf maximization) that are well-defined even for heavy-tailed fields. In general, BME is a nonlinear estimator. No constraints are imposed on the estimator being sought, non-Gaussian laws are automatically incorporated, and by taking into account physical laws, BME possesses global estimation features. These are significant improvements. Indeed, as emphasized by Stein (1999), the linear estimators commonly used in spatial statistics can be highly inefficient compared to nonlinear estimators associated with non-Gaussian random fields.

One may symbolically represent the application domains of the methods of modern spatiotemporal geostatistics as in Figure 12.15, where: *Epistemic* $\supset$ *BME* $\supset$ *Traditional kriging*. More specifically, one could write

$$\textit{Epistemic} \xrightarrow[\textit{KB-cond.} = \textit{Bayes}]{\textit{Info} = -\log f_{\mathcal{G}}} \text{BME} \xrightarrow[\substack{\mathcal{S} = \chi_{hard} \\ \textit{Single-point}}]{\mathcal{G} = \{\textit{Moments} \le 2\}} \textit{Kriging}$$

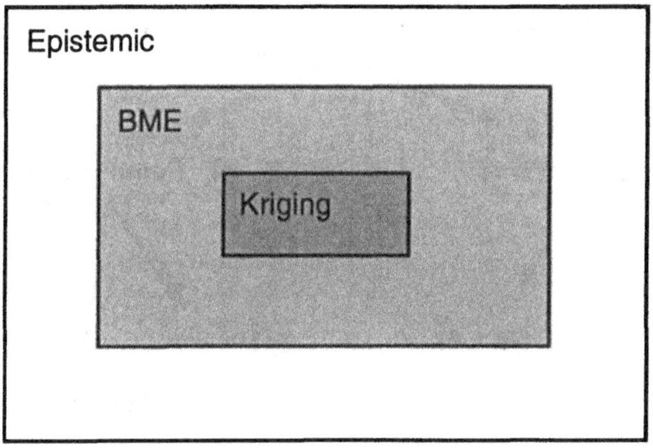

Figure 12.15. The application domains of the methods of modern spatiotemporal geostatistics.

where *KB-cond.* denotes the knowledge-base conditionalization process. Thus, BME methods are special cases of the epistemic approach discussed in Chapter 4, beginning on p. 90, if we assume that the map information measure is of the form of Equation 4.2 (p. 93) and the physical knowledge conditionalization process uses the Bayesian principle (Eq. 4.9, p. 96). Epistemic analysis may not be restricted to Bayesian conditionalization. Other forms of *KB-cond.* include material conditionals, *etc.* (p. 98). The traditional kriging techniques are, in turn, special cases of the BME concept, if the general knowledge is limited to statistical moments up to order 2, a restricted (hard) data set is considered, and a single-point estimator is sought.

Random Field Models of Modern Spatiotemporal Geostatistics

A unified framework

The mathematical apparatus of modern spatiotemporal geostatistics includes a variety of theories and models. Among them, a powerful unified framework is provided by the *generalized* S/TRF theory. Indeed, a large number of popular stochastic models, including coarse-grained random fields, wavelet random fields, and fractal random fields, can be derived from the theory of generalized S/TRF. Mathematically, generalized fields were introduced in the works of Dirac (1926–27), Sobolev (1938), and Schwartz (1950–51) on distribution theory. Studies on generalized random fields in a purely spatial domain were spearheaded by Itô (1954) and Gel'fand (1955) with important contributions by Yaglom (1957, 1987) and Matheron (1973). An extension of the generalized random field theory in the spatiotemporal domain can be found in Christakos (1991a, 1991b, 1992) and in Christakos and Hristopulos (1998).

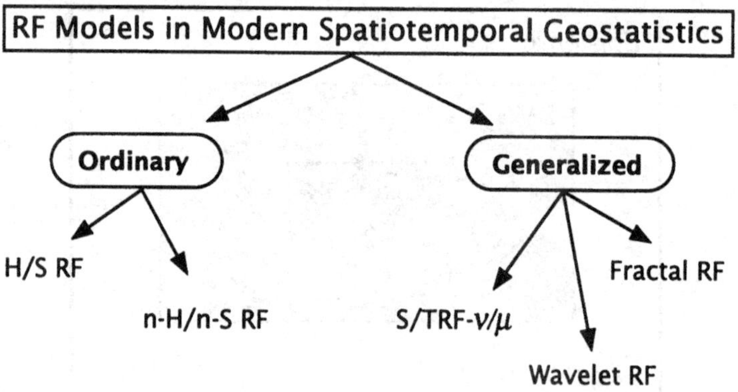

Figure 12.16. Classification of spatiotemporal random field (RF) models.

The main classes of random field (RF) models used in modern spatiotemporal geostatistics are outlined diagrammatically in Figure 12.16. This classification distinguishes between ordinary and generalized random fields. Ordinary models include homogeneous/stationary fields (H/S) and nonhomogeneous/nonstationary fields (n-H/n-S). Generalized models, on the other hand, include S/TRF-ν/μ , wavelet RF (WRF), and fractal RF (FRF). In the remainder of this section we will discuss very briefly the various parts of the classification in Figure 12.16. The interested reader is referred to the relevant literature for a more detailed presentation of the various RF models shown in the illustration.

Ordinary S/TRF models were defined in Chapter 2 (Definition 2.10, p. 60) and in references cited in that section. A *generalized* S/TRF $X[q]$ is defined by means of the following functional expression

$$X[q] = \langle X(\boldsymbol{p}),\, q(\boldsymbol{p}) \rangle = \int_{\Omega(q)} d\boldsymbol{p}\, X(\boldsymbol{p})\, q(\boldsymbol{p}) \tag{12.33}$$

where $X(\boldsymbol{p})$ is an ordinary S/TRF with point values and $q\,(\boldsymbol{p})$ is the so-called *test function* that belongs to a properly chosen space $\mathcal{D}$. The choice of $\mathcal{D}$ reflects certain characteristics of the phenomenon under study and the goals of the analysis. The set of linear functionals on $\mathcal{D}$ is called its dual space and is denoted by $\mathcal{D}'$. The generalized field (Eq. 12.33) corresponds to an average of $X(\boldsymbol{p})$ over a support $\Omega\,(q)$ and satisfies the linearity conditions $X[q_1 + q_2] = X[q_1] + X[q_2]$ and $X[\alpha\, q] = \alpha\, X[q]$, where α is an arbitrary (real or complex) number. Useful generalized fields are also defined by means of the convolution integral

$$X[q](\boldsymbol{p}) = \langle X(\boldsymbol{u}),\, q(\boldsymbol{p} - \boldsymbol{u}) \rangle = \int_{\Omega(q)} d\boldsymbol{u}\, X(\boldsymbol{u})\, q(\boldsymbol{p} - \boldsymbol{u}) \tag{12.34}$$

The $X[q](\boldsymbol{p})$ is a function of the space/time point which represents a nonlocal average of the S/TRF $X(\boldsymbol{p})$ over a window determined by the test function q. A standard example of a space $\mathcal{D}$ is that of delta functions $\delta(\boldsymbol{p})$ (Dirac,

1926–27; Schwartz, 1950–51). A practical space $\mathcal{D}$ contains functions that are continuous, integrable, and infinitely differentiable. It is also customary to require that the test functions and all their derivatives vanish outside a certain interval in $n+1$-dimensional support $\Omega(q)$ (Gel'fand and Vilenkin, 1964). Other $\mathcal{D}$-spaces are chosen on the basis of the physics of the situation (see below).

The stochastic moments of a generalized S/TRF are defined in a straightforward manner: the point values χ of the random field are replaced by the functionals $X[q]$. Several properties and classifications that are valid for ordinary fields can be extended to generalized fields, as well (see, *e.g.*, Christakos and Hristopulos, 1998). Generalized random fields share the stochastic symmetries of ordinary fields. Thus, homogeneous/stationary generalized random fields are defined by means of the second order moment functionals (in the weak sense) or the pdf (strict sense), *etc.*

EXAMPLE 12.14: Second-order moments of the generalized fields (Eq. 12.34) include the space/time mean functional

$$\overline{X}[q](\boldsymbol{p}) = \langle \overline{X}(\boldsymbol{u}),\ q(\boldsymbol{p}-\boldsymbol{u})\rangle = \int_{\Omega(q)} d\boldsymbol{u}\, \overline{X}(\boldsymbol{u})\, q(\boldsymbol{p}-\boldsymbol{u}) \tag{12.35}$$

and the (non-centered) covariance functional

$$\begin{aligned} C_x[q_1, q_2](\boldsymbol{p}, \boldsymbol{p}') &= \langle C_x(\boldsymbol{u}, \boldsymbol{v}),\ q_1(\boldsymbol{p}-\boldsymbol{u})\, q_2(\boldsymbol{p}'-\boldsymbol{v})\rangle \\ &= \int_{\Omega(q_1)} \int_{\Omega(q_2)} d\boldsymbol{u}\, d\boldsymbol{v}\, C_x(\boldsymbol{u}, \boldsymbol{v})\, q_1(\boldsymbol{p}-\boldsymbol{u})\, q_2(\boldsymbol{p}'-\boldsymbol{v}) \end{aligned} \tag{12.36}$$

Higher order moment functionals may be defined in a similar manner. □

Although the derivative of an ordinary random field may not exist in the usual sense, it may still be defined in terms of generalized random fields. The derivatives of generalized random fields in $\mathcal{D}$ always exist as generalized fields. The partial derivatives of any order of the generalized field are obtained by

$$\partial^m X[q]/\partial p_i^m = (-1^m)\, X\left[\partial^m q(\boldsymbol{p})/\partial p_i^m\right] \tag{12.37}$$

Therefore, the derivatives of the generalized field can be evaluated by means of the derivatives of the test function.

Generally, the $\mathcal{D}$-spaces of test functions q are selected on the basis of the physics of the situation and the goals of the analysis. Below we will briefly discuss three special cases of generalized S/TRF that are derived by properly specifying the desirable properties of the test function q (again, for a detailed discussion, the interested reader is referred to the relevant literature). These special cases include:

(*i.*) *coarse-graining* applications in which q is chosen so that it represents averaging effects in physical measurements,

(*ii.*) *heterogeneity* analysis in which q is chosen so that the resulting random field is spatially homogeneous/temporally stationary [this class of random fields includes fractal random fields (FRF) as a special case], and (*iii.*) *signal processing* in which the test function q (also called a *mother wavelet*) is chosen so that it characterizes the local smoothness properties of a signal. This gives rise to the wavelet random fields (WRF).

The class of coarse-grained RF

Generalized S/TRF provides the mathematical framework for a rigorous formulation of physical coarse graining. Coarse-grained fields represent averaging effects in physical measurements due to finite instrument bandwidth or effects of numerical averaging. Generalized S/TRF leads to well-defined representations of discontinuous random fields that lack well-defined values at all space/time points (*e.g.*, for intermittent processes such as rainfall, coarse-grained values can be obtained by averaging over specific space/time windows). The resulting physical processes generally depend on the scale of the coarse-graining window. Expressions such as Equations 12.33 and 12.34 provide mathematical representations of the coarse graining involved in measurements of random fields.

Test functions q with finite support provide appropriate models for the observation effect (this is the case when the coarse-graining process can significantly modify the properties of random fields; see, *e.g.*, Cushman, 1984 and Baveye and Sposito, 1984). If the apparatus function is known, the point values of the S/TRF are determined by the deconvolution of Equations 12.33 or 12.34, which can be realized by means of maximum entropy techniques. Such methods have been used successfully for enhanced information recovery by spectrum deconvolution in atomic spectroscopy (*e.g.*, Davies *et al.*, 1991; Fisher *et al.*, 1997).

The class of S/TRF-ν/μ models in heterogeneity analysis

The spatiotemporal variations of many natural processes exhibit considerable heterogeneities (complicated space/time patterns, local trends, *etc.*). In such cases, the homogeneous/stationary S/TRF model is not the best option. Instead, a more general and powerful random field model that is capable of handling complicated space/time heterogeneities is sought. It turns out that by properly selecting the test function q, such a model can be constructed on the basis of the generalized random field theory above. In particular, if the test function q is chosen so that (a) it eliminates polynomial functions of degree ν in space and μ in time representing space/time trends within each $\Omega(q)$, and (b) the expression (Eq. 12.34) is a zero-mean, spatially homogeneous/temporally stationary field, then $X(\boldsymbol{p})$ belongs to the useful class of S/TRF-ν/μ models (Christakos, 1991b). The variability of $X(\boldsymbol{p})$ is in general characterized by the two integer indices, ν for space and μ for time, which are called continuity orders. Their values determine the degree of departure from homogeneity and stationarity. The case $\nu = \mu = 0$, *e.g.*, denotes an S/TRF with homogeneous/stationary increments (RF of this type may have linear trends in space and time that are due to the mean of the increment; if the mean of the increment is zero, the mean of $X(\boldsymbol{p})$ is constant). This class of models can handle complicated space/time variabilities of any size in a mathematically rigorous and physically meaningful manner, and is considerably more general than the restricted class of homogeneous/stationary S/TRF (by standard convention, the index values $\nu = \mu = -1$ correspond to spatially homogeneous/temporally stationary fields with no trends).

There are various useful representations of the S/TRF-ν/μ $X(\boldsymbol{p})$ (Christakos and Hristopulos, 1998). An interesting representation is derived by choosing the following test function

$$q(\boldsymbol{s},\, t) = (-1)^{\nu+\mu}\, \delta^{(\nu+\mu+2)}(\boldsymbol{s},\, t) \tag{12.38}$$

where the superscript $(\nu+\mu+2)$ denotes the space/time differentiation operator $\partial^{\nu+\mu+2}[\cdot]/\partial s_1^{\nu_1}\, \partial s_2^{\nu_2} \ldots \partial s_n^{\nu_n}\, \partial t^{\mu+1}$ with $\nu + 1 = \sum_{i=1}^{n} \nu_i$. In light of Equation 12.38, Equation 12.34 leads to a continuous-domain S/TRF-ν/μ representation as follows

$$X[q](\boldsymbol{p}) = \partial^{\nu+\mu+2} X(\boldsymbol{p}) \,/\partial s_1^{\nu_1}\, \partial s_2^{\nu_2} \ldots \partial s_n^{\nu_n}\, \partial t^{\mu+1} \tag{12.39}$$

where $X[q](\boldsymbol{p}) = Y(\boldsymbol{p})$ is a zero-mean homogeneous/stationary field. A discrete domain representation is possible by means of the summation

$$X[q](\boldsymbol{p}) = \sum_{i=1}^{m} X(\boldsymbol{p}_i)\; q(\boldsymbol{p},\, \boldsymbol{p}_i) \tag{12.40}$$

where $q(\boldsymbol{p},\, \boldsymbol{p}_i)$ are local weights and the $X(\boldsymbol{p}_i)$ represent the values of the random field at space/time points $\boldsymbol{p}_i$. Several other S/TRF-ν/μ representations (continuous and discrete) involving test functions different than that of Equation 12.38 can be found in the literature.

The space/time covariance of $X(\boldsymbol{p})$ satisfies the decomposition relationship

$$c_x(\boldsymbol{p},\, \boldsymbol{p}') = k_x(\boldsymbol{r}, \tau) + \Theta_{\nu/\mu}(\boldsymbol{p},\, \boldsymbol{p}') \tag{12.41}$$

where $k_x(\boldsymbol{r}, \tau)$ is the generalized spatiotemporal covariance that depends only on the spatial and temporal lags $\boldsymbol{r} = \boldsymbol{s} - \boldsymbol{s}'$ and $\tau = t - t'$; the $\Theta_{\nu/\mu}(\boldsymbol{p},\, \boldsymbol{p}')$ denotes polynomials of degrees ν in space and μ in time.

The generalized covariance k_x can be viewed as a generalized function defined on the spaces $\mathcal{D}$ or $\mathcal{D}'$ by means of a linear functional as follows

$$\langle k_x,\, q\rangle = \int_{\Omega(q)} d\boldsymbol{p}\; q(\boldsymbol{p})\; k_x(\boldsymbol{p}) \tag{12.42}$$

with $\boldsymbol{p} = (\boldsymbol{r},\, \tau)$. The generalized Fourier transform $\tilde{k}_x(\boldsymbol{w})$, $\boldsymbol{w} = (\boldsymbol{\kappa},\, \omega)$, of $k_x(\boldsymbol{p})$ is defined as $\langle k_x, q\rangle = \langle \tilde{k}_x, \tilde{q}\rangle$; the functional $\langle \tilde{k}_x, \tilde{q}\rangle = \int_{\boldsymbol{w}} \tilde{q}(\boldsymbol{w})\tilde{k}_x(\boldsymbol{w})$ is a frequency space integral, where $\tilde{q}(\boldsymbol{w})$ is the ordinary Fourier transform of the test function. Various examples of generalized space/time covariances exist. Two such examples were given in this volume in the previous chapter (p. 227–228); other examples can be found in the relevant literature.

The class of space/time fractal RF models

It is noteworthy that members of the class of *fractal RF* (FRF) models can be derived from the generalized S/TRF theory. Let a spatiotemporal FRF $X(\boldsymbol{s}, t)$ satisfy the relationship

$$X(c^{\eta}\boldsymbol{s},\, c^{\zeta}t) = c^{H} X(\boldsymbol{s},\, t) \tag{12.43}$$

in a stochastic sense ($c > 0$; η, ζ and H are suitable scaling exponents), *i.e.*, the $X(\boldsymbol{s}, t)$ and $X(c^{\eta}\boldsymbol{s}, c^{\zeta}t)$ have the same probability law. Sometimes we write FRF-H to denote the associated fractal exponent. In many cases, the FRF-H can be characterized equivalently by the weaker condition

$$\overline{X^2(c^{\eta}\boldsymbol{s},\, c^{\zeta}t)} = c^{2H}\overline{X^2(\boldsymbol{s},\, t)} \tag{12.44}$$

(*e.g.*, when the underlying distribution is Gaussian). To show that a S/TRF-ν/μ $X(\boldsymbol{p})$ can admit FRF-H representations, one can use (*i.*) generalized $X(\boldsymbol{p})$ representations and Equation 12.43 or (*ii.*) Equations 12.41 and 12.44. These two approaches are illustrated by means of the following two examples.

EXAMPLE 12.15: Let $X(\boldsymbol{p})$ be a generalized S/TRF-ν/μ with the representation (Eq. 12.39) in the $R^1 \times T$ domain, *i.e.*, $\boldsymbol{p} = (s, t)$ and

$$\partial^{\nu+\mu+2} X(\boldsymbol{p}) / \partial\, s^{\nu+1}\; \partial\, t^{\mu+1} = Y(\boldsymbol{p}) \tag{12.45}$$

where $Y(\boldsymbol{p})$ is a Gaussian white-noise field with zero mean and $C_y(r, \tau) \propto \delta(r)\,\delta(\tau)$. A corresponding generalized covariance is

$$\kappa_x(r, \tau) = A(\nu, \mu)\; r^{2\nu+1} \tau^{2\mu+1} \tag{12.46}$$

where $A(\nu, \mu) = (-1)^{\nu+\mu} / \left[(2\nu+1)!\,(2\mu+1)!\right]$. If Equation 12.43 holds, the generalized S/TRF-ν/μ $X(\boldsymbol{p})$ with the representation of Equation 12.45 can be considered as an FRF-H with $H = (\nu + \frac{1}{2})\eta + (\mu + \frac{1}{2})\zeta$, where the range of values for ν, μ, H, *etc.*, satisfy the appropriate conditions (integrability, continuity, *etc.*). Note that one could study an FRF using S/TRF-ν/μ representations other than Equation 12.45. □

EXAMPLE 12.16: Let $X(t)$ be a generalized temporal RF of order ν, in which case Equation 12.41 reduces to

$$c_x(t, t') = \kappa_x(\tau) + \sum\nolimits_{i=0}^{\nu} b_i(t')\, t^i + \sum\nolimits_{i=0}^{\nu} b_i(t)\, t'^{\,i} \tag{12.47}$$

where

$$\kappa_x(\tau) = (-1)^{\nu+1}\, \tau^{2\nu+1} / (2\nu+1)! \tag{12.48}$$

In light of the expression

$$\overline{X^2(ct)} = c^{2H}\overline{X^2(t)} \tag{12.49}$$

and letting $t = t'$, Equation 12.47 yields $\overline{X^2(t)} = \kappa_x(0) + \text{const.} \times t^{2H}$, where $H = \nu + 1/2$. In other words, for an appropriate range of H-values the generalized RF-ν $X(t)$ can be considered as an FRF with $H = \nu + 1/2$. The determination of the H and ν values depends on the situation. Assume, *e.g.*, that $X(t)$ is a *Wiener* (*Brownian*) RF. This process is a generalized temporal RF with $\nu = 0$, which has a zero mean and covariance

$$c_x(t, t') = \alpha\,(-\tau + t + t') \tag{12.50}$$

where α is a constant and $\tau = |t - t'|$. In light of Equation 12.47, Equation 12.50 implies that $\kappa_x(\tau) = -\alpha\,\tau$, $b_0(t) = \alpha\,t$, and $b_0(t') = \alpha\,t'$; hence, $H = 1/2$. Equation 12.49 is satisfied for $H = 1/2$ and, thus, the Wiener (Brownian) process $X(t)$ is an FRF-1/2, *i.e.*, it is valid that $X(ct) = \sqrt{c}\,X(t)$. Furthermore, a Gaussian RF with $\nu = 0$ and $\kappa_x(\tau) = -a\,\tau^{2H}$ $(0 < H < 1)$ is an FRF-H, also called a *fractional Brownian* RF. □

The greater part of the above analysis can be extended to *multi-FRF* models. The latter include random fields that can be characterized by Equation 12.44. In this case, the fractal exponent is a function of the space/time point $\boldsymbol{p}$, *i.e.*, the exponent must be written as $H(\boldsymbol{p})$. The class of S/TRF-ν/μ is considerably wider than these of FRF. The former, *e.g.*, can be characterized by generalized covariances that have forms more complicated than Equations 12.46 and 12.48.

The class of wavelet RF

As we saw above, the $\mathcal{D}$-spaces of test functions q in generalized S/TRF are selected on the basis of the physics of the situation and the goals of the analysis. In several physical applications, the test functions that are selected possess properties that can be used to study certain important characteristics of the random field of interest (*e.g.*, singularities caused by physical laws, multiscale features, and sharp variations).

We start by using the fundamental generalized S/TRF concept to define the *wavelet random field* (WRF) model in the space/time domain. For this purpose, we select a test function such as

$$q(\boldsymbol{p}) = A(\boldsymbol{a})\;\psi(\boldsymbol{p},\,\boldsymbol{u},\,\boldsymbol{a}) = A(\boldsymbol{a})\;\psi\Big(\frac{s_1 - u_1}{a_1},\,...,\,\frac{s_n - u_n}{a_n},\,\frac{t - u_t}{a_t}\Big) \tag{12.51}$$

where $\psi(\boldsymbol{p},\,\boldsymbol{u},\,\boldsymbol{a})$ is called the *mother space/time wavelet*, and the coefficient $A(\boldsymbol{a})$ is chosen so that it magnifies or reduces the sensitivity of the field to different scale parameters; a_i and s_i $(i = 1,\,...,\,n)$ are the spatial scale and translation parameters, respectively, whereas the a_t and u_t are the temporal scale and translation parameters, respectively. Then, the generalized random field concept of Equation 12.34 leads naturally to the definition of a *continuous space/time* WRF model as follows

$$X[\psi](\boldsymbol{p}) = \Big\langle X(\boldsymbol{u}),\, \psi(\boldsymbol{p},\, \boldsymbol{u},\, \boldsymbol{a})\Big\rangle = A(\boldsymbol{a}) \int d\boldsymbol{u}\; X(\boldsymbol{u})\, \psi(\boldsymbol{p},\, \boldsymbol{u},\, \boldsymbol{a}) \quad (12.52)$$

The WRF above is also denoted as $X[\psi](\boldsymbol{p}) = X(\boldsymbol{p},\, \boldsymbol{a})$. Note that we do not explicitly use a particular support $\Omega(q) = \Omega(\psi)$ in the multiple integral of Equation 12.52. Since the mother wavelets are usually chosen so that they decay rapidly, the integral is assumed to cover the entire $R^n \times T$ domain. The relationship between the original S/TRF $X(\boldsymbol{p})$ and the WRF $X(\boldsymbol{p},\, \boldsymbol{a})$ in Equation 12.52 depends on the properties of the mother wavelet $\psi(\boldsymbol{p},\, \boldsymbol{u},\, \boldsymbol{a})$ (*e.g.*, some of these properties aim at detecting singularities of functions and edges of images). The notation WRF-ψ is often used to denote dependence on the specific mother wavelet ψ. In general, the wavelet is a spatially anisotropic function, *i.e.*, it has selectivity for spatial orientation. In some cases, however, $\psi(\boldsymbol{p},\, \boldsymbol{u},\, \boldsymbol{a}) = \psi(r,\, t,\, \boldsymbol{u},\, \boldsymbol{a})$ with $r = |\, \boldsymbol{s}\,|$, which means that the wavelet is spatially isotropic having no orientation selectivity.

The derivation of the statistical moments of a WRF-ψ on the basis of the generalized S/TRF theory is straightforward, as demonstrated in the following example.

EXAMPLE 12.17: In light of the generalized moment functionals (Eqs. 12.35 and 12.36), the wavelet mean and (non-centered) covariance for the WRF (Eq. 12.52) are expressed as

$$\overline{X}(\boldsymbol{p},\, \boldsymbol{a}) = \Big\langle \overline{X}(\boldsymbol{u}),\; \psi(\boldsymbol{p},\, \boldsymbol{u},\, \boldsymbol{a})\Big\rangle \quad (12.53)$$

and

$$C_x(\boldsymbol{p},\, \boldsymbol{p}',\, \boldsymbol{a},\, \boldsymbol{b}) = \Big\langle C_x(\boldsymbol{u},\, \boldsymbol{v}),\; \psi_1(\boldsymbol{p},\, \boldsymbol{u},\, \boldsymbol{a})\, \psi_2(\boldsymbol{p}',\, \boldsymbol{v},\, \boldsymbol{b})\Big\rangle \quad (12.54)$$

Higher order space/time moments can be derived in a similar fashion. □

In image processing applications, two-dimensional test functions $q(\boldsymbol{s})$, $\boldsymbol{s} = (s_1,\, s_2)$, are selected so that they provide information about shapes of objects and intersections between surfaces and between textures. These kinds of test functions are called *two-dimensional* wavelets and are denoted by $q(\boldsymbol{s}) = A(\boldsymbol{a})\psi(\boldsymbol{s},\, \boldsymbol{u},\, \boldsymbol{a})$ (*e.g.*, see Poularikas, 1996). In the last decade or so, one-dimensional WRF-ψ models have become very popular in signal processing. *One-dimensional* wavelets $q(t) = A(a)\psi(t,\, u,\, a)$ are chosen so that they characterize the local smoothness properties of a temporal signal (Daubechies, 1992). This kind of WRF-ψ is demonstrated in the following example.

EXAMPLE 12.18: If one focuses on the one-dimensional t domain and lets

$$q(t) = A(a)\; \psi(t,\, a) = A(a)\; \psi\Big(\frac{t-u}{a}\Big) \quad (12.55)$$

and $\Omega(q) = \Omega(\psi) = (-\infty,\, \infty)$, the mother wavelet (Eq. 12.55) accounts for both the location and the scale properties in terms of the parameters t and a, respectively. Then, Equation 12.52 leads to the definition of the continuous WRF-ψ on R^1 as follows

$$X[\psi](t) = \langle X(u),\ \psi(t,\ u,\ a)\rangle = A(a) \int_{-\infty}^{\infty} du\ X(u)\ \psi\Big(\frac{t-u}{a}\Big) \tag{12.56}$$

The WRF $X[\psi](t)$ is sometimes denoted $X(a,\ t)$ to emphasize the scale parameter a of the wavelet. In certain applications, these WRF constitute efficient tools for local analysis of nonstationary and fast transient signals. The choice of the mother wavelet is of utmost importance in applications. Localized mother wavelets $\psi(t,\ u,\ a)$, which are such that they decay rapidly to 0 as $t \to \pm\infty$, are of special interest. Test functions used as wavelets are known in the literature under several names: *Morlet wavelet*, *Meyer wavelet*, *Daubechies wavelet*, and the "Mexican hat" wavelet (see, *e.g.*, Poularikas, 1996). One-dimensional WRF of the form of Equation 12.55 with $A(a) = 1/\sqrt{a}$ have been studied extensively in the literature. The wavelet mean and (non-centered) covariance for the WRF (Eq. 12.56) are given by

$$\overline{X}(a,\ t) = \Big\langle \overline{X}(u),\ \psi\Big(\frac{t-u}{a}\Big)\Big\rangle \tag{12.57}$$

and

$$C_x(a,\ b;\ t,\ t') = \Big\langle C_x(u,\ v),\ \psi_1\Big(\frac{t-u}{a}\Big)\ \psi_2\Big(\frac{t'-v}{b}\Big)\Big\rangle \tag{12.58}$$

where mother wavelets of a suitable form may be used. □

In addition to its physical features, the class of mother wavelets determined by Equation 12.55 has certain interesting analytical properties, some of which we may discuss very briefly.

EXAMPLE 12.19: Let $X(t)$ be an FRF-H in the sense that $C_x(ct,\ ct') = c^{2H} C_x(t,\ t')$, where $c,\ H > 0$; the $C_x(t,\ t')$ and $\psi(t)$ satisfy some well-defined conditions (Cambanis and Houdre, 1995). Then, it is easily shown that $C_x(ca,\ cb;\ ct,\ ct') = c^{2H+1} C_x(a,\ b;\ t,\ t')$, and *vice versa*. This implies that $X(t)$ is an FRF-H if the corresponding WRF, $X[\psi](t) = X(a,\ t)$, is an FRF-$H + \frac{1}{2}$. A similar result is valid in terms of RF with homogeneous increments of order ν. □

A symbolic summary of some of the results above is presented in Figure 12.17. This figure illustrates how certain classes of S/TRF-ν/μ, FRF-H, and WRF-ψ models are derived from the generalized random field theory of modern spatiotemporal geostatistics.

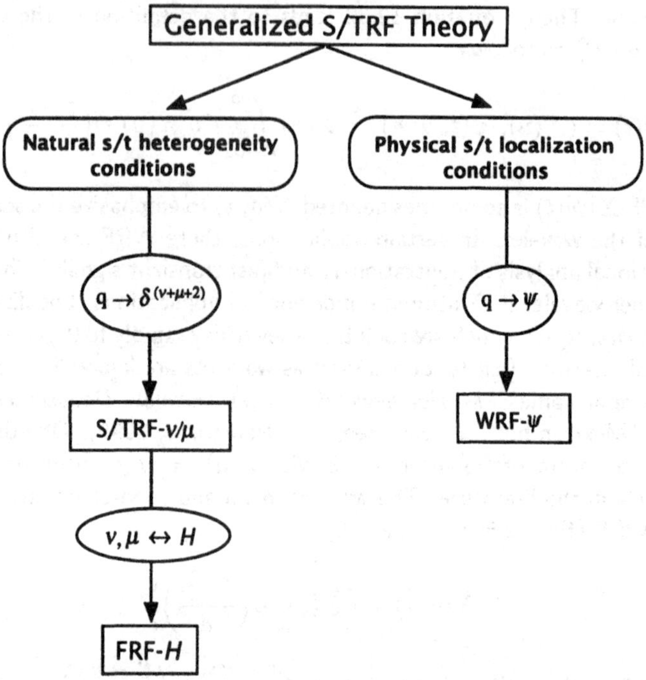

Figure 12.17. An example flowchart of S/TRF models and their relations.

The Emergence of the Computational Viewpoint in BME Analysis

Chapters 2–12 provide an introduction to the world of modern spatiotemporal geostatistics. Distinguishing characteristics of the BME model, which are of considerable importance in the scientific analysis and mapping of spatiotemporal natural variables, were discussed. These characteristics include BME's sound epistemic foundations, its considerable generalization power, its fertility, the high correspondence of BME with reality and, of course, its beauty (which, regretfully, cannot be captured by words). The generality of the BME model has been emphasized throughout the book. Several demonstrations of this generality have been offered, including the fact that most of the mapping techniques of classical geostatistics are special cases of the BME model. Indeed, BME does not discard the kriging methods, rather it shows them to be special cases of limited application within a more general concept. In this sense, the BME theory has been amply confirmed.

BME focuses on levels of spatiotemporal mapping as they relate to understanding. It produces elegant analytical formulations that are able to account for physical knowledge bases of considerable complexity. Given the latter, which reflect the complexity of real-world applications, future developments of BME analysis should focus on computational issues as well. A natural consequence of such developments will be the emergence of a powerful computational viewpoint in modern spatiotemporal geostatistics that may be called *computational* BME. With the aid of computers, the rigorous mathematical structure of the BME model can distill the complexity of physical knowledge bases into humanly manageable amounts of information in the form of space/time maps, to which scientists and engineers can apply their intuition to produce a realistic picture of the phenomenon under consideration. Computational BME is, therefore, a theme of the modern geostatistics paradigm that involves combining actual observations of real phenomena with computer *simulations* of the phenomena. These simulations (the computer's way of "experimenting") are, of course, physical model-based simulations rather than merely graphical displays of data or model-free, fuzzy-system representations.

Computational BME involves a wide range of numerical techniques, including deterministic (linear algebra, finite differences, *etc.*) and stochastic schemes of various forms (Monte Carlo, Brownian dynamics, *etc.*). In the context of BME analysis, the implementation of these techniques has two main goals, to

- solve integrodifferential and integrodifference equations with respect to the BME coefficients at the prior stage, and
- evaluate multidimensional integrals of functions of the pdf at the posterior stage.

Systematic applications of the above techniques—together with analytical methods—seek to incorporate a part of physical reality as large as possible into BME mapping. For computational BME, abstraction and construction become close partners in mapping investigations. Because of its capacity to manage large amounts of knowledge and solve large systems of mathematical equations, computational BME will divulge new aspects of BME theory and will even lead to theoretical improvements.

Modern Spatiotemporal Geostatistics and GIS Integration Technologies

Geostatistical analysis and modeling of natural processes is not merely an issue of inserting data bases into a library of "black box"-type computer programs. In many practical applications, a major challenge faced by geostatistical practitioners is how to integrate, rigorously and effectively, two aspects of the information (see Fig. 12.18) upon which they draw:

(*i.*) the powerful theoretical tools and conceptual models (mathematical equations expressing physical laws, scientific theories, phenomenological relationships, optimal estimators, *etc.*) which have usually been developed for well-defined *conceptual environments*; and

(*ii.*) the site-specific details of the *real environment* under consideration; in the case of a hazardous waste site, *e.g.*, these details may include the specific hydrogeological characteristics of the site (local drainage catchment basins, stratigraphy, faults, dikes, sills, zones of alteration, depth to water table, sources/sinks, topographic contours, *etc.*), and the goals of the study (preventing outbreaks of contamination, optimal sampling design, remediation strategy, *etc.*).

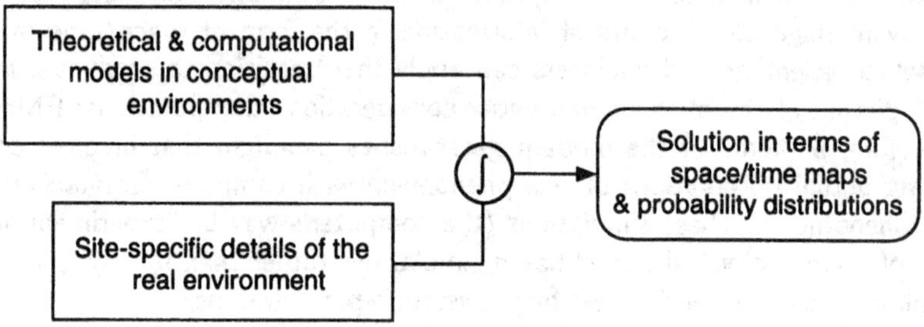

Figure 12.18. The integration issue in real-world applications.

In terms of modern spatiotemporal geostatistics, aspect (*i.*) above is part of the general knowledge base, whereas aspect (*ii.*) is part of the specificatory knowledge base. To aid in their integration, the physical knowledge processing tools of modern geostatistics can benefit considerably by technologies associated with a *geographic information system* (GIS) and *vice versa*. The unique features of GIS constitute a major technological breakthrough that includes visualization power, considerable flexibility, and the ability to analyze a variety of data sources (see, *e.g.*, Clarke, 1986, and Laurini and Thompson, 1995). A GIS is usually characterized by its set of basic *functions*. Various such sets have been proposed in the literature (*e.g.*, Chrisman, 1983; Rhind and Green, 1988). These GIS functions are usually classified into fundamental and advanced functions (Malczewski, 1999). The *fundamental* functions involve low-order geometric operations and may be considered as tools to establish relationships between spatial objects, while the *advanced* functions provide mathematical models and techniques for rigorous and efficient knowledge processing. The contribution of modern spatiotemporal geostatistics fits mainly into the framework of the advanced GIS functions which provide the system with the adequate means to incorporate the models of aspect (*i.*) above. In order to describe the application-specific details of aspect (*ii.*), GIS must employ computerized data of two types: base maps (graphic representations of geographical layout, *etc.*) and attribute data (physical measurements, demographics, *etc.*).

GIS techniques for integrating and visualizing spatial data have been used with increasing frequency during the last two decades, but significantly less work has been done in the area of temporal GIS. Most of what is available in the current GIS industry is limited to use in descriptive analysis, and is far too restricted in real analytical power (Birkin *et al.*, 1996). The kind of GIS needed for the scientific study of real-world problems are the *model-based* systems with real analytical power rather than systems that incorporate a group of techniques possessing merely descriptive capabilities. We will next consider an example of a specialized, model-based GIS.

EXAMPLE 12.20: Geographical plume analysis (Osleeb and Kahn, 1999) is an analytical tool which consists of a chemical dispersion model that is integrated with a GIS. Dispersion models take into account information about weather conditions, pollutant type and amount, plume characteristics, *etc.*, which is used to predict the spatial distribution of the pollutant concentrations. This spatial distribution is then combined with a site-specific GIS database to assess its effects on the local environment. □

Conceptual models of the natural phenomenon under consideration are often needed to guide the various stages of GIS analysis. This is the case of the following example.

EXAMPLE 12.21: Mineral deposit models are conceptual models which describe the main properties of deposits belonging to the same group (stratigraphy, dikes and fractures, genesis and deposit formation processes, *etc.*). Such models can provide the theoretical framework for guiding GIS studies of mineral potential (Bonham-Carter, 1994). □

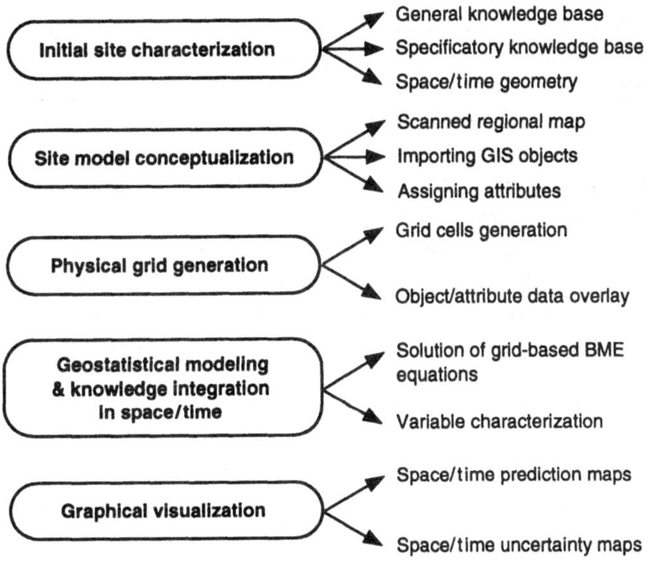

Figure 12.19. Flowchart of a physical model-based GIS approach.

Recent developments in the field of computer-programming language make it possible for GIS to incorporate the advanced functions of modern spatiotemporal geostatistics, thus leading to more accurate and informative physical model-based maps in space/time. In many environmental applications, a physical model-based GIS approach can be boiled down to a few major steps, as shown in Figure 12.19. In almost every step of the GIS approach, the techniques of modern spatiotemporal geostatistics can play a vital role in efforts to achieve realistic environmental modeling. These steps are illustrated in the following example.

EXAMPLE 12.22: Computerized hydrogeologic systems are used to integrate and simplify the process of ground-water flow and transport modeling by bringing together all of the tools needed to complete a successful study (*e.g.*, DoD-GMS, 1997). Below we present an approach based on Figure 12.19 that uses modern spatiotemporal geostatistics techniques to study the hydrogeologic properties of real sites. In the first step of the approach (initial site characterization), the concern focuses on: (1a) the development of general knowledge bases (ground-water flow equations, statistical moments of the hydrologic variables, *etc.*); (1b) the acquisition and storage of specificatory knowledge bases (borehole data display, stratigraphy representation, hard and soft attribute data, *etc.*); and (1c) the establishment of an adequate space/time geometry (coordinates, metric, local patches, *etc.*). The second step (site model conceptualization) focuses on: (2a) the scanned regional map; (2b) importing from GIS objects consisting of points, arcs, and polygons organized into coverages; and (2c) assigning attributes to the GIS objects (*e.g.*, for a drain arc, the conduction of the drain is assigned to the arc and the elevation of the drain is assigned to the endpoints of the arc). In the third step (physical grid generation): (3a) an appropriate physical grid is constructed dividing the area into cells on the basis of the conceptual model of the site (*e.g.*, the grid may be refined around the wells or other points in which a large gradient in head is expected and the grid cells outside the model domain are inactivated); and (3b) the GIS objects and attribute data of the specificatory knowledge bases of the first step above are overlaid on the conceptual model and all stresses (wells, rivers, drains, heads, etc.), recharge, and hydraulic conductivity zones are inherited by the grid cells in the appropriate format for use as inputs in BME codes. In the fourth step (geostatistical modeling and knowledge integration in space/time): (4a) the physical grid-based BME equations are solved within each grid cell subject to the constraints of the previous first three steps (which integrate the knowledge bases collected in these steps), thus leading to (4b) a complete characterization of the hydrogeologic variables of interest in terms of multivariate pdf, space/time maps, *etc.* Finally, in the fifth step (graphical visualization of the results), the solutions (hydraulic heads, contaminant concentrations, *etc.*) are plotted in the form of space/time maps. □

By way of summary, we may conclude that modern geostatisticians and GIS analysts can benefit a great deal from mutual interaction. In addition to the powerful logical concepts and sophisticated mathematical techniques of modern spatiotemporal geostatistics, the methodology can provide a valuable guide to the new realities of GIS, some of which have been described above. GIS, in turn, offers a highly efficient network of computerized technologies for representing and visualizing space/time data. Therefore, we should expect that a mathematically rigorous and physically meaningful integration of the theoretical and computational models of modern spatiotemporal geostatistics with the versatility and flexibility of GIS technologies will be a main topic of future research in our field.

13

A CALL NOT TO ARMS, BUT TO RESEARCH

"What an interior strength a man can summon if he devotes himself entirely to knowledge and creation, rather than to a vain search for honors and celebrity! What a lesson!" B. Russell for G. Frege

Unification and Distinction

A careful study of the developments in geostatistics reveals that the latter continuously undergoes two major processes of growth and advancement:

1. A *unification* process with respect to its methods.
2. A *distinction* process with respect to its mathematical structure and the way it relates to experience.

These two processes may seem to follow opposite paths, but in essence they complement each other, and are equally important for the growth of the field.

As the methods of modern spatiotemporal geostatistics lay claim to broader domains of space/time data analysis, they also become increasingly unified. For example, as we saw in Chapter 12, generalized random fields unite previously separate classes of fractal and wavelet random fields; also, BME mapping incorporates the interpolation methods of traditional geostatistics and spatial statistics as special cases. At the basis of the unified framework lies a physical geometry in which space and time constitute an integrated whole. The choice of an appropriate geometry depends on the local properties of space/time and on the physical constraints imposed by the phenomenon of interest (Chapter 2). Within the unified framework, useful classifications of the geostatistical techniques are possible, which depend on the physical knowledge sources processed, the information functions assumed, the conditionalization rules used, *etc.* (Chapters 3–8).

Considered as a theory of physical model-based estimation, geostatistical mapping has two distinct parts: a *formal* part and an *interpretive* part.

The formal part involves mathematical tools and logical rules on how to use them. This part is independent of empirical data and natural language. The interpretive part, on the other hand, provides meaningful justifications for the mathematical assumptions made in the formal part, connects the formal part to experience, and relies on a natural language which guides experimentation and involves physically testable statements. Each of these two parts of scientific inquiry has its own challenging problems and intriguing research questions to deal with. Some geostatisticians may chose to concentrate their efforts on formal geostatistics, whereas some others may find it more profitable to focus on the interpretive approach. Both groups should serve the field of geostatistics well. Below we continue our discussion of the two fundamental components of geostatistical research and development.

The Formal Part

From the formal point of view, the concentration is on the purely mathematical structure of the space/time mapping approach. The main steps of this viewpoint were discussed in Comment 6.3 (p. 134) and elsewhere in this volume. Certain assumptions are made regarding the form of the general knowledge-based probability distribution, the space/time estimation equations are formulated and solved, and specificatory knowledge-based maps are derived from a set of logical rules involving mathematical definitions, theorems, and proofs (Chapters 9–11). The formal approach is a rigorous generalization that is based on an internally consistent structure. Although the formal part of BME differs profoundly from the formal part of MMSE mapping, it nevertheless contains MMSE mapping as a limiting case, valid under specific conditions (Chapter 12). BME avoids several limitations of spatial statistics. As Stein (1999, p. 2) notes, "In practice, specifying the law of a random field can be a daunting task. . .calculating this conditional distribution may be extremely difficult. For these reasons, it is common to restrict attention to linear predictors." The formal structure of BME, however, makes it possible to derive such probability distributions in a way that guarantees consistency with physical knowledge (Chapters 5 and 6). Moreover, there is no need to restrict spatiotemporal modeling to linear predictors—nonlinear BME predictors that are more informative than the linear ones can be considered (Chapter 7). In certain cases, standard practice in spatial statistics based on the empirical covariance or variogram can be seriously flawed (see Stein, 1999, p. 221). BME mapping can overcome these kinds of problems by incorporating the covariance or variogram in terms of physical laws (see Comment 3.3 on p. 81 and Comment 5.4 on p. 110 in this volume).

The formal part presents the ambitious geostatistician with a series of challenging research problems (existence and uniqueness of the solutions of integrodifferential equations, formulation of logical conditionalization rules, permissibility of space/time correlation functions, calculation of multiple posterior integrals, *etc.*). The solution of these problems many times leads to the development of powerful means of calculation—analytical and numerical (diagrammatic representations, perturbation expansions, Monte Carlo simulations, *etc.*) The physical meaning of the mathematical terms, however, is not among the issues that are discussed in this part. For example, from the viewpoint of formal geostatistics, the shape of the general knowledge-based pdf is a mathematical assumption that is made on the basis of internal consistency rules and can be justified in terms of the predictive maps to which it leads.

Interpretive BME and the Search for "Rosebud"

Interpretive issues are relevant when we need to establish relationships between the natural world in which we use the pdf of the physical maps, and the formal mathematics which describe them, *i.e.*, to measure and test formal structures or to justify certain methodological steps of the mapping procedure. For example, if explanation in terms of epistemic ideals is an issue for a study under consideration, the principle of maximum expected information (or maximum entropy; Chapter 5) can be used to justify the association of a particular pdf with the physical map; other rules may be needed to translate mathematical expressions into testable statements, *etc.* Therefore, the interpretive part examines carefully the physical content and scientific substance of the geostatistical models.

Interpretation is an important component of applied stochastic analysis, in general. While probability theory and statistics establish the mathematical properties of stochastic concepts and tools, they do not tell us how to measure, interpret, or derive them from physical data, laws, and theories. A physically meaningful interpretation of probability, *e.g.*, cannot be obtained by means of statistical arguments, but rather by establishment of relationships between the natural world in which we use probabilities and the stochastic mathematics which describe them. This important fact which is at the heart of many physical scientists' criticism of statistical approaches has been acknowledged by a number of statisticians, as well. Dempster and Wang, among others, have suggested (see Wang, 1993, p. 87) that "although statistical model-building makes use of formal probability calculations, the probabilities usually have no sharply defined interpretation, so the whole model-building process is really a form of exploratory analysis." One may also emphasize the fundamental differences—in substance and scope—between statistical and scientific hypotheses (*e.g.*, while the former usually focus on a distinct feature of a specified population, the latter involve a deeper understanding of the underlying physical mechanisms).

The interpretive part of BME involves both ontological and epistemic investigations. We will attempt a comparative discussion of these investigations with the help of the masterpiece movie *Citizen Kane*. In this film, Kane's famous last word was "rosebud." This word gave rise to an intriguing search by reporters, movie historians, *etc.* regarding two central questions: "What is 'rosebud'?" and "What does 'rosebud' mean?" The film's narrative thus has two sides, each reflecting one of the two questions. While one concerns itself with facts, the other turns toward the meaning of Kane's life. Taking the viewpoint of scientific reasoning as we have considered it in this book, the first concern is ontological in character (*i.e.*, somewhere in the world exists an objective thing called "rosebud"), while the second is epistemological (*i.e.*, the word "rosebud" had a special meaning in the context of Mr. Kane's life). Hence, as the narrative unfolds, an attempt is made to account both for the facts and for the meaning behind the facts.

Table 13.1. Double perspectives.

Citizen Kane	BME
"Rosebud"	"Knowledge bases"
Role of "rosebud" in Kane's life	Role of "knowledge bases" in geostatistician's conceptual framework of spatiotemporal analysis and mapping

The situation with the narrative of *Citizen Kane* is analogous, *mutatis mutandis*, to the double perspective of BME (Table 13.1). The procedure leading to a meaningful and informative map involves (1) ontological investigation with questions such as "What are the objectively available knowledge bases (facts, data, laws, *etc.*)?"; and (2) epistemic investigation that seeks the answers to questions like, "What is the meaning or role of these bases in the context of scientific map building and interpretation?." Investigations (1) and (2) are closely related. To continue our metaphor, this double perspective, for both the film's narrative and the BME paradigm, expands the study domain to include the observed ("rosebud" and natural processes, respectively) as well as the observer (the examiner of Kane's existence and the geostatistician, respectively).

The Argument of Modern Spatiotemporal Geostatistics

The burden of the argument underlying the development of modern geostatistics is that geostatisticians should always try to see their work critically in the light of well-established scientific methods and sound epistemic arguments. In such a context, modern spatiotemporal geostatistics is viewed as a field of concepts, theories, and methods whose initial conditions are the knowledge bases that are currently available. The field involves depth, originality, and excellence—factors that have been and should ever be present in the evaluation of a scientific approach. But somewhere in the mass-production of recipes and provisional tricks, these requisites have been lost.

Figure 13.1. The world of the future according to the cookbook enthusiasts (from Asimov, 1986).

It has been said that if cookbook enthusiasts had their way (Fig. 13.1), they would flatten the natural hierarchy of the human mind until people could not tell the difference between the telephone directory and Homer's *Odyssey*. It can be a matter of intellectual survival. Conrad, a character in Tom Wolfe's novel, *A Man in Full* (Wolfe, 1999), was able to survive because he got a book of writings from the Stoics and critically examined their ideas. Therefore, if we wish to reclaim the true art of critical thinking from such a crippling situation, we have no other choice than to act immediately. BME analysis is but one attempt to build a framework for modern spatiotemporal geostatistics—perhaps not the best one; other frameworks already exist and in the future, perhaps, more will be created.

The important issue is that there should always be room for critical thinking. The solution to the problems faced by today's geostatistician is not a cookbook, for the very reason that reliance on such a shortcut does not promote critical thinking. Instead, the emphasis on quick and provisional fixes prevents any real development of understanding. What is needed is a larger intellectual perspective that offers a deeper theoretical comprehension of the issues, while taking into consideration more forms of physical knowledge. This is a perspective that must explicitly involve a *theory of knowledge*. A spatiotemporal map, *e.g.*, should depend on what we know about the natural phenomena it represents, as well as how we know it (*i.e.*, the critical-thinking operations through which we collect and process physical knowledge). There is no doubt in my mind that the modern geostatistician will find it more profitable in the long run to develop a sound theoretical background, rather than to rely on collecting recipes and techniques that may soon be obsolete.

Looking back on some of the important geostatistical concepts, many of them seem simple to us now. Indeed, introducing kriging into spatial estimation was brilliant, but in retrospect it is a simple concept. The same is true for conditional simulation, *etc.* So, some geostatisticians may despair, reckoning that all the good ideas in geostatistics have been discovered and that our only task is to fill in small gaps here and there. This assessment, however, is not true. The fact is, there are always opportunities around the corner; the BME approach is a case in point. As it has been developed so far, BME is the joint product of theoretical and practical reasoning by which certain of the well-documented theoretical limitations of the previous geostatistical methods can be rigorously and efficiently eliminated. The basic BME equations possess significant generalization power that takes into account a wide range of knowledge sources that could not be considered using older methods. The problem-solving power of BME comes not only from the mathematical formalisms and inference schemes it utilizes, but from the logical manner in which it processes the extended knowledge it is able to incorporate. As a result, the spatiotemporal maps obtained offer a body of information as well as a point of view.

As the domain of modern geostatistics continues to expand in search of new conquests, the return to its foundations will continue, each of the two processes nourishing the other. It has become clear that a definite epistemic outlook is a necessity for developing modern spatiotemporal geostatistics, one which takes into consideration both the internal and external aspects of obtaining and ordering physical knowledge. It is one thing to acquire knowledge bases, and quite another to organize the various knowledge bases in an appropriate manner so that when taken all together they form a realistic picture of the phenomenon under investigation. This new outlook is at the core of BME analysis, supplanting fallacies that have barred progress and becoming the direct cause of far-reaching advances. Modern spatiotemporal geostatistics requires blending sufficient skill and depth in stochastic theory and techniques with substantive knowledge and scientific content. Mathematics and statistics, because of their ability to account for structure as well as randomness, provide rigorous representations of spatiotemporal variations and play an important role in modern spatiotemporal geostatistics. The natural sciences also offer important sources of knowledge that can significantly improve the quality of the map and its scientific interpretation. Last but not least, much depends upon the insight with which ideas and data are handled before they reach the stage of mathematics. By making practical application of the epistemic postulates and conclusions, we subject them to the same sort of observational testing and control that physical models and assumptions undergo.

The above remarks may be viewed as a "call to research"—an appeal for establishment of a multidisciplinary conception of modern geostatistics aimed at novel ideas and models that consider the advances of numerous scientific disciplines in which geostatistical methods can be applied. Such a conception of modern spatiotemporal geostatistics should take advantage of the striking phenomenon of convergence in science and research nowadays. New, highly interdisciplinary subjects are emerging, the result of the realization on the part of many researchers that the problems they are confronting are shared by other researchers in disparate fields. In many cases these interdisciplinary subjects represent the frontiers of research. Like all other modern geostatistical approaches, BME at the onset poses a series of difficult problems, both theoretical and computational. Some of these problems have been mentioned in this book. Many others exist. Solving these difficulties is the task of the new class of geostatisticians who realize the importance of the foregoing considerations and who will not hesitate to confront the entrenched old guard in order to defend the bright promise of modern spatiotemporal geostatistics. The work of the modern geostatisticians and their struggle for intellectual and social acceptance will not be easy.

The Ending as a New Beginning

In order for a book of the present kind to be successful, its ending must be a new beginning. One should expect that the situation with modern spatiotemporal geostatistics in general, and BME in particular, will follow the well-established path of scientific paradigms. Most active members of the geostatistics community will have to agree on such an approach. Once such a consensus has occurred, a period will follow during which the features and implications of the modern geostatistics paradigm will be studied and its boundary conditions—*i.e.*, the limits of its applicability—will be determined and appreciated. In the process, we should not underestimate the importance of bringing together geostatisticians working in different scientific fields and providing the community of interested researchers with an interdisciplinary forum for the exchange of ideas, communication of issues that cut across disciplinary barriers, and dissemination of stochastic techniques used in the various fields.

Just as the intensity of drama feeds upon conflict, so any intellectual pursuit thrives on unresolved problems. Therefore, in addition to promoting new ideas and approaches, modern spatiotemporal geostatistics must pose challenging problems to attract new talent and ensure the continuing vitality of the field. Without the challenge of the unknown, the novel, and the exciting, any field of endeavor grows stale and begins to decline, having ceased to attract outstanding individuals. Furthermore, it should be recognized that geostatisticians may not only feel intellectually stifled, but suffer intense emotional and existential doubt, as well (the same can be said for any research scientist, in general). Some may feel that they are impoverishing their lives by dealing with the same old kinds of problems, routinely applying the same old concepts and models, and not exercising their ability to think about challenging problems or trying new avenues.

As I write the last lines of this book, the 20th century is drawing rapidly to a close. It is time for looking back to past successes and failures, assessing the current state of geostatistics, and making plans that will assure the continuing vitality of the field in the new century. I would like, therefore, to conclude this book by reiterating the view that the current situation in geostatistics requires that we take a few steps along the path of creativity and innovation. We should not be afraid to take these steps, even if this path is strewn with the bones of those consumed by the vultures of conservatism, mediocrity, and myopia. After all, life is an effort that deserves a better cause!

As far as BME is concerned, it is not conducive to its real strength that it should occupy the field of modern geostatistics alone. The intellectual pressure which creative criticism exerts will force BME proponents to examine their positions for their weaknesses and complacencies. Indeed, some of the views that have been presented in this book may well be modified by others as time progresses. The words of a 19th century English philosopher and economist seem most appropriate at this point:

> Lord, enlighten our critics, sharpen their wits, give acuteness to their perceptions and consecutiveness and clearness to their reasoning powers. We are in danger from their folly, not from their wisdom: their weakness is what fills us with apprehension, not their strength.
>
> *John Stuart Mill*

BIBLIOGRAPHY

"The fallacy does not consist in believing that in the struggle for existence the fittest to survive eliminates the less fit. The fact is obvious and stares us in the face. The fallacy is the belief that fitness for survival is identical with the best exemplification of the Art of Life." A.N. Whitehead

Aczel, J. and Z. Daroczy, 1975. *On Measures of Information and Their Characterization.* Academic Press, New York, NY.

Agterberg, F.P., 1974. *Geomathematics.* Elsevier Scientific Publ., Amsterdam, The Netherlands.

Anderson, J.R., 1985. *Psychology and Its Implications.* W.H. Freeman, New York, NY.

Anderson K.R., E.L. Avol, S.A. Edwards, D.A. Shamoo, R.C. Peng, W.S. Linn, and J.D. Hacknery, 1992. Controlled exposures of volunteers to respirable carbon and sulfuric acid aerosols. *Jour. Air and Waste Management Assoc.,* v. 42, p. 771.

Arlinghaus, S.L. (Ed.), 1996. *Practical Book of Spatial Statistics.* CRC Press, New York, NY.

Arrow, K.J. and H. Raynaud, 1986. *Social Choice and Multicriterion Decision-Making.* The MIT Press, Cambridge, MA.

Asimov, I., 1986. *A Nineteenth-Century Vision of the Year 2000.* Henry Holt & Co., New York, NY.

Aucott, W.R. and N.C. Myers, 1998. Changes in groundwater levels and storage in the Wichita well field area, south-central Kansas, 1940–98. *Water Resources Investigation,* U.S. Geological Survey/City of Wichita, KS, Rept. 98–4141, Denver, CO.

Bagrow, L., 1985. *History of Cartography, 2nd ed.* Precedent Publ., Chicago, IL.

Baldi, P. and S. Brunak, 1998. *Bioinformatics.* The MIT Press, Cambridge, MA.

Bardossy, A., U. Haberlandt, and J. Grimm-Strele, 1997. Interpolation of ground water quality parameters using additional information, *in* A. Soares, J. Gomez-Hernandez, and R. Froidevaux (Eds.), *geoENV I—Geostatistics for Environmental Applications.* Kluwer Acad. Publ., Dordrecht, The Netherlands, p. 189–200.

Batista, A.C., A.J. Sousa, M.J. Batista, and L. Viegas, 1997. Factorial kriging with external drift: A case study on the Penedono region, Portugal, *in* V. Pawlowsky-Glahn (Ed.), *Proceedings of IAMG'97—Third Annual Conference of the International Association for Mathematical Geology, Vol. 2.*

Intl. Ctr. for Numerical Methods in Engr. (CIMNE), Barcelona, Spain, p. 687–692.

Baveye, P. and G. Sposito, 1984. The operational significance of the continuum hypothesis in the theory of water movement through soils and aquifers. *Water Resour. Res.*, v. 20, no. 5, p. 521–530.

Bear, J., 1972. *Dynamics of Fluids in Porous Media.* Dover, New York, NY.

Bevington, P.R. and D.K. Robinson, 1992. *Data Reduction and Error Analysis for the Physical Sciences, 2nd ed.* McGraw-Hill, Boston, MA.

Bilonick, R.A., 1985. The space-time distribution of sulfate deposition in the northeastern U.S. *Atmospheric Environ.*, v. 19, no. 11, p. 2513–2524.

Birkin, M., G. Clarke, M. Clarke, and A. Wilson, 1996. *Intelligent GIS.* GeoInformation International, Cambridge, UK.

Blot, W.J. and J.K. McLaughlin, 1995. Geographic patterns of breast cancer among American women. *Jour. Nat. Cancer Inst.*, v. 87, no. 24, p. 1819–1820.

Bogaert, P. and G. Christakos, 1997. Spatiotemporal analysis and processing of thermometric data over Belgium. *Jour. Geophys. Research—Atmospheres*, v. 102, no. D22, p. 25,831–25,846.

Bogaert, P., M. Serre, and G. Christakos, 1999. Efficient computational BME analysis of non-Gaussian data in terms of transformation functions, *in* S.J. Lippard, A. Næss, and R. Sinding-Larsen (Eds.), *Proceedings of IAMG'99—Fifth Annual Conference of the International Association for Mathematical Geology, Vol. 1.* Tapir, Trondheim, Norway, p. 57–62.

Boltzmann, L., 1964 [1896–98]. *Vorlesungen über Gastheorie* [English transl: *Lectures on Gas Theory*]. Univ. California Press, Berkeley, CA.

Bonham-Carter, G.F., 1994. *Geographic Information Systems for Geoscientists.* Pergamon, New York, NY.

Briggs, D.J. and P. Elliott, 1995. The use of geographical information systems in studies on environment and health. *World Health Statist. Quart.*, v. 48, p. 85–94.

Bullock, A., O. Stallybrass, and S. Trombley (Eds.), 1977. *The Fontana Dictionary of Modern Thought.* Fontana Press, London, UK.

Burris, S.N., 1998. *Logic for Mathematics and Computer Science.* Prentice Hall, Upper Saddle River, NJ.

Cambanis, S. and C. Houdre, 1995. On the continuous wavelet transform of second-order random processes. *IEEE Trans. on Information Theory*, v. 41, no. 3, p. 628–642.

Carnap, R., 1966. *Philosophical Foundations of Physics.* Basic Books, New York, NY.

Casado, L.S., S. Rouhani, C.A. Cardelino, and A.J. Ferrier, 1994. Geostatistical analysis and visualization of hourly ozone data. *Atmospheric Environ.*, v. 28, no. 12, p. 2105–2118.

Cavalli-Sforza, L.L., P. Menozzi, and A. Piazza, 1994. *The History and Geography of Human Genes.* Princeton Univ. Press, Princeton, NJ.

Chalmers, A.F., 1994. *What is this Thing called Science?* Hackett Publ. Co. Indianapolis, IN.

Chandru, V. and J.N. Hooker, 1999. *Optimization Methods for Logical Inference.* J. Wiley & Sons, New York, NY.

Chavel, I., 1995. *Riemannian Geometry: A Modern Introduction.* Cambridge Univ. Press, New York, NY.

Choi, K-M., G. Christakos, and M.L. Serre, 1998. Recent developments in vectorial and multi-point Bayesian maximum entropy analysis, *in* A. Buccianti, G. Nardi, and R. Potenza (Eds.), *Proceedings of IAMG'98—Fourth Annual Conference of the International Association for Mathematical Geology, Vol. 1.* De Frede Editore, Naples, Italy, p. 91–96.

Chrisman, N.R., 1983. The role of quality information in the long-term functioning of a geographic information system, *in* Steering Committee of the 6th International Symposium on Automated Cartography, *Proceedings of AutoCarto 6, Vol. 2.* Ottawa, Canada, p. 303–321.

Christakos, G., 1990. A Bayesian/maximum-entropy view to the spatial estimation problem. *Math. Geology,* v. 22, no. 7, p. 763–776.

Christakos, G., 1991a. Some applications of the Bayesian, maximum entropy concept in Geostatistics. *Fundamental Theories of Physics,* [Invited paper]. Kluwer Acad. Publ., Dordrecht, The Netherlands, p. 215–229.

Christakos, G., 1991b. A theory of spatiotemporal random fields and its application to space-time data processing. *IEEE Trans., Systems, Man & Cybernetics,* v. 21, no. 4, p. 861–875.

Christakos, G., 1992. *Random Field Models in Earth Sciences.* Academic Press, San Diego, CA.

Christakos, G., 1998a. Spatiotemporal information systems in soil and environmental sciences. *Geoderma,* v. 85, no. 2–3, p. 141–179.

Christakos, G., 1998b. Multi-point BME space/time mapping of environmental variables, *in* V.N. Burganos, G.P. Karatzas, A.C. Payatakes, W.G. Gray, and G.F. Pinder (Eds.), *Computational Methods in Water Resources XII, Vol. 2—Computational Methods in Surface and Groundwater Transport.* Computational Mechanics Publ., Southampton, UK, p. 289-296.

Christakos, G., 1998c. While God is raining brains, are we holding umbrellas? The role of Modern Geostatistics in spatiotemporal analysis and mapping—Keynote lecture, *in* A. Buccianti, G. Nardi, and R. Potenza (Eds.), *Proceedings of IAMG'98—Fourth Annual Conference of the International Association for Mathematical Geology, Vol. 1.* De Frede Editore, Naples, Italy, p. 33–53.

Christakos, G. and D.T. Hristopulos, 1998. *Spatiotemporal Environmental Health Modelling: A Tractatus Stochasticus.* Kluwer Acad. Publ., Dordrecht, The Netherlands.

Christakos, G., D.T. Hristopulos, and P. Bogaert, 2000a. On the physical geometry hypothesis at the basis of spatiotemporal analysis of hydrologic geostatistics. *Advanc. Water Resour.,* in press.

Christakos, G., D.T. Hristopulos, and A. Kolovos, 2000b. Stochastic flowpath analysis of multiphase flow in random porous media. *SIAM—Applied Mathematics,* in press.

Christakos, G., D.T. Hristopulos, and M. Serre, 1999. BME studies of stochastic differential equations representing physical laws—Part I, *in* S.J. Lippard, A. Næss, and R. Sinding-Larsen (Eds.), *Proceedings of IAMG'99—Fifth Annual Conference of the International Association for Mathematical Geology, Vol. 1.* Tapir, Trondheim, Norway, p. 63–68.

Christakos, G. and B.R. Killam, 1993. Sampling design for classifying contaminant level using annealing search algorithms. *Water Resour. Res.,* v. 29, no. 12, p. 4063–4076.

Christakos, G. and A. Kolovos, 1999. A study of the spatiotemporal health impacts of ozone exposure. *Jour. Exposure Analysis & Environ. Epidemiol.,* v. 9, no. 4, p. 322–335.

Christakos, G. and X. Li, 1998. Bayesian maximum entropy analysis and mapping: A farewell to kriging estimators? *Math. Geology,* v. 30, no. 4, p. 435–462.

Christakos, G. and V. Papanicolaou, 2000. Norm-dependent covariance permissibility of weakly homogeneous spatial fields. *Annals of Statistics,* submitted.

Christakos, G. and M.L. Serre, 2000a. A spatiotemporal study of exposure–health effect associations. *Jour. Exposure Analysis & Environ. Epidemiol.,* v. 10, no. 2, p. 168–187.

Christakos, G. and M.L. Serre, 2000b. BME Analysis of spatiotemporal particulate matter distributions in North Carolina. *Atmospheric Environ.,* in press.

Christakos, G. and V. Vyas, 1998. A composite spatiotemporal study of ozone distribution over eastern United States. *Atmospheric Environ.*, v. 32, no. 16, p. 2845–2857.

Christophersen, O., 1997. Mortality during the 1996/7 winter. *Population Trends*, v. 90, p. 11–17.

Clarke, K.C., 1986. Advances in geographic information systems. *Computers, Environ. and Urban Systems*, v. 10, no. 3/4, p. 175–184.

Cramer, H. and M.R. Leadbetter, 1967. *Stationary and Related Stochastic Processes.* J. Wiley & Sons, New York, NY.

Cressie N., 1991. *Statistics for Spatial Data.* J. Wiley & Sons, New York, NY.

Cushman, J.H., 1984. On unifying the concepts of scale, instrumentation and stochastics in the development of multiphase transport theory. *Water Resour. Res.*, v. 20, no. 11, p. 1668–1676.

Dab, W., S. Medina, P. Quenel, Y. Le Moullec, A. Le Tertre, B. Thelot, C. Monteil, P. Lameloise, P. Pirard, I. Momas, R. Ferry, and B. Festy, 1996. Short term respiratory health effects of ambient air pollution: Results of the APHEA project in Paris. *Jour. Epidemiol. Community Health*, v. 50, p. S42–S46.

Dagan, G., 1989. *Flow and Transport in Porous Formations.* Springer-Verlag, Berlin, Germany.

Daubechies, I., 1992. *Ten Lectures on Wavelets.* SIAM, Philadelphia, PA.

David, M., 1977. *Geostatistical Ore Reserve Estimation.* Elsevier Scientific Publ., Amsterdam, The Netherlands.

Davies, S., K.J. Packer, A. Baruya, and A.I. Grant, 1991. Enhanced information recovery in spectroscopy, *in* B. Buck and V.A. Macaulay (Eds.), *Maximum Entropy in Action.* Oxford Science Publ., Oxford, UK.

Davis, J.C., 1986. *Statistics and Data Analysis in Geology, 2nd ed.* J. Wiley & Sons, New York, NY.

DoD-GMS, 1997. *Groundwater Modelling System.* Hydraulics Laboratory, U.S. Army Engr. Waterways Experiment Station, Department of Defense.

Deutsch, C.V. and A.G. Journel, 1992. *Geostatistical Software Library and User's Guide.* Oxford Univ. Press, New York, NY.

D'haeseleer, W.D., W.N.G. Hitchon, J.D. Callen, and J.L. Shohet, 1991. *Flux Coordinates and Magnetic Field Structure.* Springer-Verlag, New York, NY.

Dietrich, C.R. and G.N. Newsam, 1989. A stability analysis of the geostatistical approach to aquifer transmissivity identification. *Stochastic Hydrology & Hydraulics*, v. 3, p. 293–316.

Dirac, P., 1926–27. The physical interpretation of the quantum mechanics. *Proc. Royal Soc.*, London, Vol. A, p. 113, 621–641.

Dockery, D.W. and C.A. Pope, III, 1994. Acute respiratory effects of particulate air pollution. *Annual Review of Public Health*, v. 15, p. 107–132.

D'Or, D., 1999. Application of Bayesian maximum entropy in soil sciences. *Research Report, Unité de Biometrie et Analyse des Donnees*, Université Catholique de Louvain, Faculté des Sciénces Agronomiques, Louvain-la-Neuve, Belgium.

Doveton, J.H., 1986. *Log Analysis of Subsurface Geology: Concepts and Computer Methods.* Wiley-Interscience, New York, NY.

Dowd, P.A., 1992. A review of recent developments in geostatistics. *Computers & Geosciences*, v. 17, no. 10, p. 1481–1500.

Dunbar, R., 1996. *The Trouble with Science.* Faber and Faber Ltd., London, UK.

Dziewonski, A.M. and J.H. Woodhouse, 1987. Global images of the earth's interior. *Science*, v. 236, p. 37–48.

Ebanks, B., P. Sahoo, and W. Sander, 1998. *Characterization of Information Measures.* World Scientific, Singapore.

Eddington, A., 1959. *New Pathways in Science.* Ann Arbor Paperbacks, Univ. Michigan Press, Ann Arbor, MI.

Eerens, H., C.J. Sliggers, and K.D. van den Hout, 1993. The CAR model: The Dutch method to determine city street air quality. *Atmospheric Environ.*, v. 27B, no. 4, p. 389–399.

Einstein, A., 1994 [1954]. *Ideas and Opinions. Introduction* by Alan Lightman, Modern Library, New York, NY.

Enger, E. and B. Smith, 1995. *Environmental Science: A Study of Interrelationships, 5th ed.* Wm. C. Brown Publishers, Dubuque, Iowa.

Ewing, G.M., 1969. *Calculus of Variations with Applications.* W.W. Norton, New York, NY.

Faber, R.L., 1983. *Foundations of Euclidean and Non-Euclidean Geometry.* Marcel Dekker, New York, NY.

Feder, J., 1988. *Fractals.* Plenum Press, New York, NY.

Fisher R., M. Mayer, W. von der Linden, and V. Dose, 1997. Enhancement of the energy resolution in ion-beam experiments with the maximum entropy method. *Phys. Rev. E,* v. 55, no. 6, p. 6667–6673.

Flanders, W.D. and M.J. Khoury, 1990. Indirect assessment of confounding: Graphical description and limits on effect of adjusting for covariates. *Epidemiology,* v. 1, p. 239–246.

Foster, K.R., D.E. Bernstein, and P.W. Huber (Eds.), 1993. *Phantom Risk: Scientific Inference and the Law.* The MIT Press, Cambridge, MA.

Foster, K.R. and P.W. Huber, 1999. *Judging Science. Scientific Knowledge and the Federal Courts.* The MIT Press, Cambridge, MA.

Friedman, M., 1983. *Foundations of Space-Time Theories.* Princeton Univ. Press, Princeton, NJ.

Gandin, L.S., 1963. *Objective Analysis of Meteorological Fields.* Gidrometeorolog. Izdat., Leningrad, USSR [English transl., Israel Program of Scient. Transl., Jerusalem, Israel, 1965].

Gel'fand, I.M., 1955. Generalized Random Processes. *Dok. Akad. Nauk. SSSR,* v. 100, p. 853–856.

Gel'fand, I.M. and G.E. Shilov, 1964. *Generalized Functions, Vol. I.* Academic Press, New York, NY.

Gel'fand, I.M. and N. Ya. Vilenkin, 1964. *Generalized Random Functions, Vol. IV.* Academic Press, New York, NY.

Genz, A., 1992. Numerical computation of multivariate normal probabilities. *Jour. Statistical Computation and Simulations,* v. 2, p. 141–150.

Glattre, E., 1989. Atlas of cancer incidence in Norway 1970–1979. *Recent Results in Cancer Research,* v. 114, p. 216–226.

Gneiting, T., 1999. Radial positive definite functions generated by Euclid's hat. *Jour. of Multivariate Anal.,* v. 69, p. 88–119.

Golden, A., 1997. *Memoirs of a Geisha.* Vintage Books, New York, NY.

Goldman, A.I., 1986. *Epistemology and Cognition.* Harvard Univ. Press, Cambridge, MA.

Goovaerts, P., 1997. *Geostatistics for Natural Resources Evaluation.* Oxford Univ. Press, New York, NY.

Gregory, R.L., 1990. *Eye and Brain. The Psychology of Seeing, 4th ed.* Princeton Univ. Press, Princeton, NJ.

Hall, S.S., 1992. *Mapping the Next Millenium.* Random House, New York, NY.

Harley, R., G. Cass, J.H. Seinfeld, L. McNair, A. Russell, and G. McRae, 1992. Application of the CIT photochemical airshed model to SCAQS Data Base 1. *Southern California Air Quality Study Data Conference,* Los Angeles, CA.

Harre, R., 1989. *The Philosophies of Science.* Oxford Univ. Press, New York, NY.

Harris, E.E., 1996. *Hypothesis and Perception—The Roots of the Scientific Method.* Humanities Press Intl., Atlantic Highlands, NJ.

Heath, T.L., 1956 [1908]. *The Thirteen Books of Euclid's Elements.* Translated from the text of Heiberg, *Introduction* and *Commentary* by Sir Thomas L. Heath, Dover, New York, NY [unabridged, unaltered repub. of 2nd ed. of 1926].

Heine, V., 1955. Models for two-dimensional stationary stochastic processes. *Biometrika,* v. 42, p. 170–178.

Hill, A.B., 1965. The environment and disease: Association or causation? *Proc. Royal Soc. Med.,* v. 58, p. 295–300.

Hoel, D.G. and P.J. Landrigan, 1987. Comprehensive evaluation of humans' data, *in* R.G. Tardiff and J.V. Rodricks (Eds.), *Toxic Substances and Human Risk—Principles of Data Interpretation.* Plenum Press, New York, NY, p. 121–130.

Honda Y., M. Ono, A. Sasaki, and I. Uchiyama, 1995. Relationship between daily high temperature and mortality in Kyushu, Japan [in Japanese]. *Japanese Jour. Public Health,* v. 42, no. 4, p. 260–8.

Howson, C. and P. Urbach, 1993. *Scientific Reasoning: The Bayesian Approach.* Open Court Publ. Co., Chicago and La Salle, IL.

Hristopulos, D.T. and G. Christakos, 2000. Calculations of non-Gaussian multivariate moments for BME analysis. *Math. Geology,* submitted.

Itô, K., 1954. Stationary random distributions. *Univ. Kyoto, Memoirs College Sci.,* Series A, v. 3, no. 28, p. 209–223.

Itô, K. and G.D. Thurston, 1996. Daily PM_{10} mortality associations: An investigation of at-risk subpopulations. *Jour. Exposure Analysis & Environ. Epidemiol.,* v. 6, p. 79–95.

Janssen, L.H.J.M., E. Buringh, A. van der Meulen, and K.D. van den Hout, 1999. A method to estimate the distribution of various fractions of PM_{10} in ambient air in The Netherlands. *Atmospheric Environ.,* v. 33, p. 3325–3334.

Jaynes, E.T., 1983. Papers on probability, statistics and statistical physics, *in* R.D. Rosenkrantz (Ed.), *Synthese Library, Vol. 158.* Reidel, Dordrecht, The Netherlands.

Johnson, D.H. and D.E. Dudgeon, 1993. *Array Signal Processing.* Prentice Hall, Englewood Cliffs, NJ.

Jones, J.A.A., 1997. *Global Hydrology.* Longman, Essex, UK.

Jones, R.H. and Y. Zhang, 1997. Models for continuous stationary space-time processes, *in* T.G. Gregoire, D.R. Brillinger, P.J. Diggle, E. Russek-Cohen, W.G. Warren, and R.D. Wolfinger (Eds.), *Modelling Longitudinal and Spatially Correlated Data.* Springer-Verlag, New York, NY, p. 289–298.

Journel, A.G., 1974. Geostatistics for conditional simulation of orebodies. *Economic Geology,* v. 69, p. 673–687.

Journel, A.G., 1983. Non-parametric estimation of spatial distributions. *Math. Geology,* v. 15, no. 3, p. 445–468.

Journel, A.G., 1986. Constrained interpolation and qualitative information: The soft kriging approach. *Math. Geology,* v. 18, no. 3, p. 269–286.

Journel, A.G., 1989. *Fundamentals of Geostatistics in Five Lessons.* American Geophys. Union, Washington, DC.

Journel, A.G. and C.J. Huijbregts, 1978. *Mining Geostatistics.* Academic Press, London, UK.

Jumarie, G., 1990. *Relative Information.* Springer-Verlag, New York, NY.

Khoury, M.J. and Q. Yang, 1998. The future of genetic studies of complex human diseases: An epidemiologic perspective. *Epidemiology,* v. 9, no. 3, p. 350–354.

Kitanidis, P.K., 1986. Parameter uncertainty in estimation of spatial functions: Bayesian analysis. *Water Resour. Res.,* v. 22, p. 449–507.

Kitanidis, P.K., 1997. *Introduction to Geostatistics.* Cambridge Univ. Press, Cambridge, UK.

Kitanidis, P.K. and E.G. Vomvoris, 1983. A geostatistical approach to the inverse problem in groundwater modeling (steady state) and one-dimensional simulations. *Water Resour. Res.*, v. 19, no. 3, p. 677–690.

Klir, G. and B. Yuan, 1995. *Fuzzy Sets and Fuzzy Logic—Theory and Applications.* Prentice Hall, Upper Saddle River, NJ.

Kolmogorov, A.N., 1939. Sur l'interpolation et extrapolation des suites stationnaires. *Comptes Rendus Académie des Sciences*, Paris, v. 208, p. 2043–2045.

Kolmogorov, A.N., 1941a. The local structure of turbulence in an incompressible fluid at very large Reynolds numbers. *Dok. Akad. Nauk. SSSR,* v. 30, p. 229–303.

Kolmogorov, A.N., 1941b. Interpolation und Extrapolation von stationären zufälligen Folgen. *Izves. Akad. Nauk SSSR,* Ser. Matemat., v. 5, p. 3–14.

Kosso, P., 1998. *Appearance and Reality.* Oxford Univ. Press, New York, NY.

Kottegoda, N.T. and R. Rosso, 1997. *Statistics, Probability, and Reliability for Civil and Environmental Engineers.* McGraw-Hill, New York, NY.

Krewski, D., D. Wigle, D.B. Clayson, and G.R. Howe, 1989. Role of epidemiology in health risk assessment. *Recent Results in Cancer Research,* Springer-Verlag, New York, NY, v. 120, p. 1–24.

Laake K. and J.M. Sverre, 1996. Winter excess mortality: A comparison between Norway and England plus Wales. *Age & Ageing,* v. 25, no. 5, p. 343–348.

Lambe, T.W. and R.V. Whitman, 1969. *Soil Mechanics.* J. Wiley & Sons, New York, NY.

Langran, G., 1992. *Time in Geographic Information Systems.* Taylor & Francis, London, UK.

Lastrucci, C.L., 1967. *The Scientific Approach.* Schenkman Publ. Co., Cambridge, MA.

Laurini, R. and D. Thompson, 1995. *Fundamentals of Spatial Information Systems.* Academic Press, San Diego, CA.

Lee, Y.-M. and J.H. Ellis, 1997a. On the equivalence of kriging and maximum entropy estimators. *Math. Geology,* v. 29, no. 1, p. 131–151.

Lee, Y.-M. and J.H. Ellis, 1997b. Estimation and simulation of lognormal random fields. *Computers & Geosciences,* v. 23, no. 1, p. 19–31.

Lefohn, A.S., H.R. Knudsen, J.A. Logan, J. Simpson, and C. Bhumralkar, 1987. An evaluation of the kriging method to predict 7-hr seasonal mean ozone concentrations for estimating crop losses. *Jour. Air Pollut. Control Ass.*, v. 37, p. 595–602.

Lewis, T., 1983. *Late Night Thoughts on Listening to Mahler's Ninth Symphony.* Penguin Books, New York, NY.

Louvar, J.F. and B.D. Louvar, 1998. *Health and Environmental Risk Analysis.* Prentice Hall, Upper Saddle River, NJ.

Malczewski, J., 1999. *GIS and Multicriteria Decision Analysis.* J. Wiley & Sons, New York, NY.

Mandelbrot, B.B., 1982. *The Fractal Geometry of Nature.* W.H. Freeman, New York, NY.

Marsden, J.E. and A.J. Tromba, 1988. *Vector Calculus.* W.H. Freeman, New York, NY.

Matern, B., 1960. *Spatial Variation.* Medd. Fran. Stat. Skogsf., v. 49, no. 5, Stockholm, Sweden.

Matheron, G., 1963. Principles of geostatistics. *Economic Geology,* v. 58, p. 1246–1266.

Matheron, G., 1965. *Les Variables Régionalisées et Leur Estimation.* Masson, Paris, France.

Matheron, G., 1968. *Osnovy Prikladnoï Geostatistiki.* Mir, Moscow, USSR.

Matheron, G., 1973. The intrinsic random functions and their applications. *Adv. Appl. Prob.*, v. 5, p. 439–468.

Mellor, D.H., 1995. *The Facts of Causation.* Routledge, London, UK.

Menozzi, P., A. Piazza, and L. Cavalli-Sforza, 1978. Synthetic maps of human gene frequencies in Europeans. *Science,* v. 201 p. 786–792.

Milloy, S. and M. Gough, 1998. *Silencing Science.* Cato Institute, Washington, DC.

Milne, J.W., D.B. Roberts, S.J. Walker, and D.J. Williams, 1982. Sources of Sydney brown haze, *in* Carras and Johnson (Eds.), *The Urban Atmosphere—Sydney. A Case Study.* CSIRO, Australia.

Morgan, M.G. and M. Henrion, 1990. *Uncertainty.* Cambridge Univ. Press, New York, NY.

Nagel, E. and J.R. Newman, 1958. *Gödel's Proof.* New York Univ. Press, New York, NY.

Newton, R.G., 1997. *The Truth of Science.* Harvard Univ. Press, Cambridge, MA.

Newton-Smith, W.H., 1981. *The Rationality of Science.* Routledge & Kegan Paul, Boston, MA.

Olea, R.A., 1972. Application of regionalized variable theory to automatic contouring, *in* Univ. Kansas Center for Research, Inc./Kansas Geol. Survey, *Special Report to the American Petroleum Institute.* Research Proj. No. 131, Lawrence, KS.

Olea, R.A., 1974. Optimal contour mapping using universal kriging. *Jour. Geophys. Research,* v. 79, no. 5, p. 695–702.

Olea, R.A., 1982. Optimization of the High Plains Aquifer Observation Network, Kansas. *Groundwater Series no. 7,* Kansas Geol. Survey, Lawrence, KS.

Olea, R.A., 1999. *Geostatistics for Engineers and Earth Scientists.* Kluwer Acad. Publ., Boston, MA.

Omnés, R., 1999. *Quantum Philosophy.* Princeton Univ. Press, Princeton, NJ.

Osleeb, J.P. and S. Kahn, 1999. Integration of geographic information, *in* V.H. Dale and M.R. English (Eds.), *Tools to Aid Environmental Decision Making.* Springer-Verlag, New York, NY.

Painter, S., 1998. Numerical method for conditional simulation of Levy random fields. *Math. Geology,* v. 30, no. 2, p. 163–179.

Platt, J.R., 1964. Strong inference. *Science,* v. 146, p. 347–353.

Poincaré, H., 1952. *Science and Hypothesis.* Dover, Mineola, NY.

Popper, K.R., 1962. *The Logic of Scientific Discovery.* Hutchinson, London, UK.

Poularikas, A.D., 1996. *The Transforms and Applications.* CRC Press, Boca Raton, FL.

Quine, W.V., 1970. On the reasons for indeterminacy of translation. *Jour. Philosophy,* v. 67, p. 178–183.

Rhind, D.W. and N.P.A. Green, 1988. Design of a geographic information system for a heterogeneous scientific community. *Intern. Jour. of GIS,* v. 2, p. 171–189.

Ridley, B.K., 1994. *Time, Space and Things.* Cambridge Univ. Press, New York, NY.

Riemann Hershey, R., 1997. Using geostatistical techniques to map the distribution of tree species from ground inventory data, *in* T.G. Gregoire, D.R. Brillinger, P.J. Figgle, E. Russek-Cohen, W.G. Warren, and R.D. Wolfinger (Eds.), *Modelling Longitudinal and Spatially Correlated Data.* Springer-Verlag, New York, NY, 187–198.

Rivoirard, J., 1994. *Introduction to Disjunctive Kriging and Non-Linear Geostatistics.* Clarendon, Oxford, UK.

Robert, C.P., 1994. *The Bayesian Choice.* Springer-Verlag, New York, NY.

Rodriguez-Iturbe, I. and J.M. Mejia, 1974. The design of rainfall networks in time and space. *Water Resour. Res.*, v. 10, p. 713–728.

Rothman, K.J. and S. Greenland, 1998. *Modern Epidemiology.* Lippincott-Raven Publ., Philadelphia, PA.

Sadowski, F.G. and S.J. Covington, 1987. Processing and analysis of commercial satellite image data of the nuclear accident near Chernobyl, USSR. *U.S. Geol. Survey Bulletin 1785,* U.S. Govt. Printing Office, Washington, DC.

Saez, M., J. Sunyer, J. Castellsague, C. Murillo, and J.M. Anto, 1995. Relationship between weather temperature and mortality: A time series analysis approach in Barcelona. *Intl. Jour. Epidemiol.*, v. 24, no. 3, p. 576–582.

SANLIB99, 1999. *Stochastic Analysis Software Library and User's Guide.* Stochastic Research Group, Environmental Modeling Program, Dept. Environ. Sci. and Engr., Univ. North Carolina, Chapel Hill, NC.

Sarkar, S., 1998. *Genetics and Reductionism.* Cambridge Univ. Press, Cambridge, UK.

Schäfer-Neth, C. and K. Stattegger, 1998. Large-scale geostatistics: GSLIB and spherical coordinates, *in* IAMG'98 Poster Session, *Fourth Annual Conference of the International Association for Mathematical Geology.* A. Buccianti, G. Nardi, and R. Potenza (organizers), Naples, Italy.

Schumacher, E.F., 1977. *A Guide for the Perplexed.* Harper & Row Publ., New York, NY.

Schwartz, L., 1950–51. *Théorie des distributions,* Vols. I–II, *Actualités Scientifiques et Industrielles.* Hermann & Cie, Paris, France.

Schwedt, G., 1997. *The Essential Guide to Analytical Chemistry.* J. Wiley & Sons, New York, NY.

Serre, M.L., P. Bogaert, and G. Christakos, 1998. Computational investigations of Bayesian maximum entropy spatiotemporal mapping, *in* A. Buccianti, G. Nardi, and R. Potenza (Eds.), *Proceedings of IAMG'98—Fourth Annual Conference of the International Association for Mathematical Geology, Vol. 1.* De Frede Editore, Naples, Italy, p. 117–122.

Serre, M.L. and G. Christakos, 1999a. Modern geostatistics: Computational BME in the light of uncertain physical knowledge—Equus Beds study. *Stochastic Environ. Research & Risk Assessment,* v. 3, no. 1/2, p. 1–26.

Serre, M.L. and G. Christakos, 1999b. BME studies of stochastic differential equations representing physical laws—Part II, *in* S.J. Lippard, A. Næss, and R. Sinding-Larsen (Eds.), *Proceedings of IAMG'99—Fifth Annual Conference of the International Association for Mathematical Geology, Vol. 1.* Tapir, Trondheim, Norway, p. 93–98.

Shafer, G. and J. Pearl, 1990. *Readings in Uncertain Reasoning.* Morgan Kaufmann, San Mateo, CA.

Shannon, C.E., 1948. A mathematical theory of communication. *Bell System Tech. Jour.*, v. 7, p. 379–423; 623–656.

Shimony, A., 1985. The status of the principle of maximum entropy. *Synthese,* v. 63, p. 35–53.

Sobolev, S., 1938. Sur une theoreme de l'analyse fontionelle. *Matematiceski Sbornik,* v. 4, p. 471–496.

Stein, M.L., 1999. *Interpolation of Spatial Data: Some Theory for Kriging.* Springer–Verlag, New York, NY.

Stramel, G.J., 1966. Progress report on the groundwater hydrology of the Equus Beds area, Kansas. *Kansas Geol. Survey Bulletin 187,* Part 2, 27 p.

Swan, A.R.H. and M. Sandilands, 1995. *Introduction to Geological Data Analysis.* Blackwell Science, Cambridge, MA.

Tanaka, K., 1997. *An Introduction to Fuzzy Logic for Practical Applications.* Springer-Verlag, New York, NY.

Thiebaux, H.J. and M.A. Pedder, 1987. *Spatial Objective Analysis.* Academic Press, San Diego, CA.

Tilman, D. and P. Kareiva (Eds.), 1997. *Spatial Ecology.* Princeton Univ. Press, Princeton, NJ.

Trebicki, J. and K. Sobczyk, 1996. Maximum entropy principle and non-stationary distributions of stochastic systems. *Probabilistic Engr. Mech.*, v. 11, p. 169–178.

USEPA, 1996. *Air Quality Criteria for Particulate Matter* [3 Vols.]. U.S. Environmental Protection Agency, Washington DC, Rept. no. EPA/600/P–95/001aF.

USEPA, 1997. *National Ambient Air Quality Standards for Particulate Matter* [Final Draft]. U.S. Environmental Protection Agency, Washington, DC, Federal Register 40 CFR, Part 50.

U.S. Supreme Court, 1993. *W. Daubert et al. vs. Merrell Dow Pharmaceuticals, Inc.* 509 U.S. 579.

van Bemmel, J.H. and M.A. Musen (Eds.), 1997. *Handbook of Medical Informatics.* Springer-Verlag, Heidelberg, Germany.

Vlastos, G. (Ed.), 1971. *The Philosophy of Socrates.* Anchor Books, Garden City, NY.

von Foerster, H., 1962. Circuitry of clues to Platonic ideation, *in* C.A. Musés (Ed.), *Aspects of the Theory of Artificial Intelligence.* Plenum Press, New York, NY, p. 43–81.

Wackernagel, H., 1995. *Multivariate Geostatistics.* Springer-Verlag, Berlin, Germany.

Wallace, B., 1992. *The Search for the Gene.* Cornell Univ. Press, Ithaca, NY.

Wang, C., 1993. *Sense and Nonsense of Statistical Inference.* Marcel Dekker, New York, NY.

Weber, H. (Ed.), 1953. *Collected Works of Bernhard Riemann.* Dover, New York, NY.

Whitehead, A.N., 1969. *The Function of Reason.* Beacon Press, Boston, MA [6th printing].

Whittle, P., 1954. On stationary processes in the plane. *Biometrika,* v. 41, p. 434–449.

Whittle, P., 1963. *Prediction and Regulation.* English Univ. Press, UK.

Wiener, N., 1949. *Time Series.* The MIT Press, Cambridge, MA.

Wold, H., 1938. *A Study in the Analysis of Stationary Time Series.* Almqvist & Wiksell, Stockholm, Sweden.

Wolfe, T., 1999. *A Man in Full.* Bantam Books, New York, NY.

Woodbury, A.D. and T.J. Ulrych, 1998. Minimum relative entropy and probabilistic groundwater inversion. *Stoch. Hydrology & Hydraulics,* v. 12, p. 317–358.

Yaglom, A.M., 1955. Correlation theory of processes with stationary random increments of order n. *Mat. USSR Sb.*, v. 37–141 [English transl. in *Am. Math. Soc. Trans.*, 1958, Ser. 2, p. 8–87].

Yaglom, A.M., 1957. Some classes of random fields in n-dimensional space related to stationary random processes. *Theory of Prob. and its Appl.,* v. II, no. 3, p. 273–320.

Yaglom, A.M., 1987. *Correlation Theory of Stationary and Related Random Functions,* I. *Springer Series in Statistics,* Springer-Verlag, New York, NY.

Yaglom, A.M. and M.S. Pinsker, 1953. Random processes with stationary increments of order n. *Dokl. Acad. Nauk USSR,* v. 90, p. 731–734.

Yamartino, R.J., J.S. Scire, G.R. Carmichael, and Y.S. Chang, 1992. The CALGRID mesoscale photochemical grid model—I. Model formulation. *Atmospheric Environ.*, v. 26A, p. 1493–1512.

INDEX